Hierarchical Methods

Fundamental Theories of Physics

An International Book Series on The Fundamental Theories of Physics:
Their Clarification, Development and Application

Volume 123

Hierarchical Methods

Hierarchy and Hierarchical Asymptotic Methods in Electrodynamics, Volume 1

by

Victor V. Kulish

National Aviation University, Kiev, Ukraine and
Sumy State University, Sumy, Ukraine

SPRINGER-SCIENCE+BUSINESS MEDIA, B.V.

A C.I.P. Catalogue record for this book is available from the Library of Congress.

ISBN 978-94-017-4058-6   ISBN 978-0-306-48061-4 (eBook)
DOI 10.1007/978-0-306-48061-4

Printed on acid-free paper

Contents

Acknowledgments   xv

Preface   xvii

1. GENERAL IDEAS, CONCEPTS,
   DEFINITIONS, AND OTHER
   PRELIMINARY INFORMATION   1

   1   ESSENCE OF THE HIERARCHICAL
       ANALYTICAL–NUMERICAL METHODS   1

       1.1   One Illustrative Example
             of the Multi-Frequency Oscillation–Wave Systems   1

       1.2   Essence of the Hierarchical Approach   4

       1.3   Classification of the Hierarchical
             Methods Discussed in the Book   7

   2   BASIC CONCEPTS AND DEFINITIONS
       OF THE OSCILLATION THEORY
       OF WEAKLY NONLINEAR SYSTEMS   8

       2.1   Nonlinear Oscillations and Nonlinear Systems   9

       2.2   Hidden and Explicit Oscillation Phases   12

       2.3   Resonances   13

       2.4   Slowly Varying Amplitudes and
             Slowly Varying Initial Phases.
             Complex Amplitudes   20

       2.5   Harmonic and Non-Harmonic Oscillations   21

   3   BASIC CONCEPTS AND DEFINITIONS
       OF THE THEORY OF WAVES
       IN WEAK NONLINEAR SYSTEMS   23

       3.1   Definition of the Wave   23

       3.2   The Phase and Group Wave Velocities   24

       3.3   The Phase of a Wave   26

	3.4	Transverse and Longitudinal Waves	26
	3.5	Surface and Volumetric Waves	27
	3.6	The Concept of Dispersion	27
	3.7	Waves with Negative, Zero and Positive Energy	28
4		OTHER BASIC DEFINITIONS AND EQUATIONS OF PHYSICS	29
	4.1	Linear Velocity and Linear Acceleration	29
	4.2	Linear Momentum and Force	29
	4.3	Curvilinear Motion	30
	4.4	Rotation	30
	4.5	Energy. Field of Forces	31
	4.6	Motion Integrals	32
	4.7	Canonical Variables. Hamiltonian Equations	33
	4.8	Electromagnetic Field	33
	4.9	Lagrange and Euler Variables	34
5		CONCEPT OF SYSTEM	35
	5.1	What is a System?	35
	5.2	State of the System	36
	5.3	General Classification of the Systems	36
	5.4	Some General Properties of Complex Systems	37
6		BASIC POSTULATES AND PRINCIPLES OF THE GENERAL SYSTEM THEORY	38
	6.1	Principle of Physicality	38
	6.2	Integrity postulate	39
	6.3	Autonomy Postulate	39
	6.4	Principle of Ability to Model	40
	6.5	Complementarity Postulate	40
	6.6	Action Postulate	41
	6.7	Uncertainty Postulate	41
	6.8	Purposefulness Principle	42
	6.9	Entropy and Negentropy	42
	6.10	Complexity	45
2.		HIERARCHY AND HIERARCHICAL SYSTEMS	49
1		BASIC CONCEPTS OF THE THEORY OF HIERARCHICAL DYNAMICAL SYSTEMS	49
	1.1	Why Does Nature Need Hierarchy?	49
	1.2	Two Approaches to the Theory of Hierarchical Systems	50

	1.3	Self-Modeling Principle	51
	1.4	Main Idea of the Hierarchical Method	52
	1.5	Again: What is Hierarchy Originally? The Structural Hierarchy	53
	1.6	Dynamical Hierarchy	54
	1.7	Hierarchical Series in Dimensionless Form	58
2		HIERARCHICAL PRINCIPLES. HIERARCHICAL DESCRIPTION	58
	2.1	Hierarchical Principles	58
	2.2	Dynamical Equation of the Zeroth Hierarchical Level	60
	2.3	Structural and Functional (Dynamical) Operators	62
	2.4	Classification of the Hierarchical Problems	64
	2.5	Hierarchical Tree	66

3. HIERARCHICAL ASYMPTOTIC METHODS. GENERAL IDEAS — 71

	1	DETERMINED HIERARCHICAL SYSTEMS	72
	1.1	Determined Hierarchical Systems	72
	1.2	Averaging Operators	76
2		STOCHASTIC HIERARCHICAL SYSTEMS	78
	2.1	Factors of Stochasticity in Dynamical Systems	78
	2.2	Example of Hierarchical Model of Stochastic System	79
3		WAVE RESONANT HIERARCHICAL SYSTEMS	81
	3.1	Standard Hierarchical Equations in the Case of Hierarchical Wave Problem	82
	3.2	Classification of Problems	85
	3.3	Krylov–Bogolyubov Substitution	87
	3.4	Case b)	88
4		VAN DER POL'S METHOD	90
	4.1	A Few Introductory Words	90
	4.2	Van der Pol's Method's Variables	90
	4.3	Truncated Equations and Their Hierarchical Sense	92
5		METHODS OF AVERAGING. STANDARD VERSION	95
	5.1	Bogolyubov's Standard System	95
	5.2	The Problem of Secular Terms	99

6   METHODS OF AVERAGING.
      TWO-LEVEL SYSTEMS WITH
      SLOW AND FAST VARIABLES   100
      6.1   Two-Level Systems with Slow and
             Fast Variables. General Case   100
      6.2   Two-Level Systems with
             Fast Rotating Phases   102

4. HIERARCHICAL SYSTEMS
   WITH FAST ROTATING PHASES   107
   1   HIERARCHICAL OSCILLATIONS   107
        1.1   A Few Introductory Words   107
        1.2   Formulation of the Hierarchical
               Single-Particle Electrodynamic Problem   109
        1.3   Classification of Oscillatory Phases
               and Resonances. Hierarchical Tree   112
        1.4   Reducing Hierarchical Multi-Level
               Standard System to the Two-Level Form.
               The Scheme of Hierarchical Transformations   117
   2   THE CASE OF SIMPLEST TWO-LEVEL
        SYSTEM WITH ONE ROTATING
        SCALAR PHASE   119
        2.1   Formulation of the Problem   119
        2.2   Algorithm of Asymptotic Integration   120
        2.3   Accuracy of Approximate Solutions   124
        2.4   Asymptotic Integration of Initial Equations
               by Means of Successive Approximations 127
        2.5   Peculiarities of Asymptotic Hierarchical
               Calculational Schemes Based on
               the Fourier Method   128
   3   CASE OF TWO FAST ROTATING SCALAR PHASES   130
        3.1   Formulation of Problem   130
        3.2   Solutions. Non-Resonant Case   131
        3.3   Solutions. Resonant Case   134
   4   THE CASE OF MANY ROTATING SCALAR PHASES 139
        4.1   Formulation of the Problem   139
        4.2   Algorithm of Asymptotic Integration 140
   5   ALGORITHM FOR SEWING TOGETHER
        RESONANT AND NON-RESONANT SOLUTIONS   143
        5.1   Essence of the Problem   143

| | 5.2 | Sewing Together of Resonant and Non-Resonant Solutions | 144 |

| | 5.3 | Example for the Solution 'Sewing': The 'Stimulated' Duffing Equation | 146 |

5. HIERARCHICAL SYSTEMS WITH FAST ROTATING PHASES. EXAMPLES OF PRACTICAL APPLICATIONS — 157

| 1 | GENERAL PROPERTIES OF THE MODEL: 'A CHARGED PARTICLE IN THE FIELD OF A STANDING ELECTROMAGNETIC WAVE' AND EXAMPLES OF ITS PRACTICAL REALIZATION | 158 |

| | 1.1 | Systems for Transformation of Optical Signals into Microwave Signals as a Convenient Illustrative Examples | 158 |

| | 1.2 | Formulation of the Problem of Electron Motion in the Field of Two Oppositely Propagating Electromagnetic Waves | 164 |

| 2 | MODEL 'AN ELECTRON IN THE FIELD OF TWO OPPOSITELY DIRECTED ELECTROMAGNETIC WAVES' AS A TWO-LEVEL HIERARCHICAL OSCILLATATIVE SYSTEM | 167 |

| | 2.1 | Reducing Initial Motion Equations to the Standard Forms with Two Rotation Phases | 167 |

| | 2.2 | Zeroth Hierarchical Level. Parametrical Resonance | 170 |

| | 2.3 | Passage to First Hierarchical Level. Nonlinear Pendulum | 171 |

| | 2.4 | Nonlinear Pendulum. The Miller–Gaponov Potential | 176 |

| | 2.5 | Nonlinear Pendulum. The First Motion Integral | 178 |

| | 2.6 | Nonlinear Pendulum. Exact Solutions and Analysis | 178 |

| | 2.7 | Full Solutions of the Initial System | 183 |

| 3 | MODEL 'AN ELECTRON IN THE FIELD OF TWO OPPOSITELY DIRECTED ELECTROMAGNETIC WAVES' AS A THREE-LEVEL HIERARCHICAL OSCILLATATIVE SYSTEM | 185 |

| | 3.1 | Transition to the Second Hierarchical Level | 185 |

3.2   Duffing Oscillator   188

4   MODEL 'AN ELECTRON IN THE FIELD OF
       THREE ELECTROMAGNETIC
       WAVES' AS A FOURTH-LEVEL HIERARCHICAL
       OSCILLATATIVE SYSTEM   192
       4.1   Stimulated Oscillation of a Charged Particle   192
       4.2   Stimulated Oscillations of an Electron Ensemble   196

6. HIERARCHICAL SYSTEMS WITH
    PARTIAL DERIVATIVES. METHOD OF
    AVERAGED CHARACTERISTICS   207
    1   SOME PRELIMINARY INFORMATION   208
         1.1   Motion Equations   208
         1.2   Field Equations   210
         1.3   Some General Information about Equations with
                Partial Derivatives   211
    2   METHOD OF AVERAGED CHARACTERISTICS   213
         2.1   Concept of the Standard Form   213
         2.2   General Scheme of the Method   214
    3   CHARACTERISTICS AND
         THE METHOD OF CHARACTERISTICS   222
         3.1   Method of Characteristics. The Scalar Case   222
         3.2   Method of Characteristics. The Vector Case   227
    4   EXAMPLE: APPLICATION OF
         THE METHOD OF AVERAGED CHARACTERISTICS
         FOR A SIMPLEST SYSTEM WITH
         OSCILLATATIVE RIGHT PARTS   231
         4.1   Initial Equations   231
         4.2   Characteristics   232
         4.3   Passage to the First Hierarchical Level   232
         4.4   Back Transformations   234
    5   HIERARCHICAL METHOD OF AVERAGED
         QUASI-HYDRODYNAMIC EQUATION   236
         5.1   Averaged Quasi-Hydrodynamic Equation   237
         5.2   Back Transformations   240
    6   THE METHOD OF AVERAGED
         CURRENT–DENSITY EQUATION   240
         6.1   Averaged Current–Density Equation   241
         6.2   Back Transformations   244

7   HIERARCHICAL METHOD OF
     THE AVERAGED KINETIC EQUATION   244
     7.1   Averaged Kinetic Equation   244

7.   EXAMPLE: APPLICATION OF
     THE METHOD OF AVERAGED
     CHARACTERISTICS IN NONLINEAR
     THEORY OF THE TWO-STREAM
     INSTABILITY   249
     1   PROBLEM OF MOTION
          OF A TWO-VELOCITY ELECTRON BEAM
          IN GIVEN ELECTROMAGNETIC FIELDS   252
          1.1   Statement of the Motion Problem   253
          1.2   Averaged Characteristics for the Motion Problem 254
          1.3   Back Transformations   260
          1.4   Integration of the Averaged Quasi-Linear
                 Equation for the Beam Velocity   261
     2   FIELD PROBLEM. HIERARCHICAL
          ASYMPTOTIC INTEGRATION OF
          THE CONTINUITY EQUATION   262
          2.1   Continuity Equation of
                 the Two-Velocity Electron Beam   263
          2.2   Averaged Characteristics and
                 the Averaged Quasi-Linear Equation   264
          2.3   Back Transformation   268
          2.4   Characteristics of
                 the Averaged Continuity Equation   270
     3   FIELD PROBLEM. APPLICATION OF
          THE METHOD OF AVERAGED
          CHARACTERISTICS FOR ASYMPTOTIC
          INTEGRATION OF
          THE MAXWELL's EQUATIONS   271
          3.1   Averaged Maxwell's Equations   271
          3.2   Back Transformations   275
          3.3   Solving the Averaged Quasilinear
                 Equation for the Electric Field   276
          3.4   Truncated Equations for the Harmonic
                 Amplitudes of a Space Charge Wave   277
          3.5   Some Commentaries for the Obtained Results 279

8.   HIERARCHICAL SYSTEMS
     WITH PARTIAL DERIVATIVES.
     SOME OTHER ASYMPTOTIC METHODS   283

1   MAIN IDEAS OF THE METHOD OF
    SLOWLY VARYING AMPLITUDES   284
    1.1   General Calculational Scheme of
          the Slowly Varying Amplitudes Method   285
    1.2   Simplified Version of the Slowly Varying
          Amplitude Method. Example: Effect of
          Parametric Amplification of a Wave   288

2   TRADITIONAL VARIANT OF
    THE SLOWLY VARYING AMPLITUDES
    METHOD. RIGOROUS VERSION   293
    2.1   Case of Spatially One-Dimensional Model   294
    2.2   Classification of Transversely Inhomogeneous
          Models   306
    2.3   Model with Moderate Inhomogeneity   306
    2.4   Method of Parabolic Equation   308

3   MODERNIZED VERSION OF THE SLOWLY
    VARYING AMPLITUDE METHOD   309
    3.1   Field Problem   309
    3.2   Current Density Problem   311
    3.3   Current Density Problem in Framework of
          the Kinetic Approach   313

4   METHOD OF HIERARCHICAL
    TRANSFORMATION OF COORDINATES   314
    4.1   Main Idea of the Hierarchical
          Transformations   314
    4.2   Hierarchical Equations   317
    4.3   Averaged Operator $\bar{\bar{\nabla}}$ 320

5   MITROPOL'SKII METHOD   322
    5.1   Reduction of a Partial Differential Equation
          to the Standard Form with
          Fast Rotating Phases   322
    5.2   Basic Solutions   323
    5.3   Truncated Equations   324

6   EXAMPLES OF REDUCING OF
    THE MAXWELL EQUATIONS TO
    THE STANDARD FORM FOR
    THE METHOD OF SLOWLY VARYING
    AMPLITUDES   326
    6.1   Kinetic Version   326
    6.2   Quasi-Hydrodynamic Case   328

7   EXAMPLE: THE TWO-STREAM
      INSTABILITY IN A TWO-VELOCITY
      ELECTRON BEAM. THE METHODS OF
      AVERAGED QUASI-HYDRODYNAMIC
      EQUATION AND SLOWLY VARYING AMPLITUDES   329

      7.1   Statement of the Problem   329
      7.2   Motion Problem. The Averaged
              Quasi-hydrodynamic Equation   330
      7.3   Motion Problem. The Back Transformations   340
      7.4   Field Problem. The Method of
              Slowly Varying Amplitudes   341

Appendices   349
Results of calculations in the second approximation   349

Index   351

Acknowledgments

The author thanks his son Vol.V. Kulish, all friends, colleagues, present and former students, for their efforts in helping to make this book an accomplished reality.

The book had been written with the support of Grant No. 1457 given by the US Government via the Scientific Research Center of Ukraine.

Preface

Everybody is current in a world surrounded by computer. Computers determine our professional activity and penetrate increasingly deeper into our everyday life. Therein we also need increasingly refined computer technology. Sometimes we think that the next generation of computer will satisfy all our dreams, giving us hope that most of our urgent problems will be solved very soon. However, the future comes and illusions dissipate. This phenomenon occurs and vanishes sporadically, and, possibly, is a fundamental law of our life. Experience shows that indeed 'systematically remaining' problems are mainly of a complex technological nature (the creation of new generation of especially perfect microschemes, elements of memory, etc.). But let us note that amongst these problems there are always ones solved by our *purely intellectual efforts* alone. Progress in this direction does not require the invention of any 'superchip' or other similar elements. It is important to note that the results obtained in this way very often turn out to be more significant than the 'fruits' of relevant technological progress.

The *hierarchical asymptotic analytical–numerical methods* can be regarded as results of such 'purely intellectual efforts'. Their application allows us to simplify essentially computer calculational procedures and, consequently, to reduce the calculational time required. It is obvious that this circumstance is very attractive to any computer user. The situations discussed are typical for various physical areas (especially for applied physics which has to do with fast oscillatative nonlinear problems) and engineering. The models often considered can not be solved by conventional numerical (or analytical) methods straightforwardly. It is necessary to have new algorithmic ideas. All known *analytical–numerical methods* are intended for similar situations. The point is the special analytical transformation of the initial differential equations to their so called *shortened form*. Such shortened equations can be solved by use some well known conventional numerical or analytical methods. The hierarchical version of the analytical–numerical methods is a specific case in which the special *hierarchical transformations* are used. In situations the hierarchical transformations are constructed on the basis of some asymptotic calculational procedures, these methods are treated in the same way as the hierarchical asymptotic analytical–numerical methods.

The main purpose of this book is to expund the basic principles and peculiarities of hierarchical analytical–numerical calculational procedures. Electrodynamics seems to the a convenient field for vivid illustration of the essence and specific features of the latter.

The book is intended, first, for students at undergraduate, graduate, and postgraduate levels who specialize in computer calculations in physical electronics, plasma physics, radiophysics, optical and electronic engineering, acceleration and various space technologies, etc.. However, the book will be of interest for mature experts also, because a significant part of it is devoted to a new hierarchical point of view as well at some widespread calculational problems and electrodynamics, as a whole. Besides that we give a number of new illustrative examples in applied relativistic electrodynamics (free electron lasers and EH-accelerators, mainly), which could be interesting to them also.

The book has a peculiar 'hierarchical' structure with three levels of complexity. The first level requires knowledge of integral and differential calculus, series, and general physics at bachelor degree level. The second level requires a basic knowledge of linear algebra and linear differential equations theory (graduate level). The third level requires some preliminary knowledge of nonlinear equations and plasma-like systems (postgraduate level). The book is structured logically in such a manner that each level forms a self-consistent educational system, i.e., each lower level does not depend on the higher levels. Thus a reader at each educational level can find his own closed set of topics.

Concerning the concept of hierarchy, in itself (the use of which is the most characteristic feature of the class of numerical-analytical methods discussed), we should recognize that it is the most unusual and mysterious thing in the modern science. At the same time, it is a trivial phenomenon of everyday life, too. Indeed, one can be convinced that there is a hierarchy in everyday life everywhere. We can affirm once more that a person lives in a completely hierarchical world. Here each occupies a fixed place in a social hierarchy and the results of all events and situations depend essentially on its level in relevant hierarchical system (or, as an ancient Latin proverb says, '*Quod licet Jovi — non licet bovi*'). The hierarchical order of the regional, state, national, and family levels reminds us of this systematically. In places of worship each recalls that the Supreme Being has created our world with some hierarchy and that the arrangements of the Universe, of the human nature, of man's social order follow similar patterns, etc.. It seems incredible, but we have found some similarity between modern physics and engineering. In particular, there is hierarchy in condensed matter physics, cybernetics, and coding and systems theories, etc. [1–9]. Recently the hierarchy has been studied in the theory of oscillations and waves [3, 10, 11], too. (It should be mentioned that the oscillatory wave like dynamical systems namely are the main objects of interest in this book).

The characteristic feature of the hierarchical system is that some of their specific properties allow us *to use the concept of hierarchy* to elaborate various asymptotic calculational procedures. That is why it is natural that in spite of the explicit calculational trend of the book we discuss some of the main ideas of the general theory of dynamical systems, too (see Chapters 1 and 2).

In what follows let us especially draw the reader's attention to the following topic. There is a widely accepted opinion that the basis of modern hierarchical theory in physics has formed during the last twenty years [1–9] and this theory has philosophical–cognitive significance mainly. However, it is a relatively little known fact that some calculational aspects of the hierarchical approach really have bean practically from used at least the *third decade of the twentieth century* (!) in electrodynamics, and electrical and radio engineering. It is important especially to note that specific philosophy, terminology, and concepts which are characteristic for the modern hierarchical theory, are not used here traditionally. That is why this observation remains some way removed from the attention of the 'traditional hierarchical' experts.

It is interesting that researchers (for instance Van der Pol [12], Krylov and Bogolyubov [13, 14], Leontovich [15], and many others) have, as a rule, no interest in the philosophy of the problems concerned. Their main interest had been in the realization of hierarchical like calculational procedures in practice, e.g., to develop new types of highly effective asymptotic calculational methods. In this connection we may mention the Van der Pol method [12, 16], the set of averaging methods [11, 13, 14, 16, 17], the slowly varying amplitude method [11, 15, 18], various types of methods based on different types of transformations (conformal mapping, convolution, operational calculus, and so on), etc.. From the modern ('new') point of view, these methods can be regarded as the simplest (two-level) versions of the *hierarchical asymptotic analytical–numerical methods*, because specific transformations used here have explicit *hierarchical* nature.

The version of hierarchical theory [10, 11, 19–25], which is described
below, essentially differs from the above mentioned 'traditional' [1–9]
one. Firstly, because a specific set of the so called *hierarchical principles*
is put in its basis. Secondly, because the calculational approaches such
as Van der Pol's [12, 16], averaging [11, 13, 14, 16, 17], and slowly varying
amplitude methods [11, 15, 18] were used for factorization of these principles (in other words, for their mathematical description) [10, 11, 19–25].

It should be mentioned that further application of the proposed 'new
hierarchical ideology' [10, 11, 19–25] for modern electrodynamics nonlinear oscillatative wave resonant models, in turn, yielded to some nontrivial sequences. First, this stimulates the 'hierarchical' grasping by the
mind of some well known electrodynamic problems, which were studied
earlier in the framework of traditional approaches [26–35]. Results of
such a 'work of thought' turn out to be unexpected. Including that we
were surprised by the formal schemes of wave electronic devices with
long-time interaction, on the one hand, and the Universe, social and
biological systems, on other hand, possess the same formal hierarchical
structure [10, 23]. Unfortunately, fundamental reasons for this resemblance are not yet understood. However, this indicates that (possibly)
the proposed 'new hierarchical ideology' [10, 11, 19–25], which was elaborated in the framework of the nonlinear electrodynamic problems, possesses a really more universal nature than it seems at first sight. It also
means that we can hope that the general ideas of this ideology can be
used successfully in other branches of science, too. For instance, in unified field theory, astrophysics, the general theory of dynamical social,
economical, and biological systems, cybernetics, etc.. So the basic interest of the present book should not be limited to the calculational electrodynamic problems only. Electrodynamics here predominantly gives
convenient examples which demonstrate the essence and peculiarities of
the application of hierarchical calculational algorithms.

It is important to also note another characteristic point of the problem discussed. We take in view the elaboration of a number of new
hierarchical calculational procedures because we use the proposed new
hierarchical paradigm for practical purposes [10, 11, 19–25]. These are
the method of averaged characteristics, the methods of averaging kinetic
and quasi-hydrodynamic equations, the method of hierarchical transformation of coordinates, etc. [10, 11, 19–25].

The book consists of two volumes. The first of them is devoted to the
description of various versions of the hierarchical asymptotic analytical–
numerical methods. Detailed illustrations of characteristic peculiarities
of application of these methods in real nonlinear problems of the relativistic electrodynamics (the theory of the Undulative Induction Accel-

erators (UNIACs or EH-accelerators) and Free Electron Lasers (FELs)) are given in Volume II.

In writing this book the author has striven to make it open for discussion, and in doing so he would also hope that experts interested in electrodynamics and other areas of research would be interested in the further development of the hierarchical ideology proposed.

References

[1] J.S. Nicolis. Dynamics of hierarchical systems. an evolutionary approach. Springer-Verlag, Berlin-Heidelberg-New York-Tokyo, 1986.

[2] R. Rammal, G. Toulouse, M.A. Virasoro. Ultrametricity for physicists. *Reviews of Modern Physics*, 58(7):765–788, 1986.

[3] H. Haken. *Advanced Synergetic. Instability Hierarchies of Self-Organizing Systems and Devices*. Springer-Verlag, Berlin-Heidelberg-New York-Tokyo, 1983.

[4] H. Kaivarainen. Hierarchical concept of matter and field. Earthpuls Press, 1997.

[5] B.M. Vladimirskij, L.D. Kislovskij. *The outer space influences and biosphere evolution*, volume 1. In Series 'Astronautics, Astronomy', Znanije, Moscow, 1986.

[6] V.V. Druzshynin, D.S. Kontorov. *System Techniques*. Radio i Sviaz, Moscow, 1985.

[7] V. Ahl, T.F.H. Allen. Hierarchy theory. New York: Columbia Univ. Press, 1996.

[8] T.C. Marshall. *Free electron laser*. Mac Millan, New York, London, 1985.

[9] Interpretation the hierarchy of nature: from systematic pattern to evolutionary process theories. Academic Press, 1994.

[10] V.V. Kulish. Hierarchical oscillations and averaging methods in nonlinear problems of relativistic electronics. *The International Journal of Infrared and Millimeter Waves*, 18(5):1053–1117, 1997.

[11] V.V. Kulish. *Methods of averaging in non-linear problems of relativistic electrodynamics*. World Federation Publishers, Atlanta, 1998.

[12] B. Van der Pol. *Nonlinear theory of electric oscillations. Russian translation*. Svyazizdat, Moscow, 1935.

[13] N.M. Krylov, N.N. Bogolyubov. *Application of methods of nonlinear mechanics to the theory of stationary oscillations*. Ukrainian Ac. Sci. Publishers, Kiev, 1934.

[14] N.M. Krylov, N.N. Bogolyubov. Introduction to nonlinear mechanics. Kiev: Ukrainian Ac. Sci. Publishers, 1937 (in Ukrainian); English translation: New Jersey: Princeton, Princeton Univ. Press., 1947.

[15] M.A. Leontovich. To the problem about propagation of electromagnetic waves in the earth atmosphere. *Izv. Akad. Nauk SSSR, ser. Fiz., Bull. Acad. Sci. USSR, Phys. Ser.*, 8:6–20, 1944.

[16] N.N. Moiseev. *Asymptotic methods of nonlinear mechanics.* Nauka, Moscow, 1981.

[17] E.A. Grebennikov. *Averaging method in applied problems.* Nauka, Moscow, 1986.

[18] M.I. Rabinovich. On the asymptotic in the theory of distributed system oscillations. *Dok. Akad. Nauk. SSSR*, 191:1253–1268, 1971. ser. Fiz., Sov. Phys.-Doklady.

[19] V.V. Kulish. Nonlinear self-consistent theory of free electron lasers. method of investigation. *Ukrainian Physical Journal*, 36(9):1318–1325, 1991.

[20] V.V. Kulish, A.V. Lysenko. Method of averaged kinetic equation and its using in nonlinear problems of plasma electrodynamics. *Fizika Plazmy (Sov. Plasma Physics)*, 19(2):216–227, 1993.

[21] V.V. Kulish, S.A. Kuleshov, A.V. Lysenko. Nonlinear self-consistent theory of superheterodyne and free electron lasers. *The International Journal of Infrared and Millimeter Waves*, 14:3, 1993.

[22] V.V. Kulish. Hierarchical approach to nonlinear problems of electrodynamics. *Visnyk Sumskoho Derzshavnoho Universytetu*, 1(7):3–11, 1997.

[23] V.V. Kulish, P.B. Kosel, A.G. Kailyuk. New acceleration principle of charged particles for electronic applications. hierarchical description. *The International Journal of Infrared and Millimeter waves*, 19(1):3–93, 1998.

[24] V.V. Kulish. Hierarchical method and its application peculiarities in nonlinear problems of relativistic electrodynamics. general theory. *Ukrainian Physical Journal*, 43(4):483–499, 1998.

[25] V.V. Kulish. *Hierarchical theory of oscillations and waves and its application for nonlinear problems of relativistic electrodynamics. In Causality and locality in modern physics.* Kluwer Academic Publishers, Dordrecht/Boston/London, 1998.

[26] S.A. Przybylki. *Cache & memory hierarchy: A performance directed approach.* Morgan Kaufmann Publishers, New York, 1990.

[27] C. Brau. Free electron laser. Boston: Academic Press, 1990.

[28] P. Luchini, U. Motz. Undulators and free electron lasers. Oxford: Clarendon Press, 1990.

[29] A.N. Kondratenko, V.M. Kuklin. *Principles of plasma electronics.* Energoatomizdat, Moscow, 1988.

[30] A.A. Ruhadze, L.S. Bogdankevich, S.E. Rosinkii, V.G. Ruhlin. *Physics of high-current relativistic beams.* Atomizdat, Moscow, 1980.

[31] R.C. Davidson. Theory of nonlinear plasmas. Mass: Benjamin, Reading, 1974.

[32] A.G. Sitenko, V.M. Malnev. *Principles of plasma theory.* Naukova Dumka, Kiev, 1994.

[33] A.P. Sukhorukov. *Nonlinear wave-interactions in optics and radiophysics.* Nauka, Moscow, 1988.

[34] N. Bloembergen. *Nonlinear optics.* Benjamin, New York, 1965.

[35] J. Weiland, H. Wilhelmsson. Coherent nonlinear interactions of waves in plasmas. Oxford: Pergamon Press, 1977.

Chapter 1

GENERAL IDEAS, CONCEPTS, DEFINITIONS, AND OTHER PRELIMINARY INFORMATION

1. ESSENCE OF THE HIERARCHICAL ANALYTICAL–NUMERICAL METHODS

1.1 One Illustrative Example of the Multi-Frequency Oscillation–Wave Systems

We begin the discussion of the concept 'hierarchical analytical–numerical methods' with a certain physical example. Below in this Volume (and especially in Volume II) we will illustrate widely relevant features of the studied calculational technologies with the simplest model of an electronic system that is referred in a special literature to as *free electron laser* (FEL) [1–4]. It is very convenient example for illustration of the essence and application possibilities of the methods discussed. Fig. 1.1.1 can explain some general basic operational principle of a FEL amplifier.

The FEL amplifier discussed comprises a pumping system (for instance, some periodic reversed (in the transverse plane) system of permanent magnets (*H-ubitron pumping system*), see item 1 in Fig. 1.1.1), as one of the essential elements. Electrons 2 of the relativistic electron beam 3 move through the working bulk of system 1 on some undulatory trajectory 3. Simultaneously, electromagnetic signal wave 4 propagates along the longitudinal direction (i.e., along the axis z) of electron 2 motion. Signal wave 4 is amplified by virtue of realization of some specific physical mechanism (see in detail in Chapter 10, Volume II).

The corresponding initial system of differential equations is found to be nonlinear in the considered case. Besides that, it contains some periodic functions in the equation's right hands sides (see Chapters 10 and 11, Volume II). It is important that the scale of their periods might

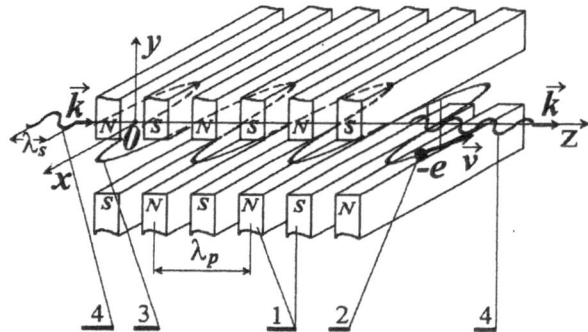

Figure 1.1.1. Simplest scheme of the free electron laser amplifier (FEL amplifier). Here 1 are the magnet poles of pumping system, 2 is an electron $(-e)$, 3 is the electron trajectory, 4 is the electromagnetic amplified signal wave, λ_p is the spatial period of the magnet pumping system 1, $\lambda_s = 2\pi/|\vec{k}|$ is the spatial period of signal wave 4, $\vec{v}$ is the vector of velocity of the electron, $\vec{k}$ is the wave vector of the signal wave.

be essentially different from some characteristic scale parameter (the longitudinal size of working bulk of the system L, for instance). *The simultaneous presence of the periods of essentially different scale is the most characteristic mathematical peculiarity of the general FEL problem.*

What are these different periods in our specific physical example? The first one is connected with the above mentioned spatial period of electron oscillations that occur under the influence of the periodic-reversed magnetic (H-ubitron) field (see Fig. 1.1.1). The second is characterized by the period of electron oscillations under the influence of the field of signal electromagnetic wave. Apart from that the electron can make the cyclotron oscillations in the case when longitudinal focused magnetic field is put along the work axis of the system. Various collective type of oscillations can also take place within the electron beam, which could be regarded as a plasma-like system, and so on [2–4]. Generally speaking, we can ascertain that in real systems there exist many real physical mechanisms which can cause electron oscillations with essentially different periods.

Let us limit ourselves to the discussion of a simplest FEL model, where only the accounting of the pumping and the signal oscillation periods are important. In this case we can separate three characteristic scale parameters. The first one is connected with the above mentioned spatial period of the pumping system λ_p (see Fig. 1.1.1). Really this period might be equal, for instance, to a few centimeters [2–4]. The second scale parameter is determined by the spatial periods of the signal wave

λ_s. In experiments these periods are usually characterized by values $\sim (1 \div 10)\mu$ [2–4]. And finally we introduce some *dynamical* parameter Λ which characterizes the *velocity of change* of electron energy, wave amplitudes, etc., caused by the interaction process. It may determine, for example, the spatial scale of changing these parameters (for instance, this could be some inverse gain factors). In a real experiment the following strong inequality is usually satisfied: $\Lambda \gg \lambda_{p,s}$ [2–4]. Or, in other words, essentially different spatial scales, indeed, characterize the above discussed oscillation periods. Let us assume for simplicity $\Lambda \leqslant L$ (where L, as before, is the longitudinal size of the system working bulk).

Dynamical objects of the discussed class are referred in literature to as *multi-frequency nonlinear resonant oscillation–wave systems* (MFN-ROWS). Thus we might be sure that the FEL model discussed could be regarded as the simplest example of the MFNROWS. The integration procedure of corresponding nonlinear differential equation system that describes the MFNROWS represents the essence of the so called *multi-frequency nonlinear resonant oscillation–wave problem*.

At first sight, any standard numerical method like to the Runge–Kutta or Adams method [5, 6] could, in principle, be used for solution of this problem. However, such a calculational method of 'straight numerical integration' really turns out to be too difficult in practice. It is clear especially in the case of the multi-frequency nonlinear resonant oscillation–wave problem [6]. But let us discuss this topic in more detail.

It is obvious that we should provide corresponding calculational step Δz for some chosen difference scheme for the sake of guaranteeing the required calculational precision. At the same time, it is quite clear that in the FEL case discussed the calculational step Δz should be smaller than the smallest period of the oscillations. In practice it means that

$$\Delta z \ll \lambda_s \ll \lambda_p \ll L. \tag{1.1.1}$$

The simplest analysis shows that the general calculational situation in such a case really becomes very complicated.

Let us give some evident numerical estimation for the illustration of these complications originating from the requirements concerning the calculational precision. Therein we take into consideration the above given typical magnitudes for characteristic FEL parameters ($\lambda_n \sim 10$cm, $\lambda_s \sim (1 \div 10) \quad \mu = (1 \div 10) \cdot 10^{-4}$cm). Besides that, let us remind that the total length of the FELs $L \gg \lambda_p$ in experimental practice is approximately equal to $(1 \div 10) \; m = (1 \div 10) \cdot 10^2 \; cm$ [2–4]. Then, let us rewrite the precision condition (1.1.1) in the following form:

$$\Delta z/L \ll \lambda_s/L \ll \lambda_p/L \ll 1, \tag{1.1.2}$$

or, taking into account the estimations chosen above for the FEL parameters, it is not difficult to obtain the estimation for a number n of calculational steps:

$$n = L/\Delta z >> L/\lambda_s = (1 - 10) \cdot 10^6 \qquad (1.1.3)$$

Inequality (1.1.3) means that we should take many more than 10^6 steps to perform the complete cycle of calculations! The practical (and quite grievous) consequences of the latter estimation for practice are evident. Hence we should look for some other approaches for solution of the problems of discussed kind. The *hierarchical asymptotic analytical–numerical methods* (hierarchical methods) can be regarded as one of the most promising ways for this. As experience shows (see, for example, Volume II), they are indeed the most efficient ways of overcoming the type of difficulties discussed.

The essential point of the hierarchical approach is the *use of the property of hierarchy* for separation of the corresponding problem scale parameters. These scale parameters are used as correspondingly small (or large) parameters for construction of the required asymptotic solutions of the problem. A number of hierarchical methods for constructing asymptotic solutions on such a basis are described below in Chapters 2–4, 6, 8.

1.2   Essence of the Hierarchical Approach

The main idea of any analytical–numerical method is some specific preliminary treatment (transformation) of the initial equations. In the particular case of the hierarchical methods a number of hierarchical schemes are proposed for such treatment. The terminal purpose of all these transformations is reduction of the initial equations into some 'smooth' form, i.e., to equations which do not contain any fast oscillations. In the theory of hierarchical methods (see below Chapters 2–3) these equations are called *truncated (shortened) forms.* This means that we should obtain some simplified set of equations in the final step of the transformation procedures. The characteristic feature of the latter is that the truncated equations are written with respect to some *slowest dynamical variables.* Because of this we obtain the situation in which the final 'smooth' (truncated, shortened) forms indeed do not comprise any fast oscillations. We regard such equations as the *equations of higher hierarchical level.* It is important that these equations are *characterized by the highest hierarchical scale parameters.*

The main merit of these truncated equations (i.e., equations of highest hierarchical level) is their evident simplicity. Hence in contrast to the situation with the initial system, we can apply conventional numerical (or exact or asymptotic analytical) methods for the solution of the truncated equations obtained.

Let us point out that in the case discussed all information about all faster motions (oscillations) turns out to be 'enciphered' in the corresponding *transformation relations*. The characteristic feature of the hierarchical methods considered in the book is that such transformation relationships can very often be found in simple analytical forms.

The 'truncated' solution obtained cannot be used as a solutions for the initial problem. For this we should perform preliminary, so called *inverse transformations*. The latter is based on the joint use of the transformation relationships mentioned and 'truncated' solutions. In this way we can finally obtain the complete solutions of the initial problem. Therein these solutions contain all groups of the above mentioned oscillation periods of different hierarchies.

Thus any calculational algorithm of the hierarchical type consists of the three following stages:

a) accomplishing corresponding hierarchical transformation of the initial equations up the hierarchical levels, i.e. these are transformations from the lowest hierarchical level to the highest one (obtaining the truncated equations);

b) solving the equations of the highest hierarchical level (i.e., truncated equations);

c) performing the inverse transformations of solutions for the highest hierarchical level to the lowest (initial) level.

All these stages are illustrated in more detail below in Fig. 1.1.2.

In what follows let us illustrate once more this calculational of idea by the simplest example of a FEL model represented in Fig. 1.1.1. But this time we will try to do it mathematically.

In the view of what has been said the corresponding initial equation set for the electron motion in the FEL working bulk must contain some periodic functions. In turn, these periodic functions should comprise various oscillation phases and their harmonics (see below about the concepts 'phase' and 'harmonics'). These phases are determined by electron oscillations under common action of periodic pumping field (of the spatial period λ_p) and the periodic signal–wave field (spatial period λ_s). Therein the initial motion equations can be represented by some nonlinear vector equation (see Fig. 1.1.2)

$$\frac{dz}{dt} = Z\left(z, t, p_p, p_s\right), \qquad (1.1.4)$$

where z is the corresponding n-dimension vector (electron coordinates, momentum, energy and so on), $Z\left(\ldots\right)$ is the corresponding vector-

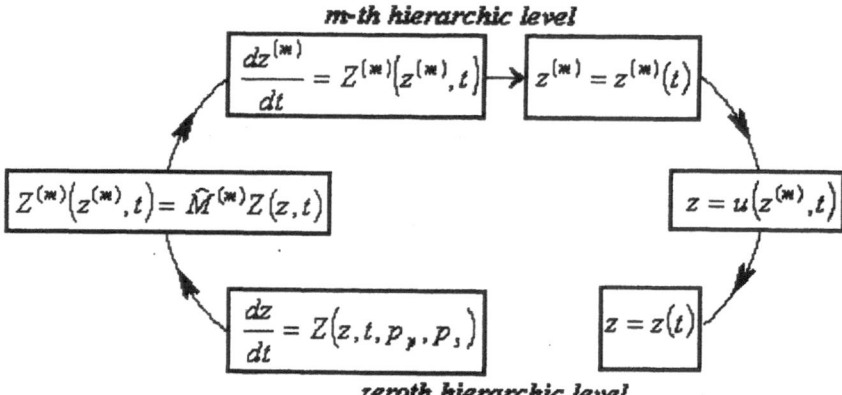

Figure 1.1.2. The illustration scheme of the principle of hierarchical calculational procedure.

function, p_p and p_s are the phases of the electron oscillations in pumping and signal fields that correspond to the mentioned above spatial periods λ_p, and λ_s, and t is the laboratory time (for more details see paragraphs 1.2 and 1.3). Within the framework of the hierarchical calculational ideology we can regard the initial equation (1.1.4) as the *standard equation of zeroth hierarchical level.* Then we transform the variables in equation (1.1.4) (as well as this equation, in itself) hierarchically into variables (equation) of the *m-th hierarchical level.* For this it is necessary to use some specific transformation relationships similar to (see Fig. 1.1.2):

$$z = u\left(z^{(m)}, t\right);\qquad (1.1.5)$$

$$Z^{(m)}\left(z^{(m)}, t\right) = \hat{M}^{(m)} Z\left(z, t\right),\qquad (1.1.6)$$

where $z^{(m)}$ and $Z^{(m)}$ are new vectors and a new vector-function for the *m-th* hierarchical level (in the considered particular FEL example the total number of hierarchical levels $m + 1$ equals 2 or 3, — see Volume II, Chapter 9 for more details), $u\left(z^{(m)}, t\right)$ and $\hat{M}^{(m)}$ are some vector-function and operator, whose construction methods are given below in Chapters 2–4. As a result of accomplished transformations the initial equation (1.1.4) can be rewritten in the form of a truncated equation of the *m-th* hierarchical level (see Fig. 1.1.2):

$$\frac{dz^{(m)}}{dt} = Z^{(m)}\left(z^{(m)}, t\right),\qquad (1.1.7)$$

It is clear that the above discussed calculational problem, concerning the difficulties related with in choosing the integration step Δz (see (1.1.2) and corresponding commentaries), is actually solved in this case. Indeed, the right part of the vector equation (1.1.7) is 'smooth' and consequently essentially simpler for solving. Any standard numerical (or analytical) method can be used for its integration. The inverse transformation of the obtained truncated solutions into solutions for the initial (zeroth) hierarchical level is realized by means of the relationships like to (1.1.5) (see Fig. 1.1.2).

Thus summarizing we can say that the essence of the hierarchical approach consists in accomplishing a specific cyclical calculational procedure of the type: '*an equation of the zeroth hierarchical level* → *the smoothed (truncated) equation of the upper hierarchical level* → *truncated solutions of this equation* → *transformation of the obtained truncated solutions on the zeroth hierarchical level*'.

1.3 Classification of the Hierarchical Methods Discussed in the Book

Accordingly with what was said in the Preface there are a lot of methods which can be classified as hierarchical. In this book we confine ourselves by discussing only those asymptotic methods which are effective for solving weak nonlinear wave oscillatory (including resonant) multifrequency electrodynamical problems. Let us classify these methods. The results of such a classification are presented in Fig. 1.1.3.

As can be easily seen, all methods discussed can be divided into those which are destined for obtaining solutions of exact differential equations, and whose objects of description are the systems with partial derivatives. The Van der Pol, Bogolyubov, and Bogolyubov–Zubarev methods are related to the methods of the first group (see Chapters 3, 4 and 5). These methods are widely known in nonlinear mechanics [7–9]. Besides that, their hierarchical versions (which are foreseen for situations when the number of hierarchical levels is more than two) are set forth, too.

An essentially different situation takes place for the methods of the second group. The latter are expounded in Chapters 6–8. Here only the method of slowly varying amplitudes can be considered as traditional. The other four methods described (the methods of averaged kinetic and quasi-hydrodynamic equations, hierarchical transformation of coordinates, and the method of averaged characteristics, namely) are relatively new, and therefore little known for the time being for wide circle of experts. Nevertheless, as will be shown below in the corresponding examples (see Chapters 7, 8 and Volume II), they can be very

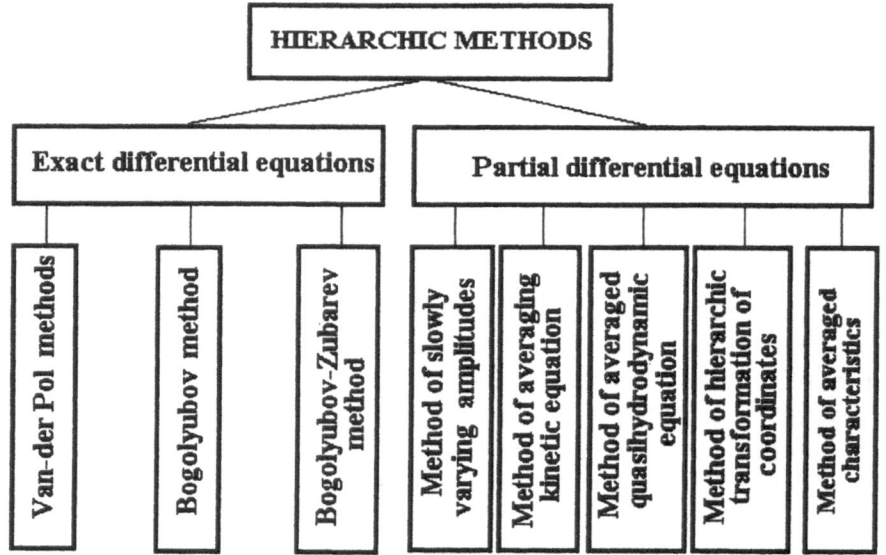

Figure 1.1.3. Classification of the hierarchical methods studied in the book.

effective for solving of a number of complex nonlinear wave resonant electrodynamic problems, especially in the case of plasma-like systems.

2. BASIC CONCEPTS AND DEFINITIONS OF THE OSCILLATION THEORY OF WEAKLY NONLINEAR SYSTEMS

The nonlinear oscillation problem is the most widespread in physics and engineering. In general, one might be convinced that the entire surrounding world has an explicit oscillatory nature. Indeed, Earth rotates around its own axis and around the Sun simultaneously; most existing machines (e.g., your car) contain rotating parts, within all electronic devices we can see oscillation or rotational electron motion, etc.. According to de Broglie's hypothesis, all surrounding matter has a wave nature (waves are complex simultaneous oscillations in space and time — see the next Subsection for more details). Therefore methods of solving the oscillation and wave problems (including the hierarchical ones) are quite popular in most areas of science. Then we take into consideration that further in Volume II we will illustrate the applied aspects of the hierarchical methods mostly by various examples, which have to do with nonlinear oscillations and waves in electrodynamical systems. That is why it seems expedient to give some preliminary information about the general theory of oscillations. This mainly concerns the basic ideas, definitions, and concepts used in the following Chapters.

Let us start with the simplest concepts 'nonlinear oscillations' and 'nonlinear oscillatative systems'.

2.1   Nonlinear Oscillations and Nonlinear Systems

Conventionally, the *mathematical pendulum* is the simplest and most evident example of the *nonlinear oscillatative system*. The scheme of the mathematical pendulum is illustrated in Fig. 1.2.1.

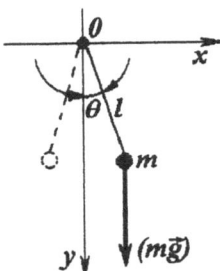

Figure 1.2.1. The simplest version of the mathematical pendulum. m is the mass of the material point, l is the length of the pendulum, q is the angle of deviation, g is free fall acceleration.

Here the material point of mass m oscillates near the vertical axis y. It is not difficult to obtain the differential equation for describing the particle's motion. As is known from any standard course of general physics, the generalization of the second law of dynamics (see below subsection 1.4) for rotational motion can be written in the form:

$$M_z = I_z \varepsilon_z, \tag{1.2.1}$$

where M_z is the z-component of moment of the force vector $\vec{F} = m\vec{g}$

$$\vec{M} = \left[\vec{r}\vec{F} \right] = \{M_x, M_y, M_z\} \tag{1.2.2}$$

$\vec{g}$ is the free fall acceleration vector, $\vec{r} = \{x, y, z\}$ is its space radius-vector, m is the material point mass, ε_z is the z-component of the angular acceleration vector

$$\vec{\varepsilon} = \frac{d^2\vec{\theta}}{dt^2}, \tag{1.2.3}$$

I_z is the moment of inertia with respect to the z-axis, $\vec{\theta} = \vec{n}\theta$ is the vector of the turning angle, $\vec{n}$ is the unit vector along the rotation axis ($\vec{n}$ is collinear to the z-axis normal to the drawing plane), θ is the magnitude of turning angle, t is the laboratory time. The corresponding set of dynamical equations (1.2.1)–(1.2.3) can be rewritten in the form:

$$I_z = ml^2; \quad \varepsilon_z = \frac{d^2\theta}{dt^2}; \quad M_z = -mgl\sin\theta, \qquad (1.2.4)$$

where l is the pendulum length. After corresponding transformations equation system (1.2.4) can be reduced to the *equation of nonlinear pendulum*

$$\frac{d^2\theta}{dt^2} + \omega_0^2 \sin\theta = 0, \qquad (1.2.5)$$

where the physical meaning of the constant

$$\omega_0 = \sqrt{g/l} \qquad (1.2.6)$$

will be clarified below. The expression (1.2.5) represents the wide class of mathematical objects which are called *nonlinear equations*. Generalizing, we can say that all equations which contain some nonlinear functions in their right part (in the case above we have the trigonometric sine), could be classified as nonlinear equations. Besides the differential equations the integral and mixed integral–differential nonlinear equations are known also. However, we will take an interest in the differential nonlinear equations only.

All physical systems whose dynamics are described by nonlinear equations of any type are called *nonlinear dynamical systems*. The nonlinear pendulum (see Fig. 1.2.1 and equation (1.2.5)) represents their simplest partial type which are called *nonlinear oscillation dynamical systems*.

In the theory of nonlinear systems, the concepts of *linear equation* and *linear system* are also used. They play a very important role in the theory of nonlinear *oscillation dynamical systems*. In particular, these objects are used as the so called *comparison equations* and *comparison systems* (generating equations or systems, also). The concepts of *weakly nonlinear equations* and *weak nonlinear systems* can also be defined by means of the latter. Further, we illustrate this in more detail.

Let us consider the pendulum to do *small oscillations* $\theta << 1$. In what follows we expand the sine in (1.2.5) in power a series in θ:

$$\sin\theta \approx \theta - \frac{1}{3!}\theta^3 + \cdots \qquad (1.2.7)$$

Accounting the first term in (1.2.7) only, we reduce the equation (1.2.5) to the form of *linear differential equation* (differential equation of linear oscillations):

$$\frac{d^2\theta}{dt^2} + \omega_0^2\theta = 0. \qquad (1.2.8)$$

The solution to (1.2.8) can be obtained easily using Euler's substitution. Simple calculations yield:

$$\theta = \theta_0 \sin(\omega_0 t + \varphi_0), \qquad (1.2.9)$$

where θ_0 is the angular *amplitude*, $p = (\omega_0 t + \varphi_0)$ is the *phase of linear oscillation*, and φ_0 is *the initial phase of oscillations*. The physical meaning of the constant ω_0 is obvious: it is the *cyclic frequency* of linear oscillations of the angular coordinate θ. It is related with to *period of oscillation* T_0

$$\omega_0 = 2\pi/T_0.$$

In what follows we take into account the second term in the expansion (1.2.7). After routine calculations it is not difficult to obtain the following nonlinear differential equation:

$$\frac{d^2\theta}{dt^2} + \omega_0^2\theta - \mu\theta^3 = 0, \qquad (1.2.10)$$

where $\mu = \omega_0^2/6$. The nonlinear pendulum equation (1.2.5) as well as (1.2.6) are very popular in the literature dedicated to nonlinear oscillations. Equations that are similar (with respect to mathematical structure) to (1.2.10) are called *Duffing equations* [10]. Analytical solutions for the latter are also well known (see, for instance, [10]). These equations are interesting because they describe the simplest cases of the *weak nonlinear oscillations*. This type of oscillations is characteristic for the *weak nonlinear dynamical systems*.

Let us further pay some more attention for the *weak nonlinear oscillations* and systems. The equation (1.2.9) had obtained supposing the series (1.2.7) converges. In the given case this condition can be written in the obvious form:

$$\theta \gg \frac{\theta^3}{6}. \qquad (1.2.11)$$

Taking this into consideration we can determine the weak nonlinear systems as the *systems for which some condition like (1.2.11) holds*. In other cases, we can say about the *moderately nonlinear systems* ($\theta \sim \theta^3/6$) or the *essentially nonlinear systems* ($\theta \ll \theta^3/6$). In the book, we confine ourselves to studying the weak nonlinear systems only. Wide literature is now dedicated to studying the moderately nonlinear and essentially nonlinear systems (see, for instance [11]). However, it should be mentioned that their theory requires essentially different calculational and ideological approaches, which are not the object of our interest here.

Let us also point out the following important circumstance. The characteristic feature of the theory of *weak nonlinear oscillations* is the wide use of concepts of *oscillation amplitude* and *oscillation phase* . But, strictly speaking, these concepts had been introduced for the case of linear oscillations only (see (1.2.9) and corresponding commentaries). Hence simple dissemination of such 'linear' terminology under the weak nonlinear theory is not a rigorous procedure. However, we have not any contradiction in this case. The point is that the physical behavior of the weak nonlinear systems at *relatively small observational intervals* (for instance, during one or a few oscillations) turns out to be almost the same as a 'purely linear' equivalent system. The essential difference appears only in the course of a large number of oscillations. In other words, we can regard the weak nonlinear system as some *slowly evolving (equivalent) linear system*. It means that the system amplitude and frequency in this case are *slowly varying parameters (variables) of the weak linear system*. Correspondingly, the phase of oscillations in this situation can be regarded as *fast changing* on background of the *slowly varying parameters (variables)*. This physical peculiarity we will widely use further to construct various hierarchical calculational algorithms with fast rotating phases.

2.2   Hidden and Explicit Oscillation Phases

The concept of oscillation phase (including fast rotating phase) is widely used in the theory of weak nonlinear oscillations in a number of effective hierarchical analytical–numerical calculational algorithms. Therefore, let us further discuss shortly some general peculiarities of the 'phase' concept.

We begin with the concept of *hidden oscillation phase*. Its simplest illustration could be given by using the example above discussed of the *mathematical pendulum* (see Fig. 1.2.1). The oscillation phase in this case can be represented in the form:

$$p = \omega_0 t + \varphi_0$$

(see (1.2.9) and corresponding commentaries). The main feature of the hidden phases is that they are not related to any evident periodicities of the acting system forces. Indeed, the main force acting (Earth's gravitation) in the case mentioned of the *mathematical pendulum* is not periodic. So the oscillation period of the latter,

$$T_0 = 2\pi \Big/ \omega_0 = 2\pi \sqrt{l/g}, \qquad (1.2.12)$$

does not depend on any external periodicity.

In contrast, the *explicit oscillation phases* are always determined by some external periodicities. We may take the free electron laser (FEL) as a convenient illustration example (see Fig. 1.1.1). The periodicities of the pumping magnetic and signal electromagnetic fields (and their corresponding acting forces) determine the periodicity of the electron motion. The *electron oscillation phases* in this case formally coincides with the *field oscillation phases*. Thus periodicity of the considered oscillation system (*FEL*) is expressed evidently, in contrast to the case of hidden phases.

Let us note that in the general case oscillation periods in the weak nonlinear oscillations always depend on the time t or the coordinate $\vec{r}$ or on both of them simultaneously. As we will see later, this circumstance is very important for constructing corresponding hierarchical procedures, as well as for physical analysis, too.

Then it should also be noted that according to the definitions given above of slowly and fast varying values, the *slow* and *fast oscillation phases* could be determined also. This means that some *hierarchy* with respect to the changing velocity of the oscillation phases can take place in the case of multi-frequency systems. The systems of this kind we classify as the *hierarchical oscillation systems*.

2.3   Resonances

In what follows let us discuss the concept of '*resonance*'. We distinguish the *proper* and *stimulated (induced) oscillations*. The proper oscillations occur in the case of a system with no *external* influence. One can be convinced that the oscillations of a mathematical pendulum are proper, because the acting force (Earth gravitation) is a part of the system considered. I.e., this force can be treated as an *internal* one. The stimulated (induced) oscillations take place in the situations some external periodic force exerts influence on the system. According to the definition given, we can regard the model of FEL (see Fig. 1.1.1) as a system with stimulated oscillations. The electron oscillations under action of the magnetic pumping system could be treated as proper oscillations. On the other hand, the action of an electromagnetic signal wave on electron motion could be regarded as a stimulated (induced) influence. The stimulated (induced) system oscillations are connected with this influence.

Thus the dividing oscillations into proper and stimulated, strictly saying, are relative. This can be explained by it depending on the defining elements (and forces) which can be included in the system considered, and which are considered as external ones.

The interesting feature of the stimulated oscillations is the phenomenon of *resonance*. It could seem incredible that until today there is no universally recognized and rigorous definition of the concept of resonance. Often any sharp increasing the oscillation amplitude (for the changing the frequency of an external force) is called the resonance (see, for instance, [12]). However, in electrodynamics there are number of rather 'strange' phenomena (such as the plasma beam *instability* or the *two-stream instability* — see Chapter 7, and Chapter 11 in Volume II). They are also characterized by rather sharp increase of the oscillation amplitude, but have no 'truly resonant nature'. Similar phenomena are also known in hydrodynamics (the Helmholtz instability [13]) and other areas of science. The point is that all forces acting in phenomena of this type have an explicitly expressed proper nature. As a rule, in electrodynamics such processes are called *instabilities* [14]. Therefore following the tradition [14] we *consider resonance to be a specific instability*. But this does not mean that any instability can be treated as a resonance. The distinguishing feature of the resonance is that here one of *acting stimulated force always is external* one.

In this book we accept the *definition of resonance* as a sharp increasing oscillation amplitude with the following condition held

$$\frac{dp_p}{dt} \approx \frac{dp_s}{dt}, \quad \text{or} \quad \left| \frac{d\left(p_p - p_s\right)}{dt} \right| << \left| \frac{dp_p}{dt} \right|,$$

$$\text{or} \quad \left| \frac{d\left(p_p - p_s\right)}{dt} \middle/ \frac{d\left(p_p + p_s\right)}{dt} \right| << 1, \tag{1.2.13}$$

where the phases $p_{p,s}$ correspond to proper and stimulated oscillations. Thus the *characteristic feature of resonance* is a closeness of velocities of phases' change of the proper and stimulated oscillations. This definition is essential in the theory of oscillations used later in the book.

The physical meaning of definition (1.2.13) can be illustrated by the FEL model (see Fig. 1.1.1). In the given case the self-evident definition of the proper phase (i.e., the evident phase of oscillations of some electron under magnetic pumping field) is

$$p_p = k_p z + \varphi_{0p}, \tag{1.2.14}$$

where $k_p = 2\pi/\lambda_p$ is the *pumping wave* number, λ_p is the spatial period of oscillations of the pumping field (see Fig. 1.1.1), $z = z(t)$ is the electron longitudinal coordinate, φ_{0p} is its initial pumping oscillation phase.

In what follows let us determine the phase of electron stimulated oscillations. We use for this the conventional definition for the phase of plane monochromatic signal wave. Then explicit expression for the electron phase of stimulated oscillations formally coincides with the signal wave phase:

$$p_s = \omega_s t - k_s z + \varphi_{0s}, \qquad (1.2.15)$$

where $\omega_s = 2\pi/T_s$ is the *cyclic frequency*, T_s is *wave period*, $k_s = 2\pi/\lambda_s = \omega_s/c$ is signal *wave number*, λ_s is signal *wave length*, c is light velocity in vacuum $\varphi_{0s} = p_s\,(t = 0, z = 0)$ is initial wave phase. The difference is only the longitudinal coordinate z is slowly varying function $z\,(t)$ (in contrast to the case of plane electromagnetic wave in itself, where variables z and t are independent values). Substituting (1.2.14) and (1.2.15) into (1.2.13), we obtain classical the *resonant condition* for the FELs with pumping by undulative magnetic field (H-ubitron pumping) [1–4]:

$$\omega_s \approx \frac{2\pi v_z}{\lambda_p\,(1 - v_z/c)}, \qquad (1.2.16)$$

where $v_z = dz/dt$ is the velocity of the electron on longitudinal direction. In the essentially relativistic case ($v_z \approx c$) the formula (1.2.16) can be rewritten as [2–4]:

$$\omega_s \approx \frac{2\pi v_z\,(1 + v_z/c)}{\lambda_p\,(1 - v_z^2/c^2)} \approx 4\pi v_z \gamma^2/\lambda_p, \qquad (1.2.17)$$

where $\gamma = \sqrt{1 - v_z^2/c^2} = \mathcal{E}/mc^2$ is the relativistic factor, $\mathcal{E}$ is the energy, m is the rest mass of the electron. Obtaining (1.2.17), we supposed $\left(1 + v_z^2/c^2 \sim 2\right)$. The formulas (1.2.16), (1.2.17) explain, necessity to use essentially relativistic electron beams ($\gamma \gg 1$) in FELs. Indeed, it can easily be seen from (1.2.17), (1.2.18) that the possibility of obtaining high frequencies ω_s (including the optical range frequency) determined by the range of electron energy $mc^2\gamma$. Let us accomplish some estimations for illustration of the latter affirmation. Therein we will use corresponding results of numerical estimations, performed earlier in this Section (see Subsection 1.1). Including we accept: $\lambda_n \sim 10\,\mathrm{cm}$, $\lambda_s = (1–10)\,\mu$ (that corresponds to the *cyclic frequency* $\omega_s = 2\pi c/\lambda_s \sim (2 \cdot 10^{14} – 2 \cdot 10^{15})\mathrm{s}^{-1}$). According to (1.2.17) it means that the ratio $\lambda_p\lambda_s = \lambda_p\omega_s/2\pi \sim \left(3 \cdot 10^4 – 3 \cdot 10^3 \gg 1\right)$ can be attained in the case of relativistic electron beam ($\gamma \sim (1.3–0.4) \cdot 10^2 \gg 1$) only [2–4].

The resonances of the type (1.2.16), (1.2.17) are classified in theory of nonlinear oscillations as the *parametric resonances* [15–17]. The peculiarity of the parametric resonances is that here both resonant phases $p_{s,p}$ are represented by some explicit phases only. The other possible variant in which one of phases is represented by a hidden phase in the resonant condition is called the *quasilinear resonance* [15–17].

We discussed before the so called *paired (two-multiple) resonances* only , when two different phases take part in resonant process (1.2.13). It should be mentioned that, in principle, the three-multiple, four-multiple, and so on, resonances can be realized, too [15, 16, 18]. For instance, the well known *four-wave parametric resonances* in plasmas, nonlinear optic mediums, etc. [19–21] can be treated as physical examples of resonances of the three-multiple type, and so on. In a general case of a m-multiple resonance the condition (1.2.13) can be generalized in the following form

$$\left| \frac{\sum\limits_{j=1}^{m} [d\,(n_j p_j)]}{dt} \middle/ \frac{d\,(n_j p_j)}{dt} \right| \leqslant \varepsilon_j << 1, \qquad (1.2.18)$$

where $n_j = \pm 1, \pm 2, \cdots$ are the numbers of *oscillation harmonics*, p_j are the oscillation phases, and $\varepsilon_j \sim \varepsilon << 1$ is the *small parameter of the problem*. One can easily be convinced that the definition (1.2.18) in particular case of two-multiple resonance ($m = 2, n_1 p_1 = p_s, n_2 p_2 = -p_p$) reduces to the simplest form (1.2.13).

In view of (1.2.18) the quantity

$$\sum_{j=i}^{m} n_j p_j = \theta \qquad (1.2.19)$$

can be regarded as a slowly varying function. In literature, it is called the *slowly varying combination phase*. Similarly, we can also introduce the *fast varying combination phase*:

$$\sum_{j-1}^{m} n'_j p_j = \psi, \qquad (1.2.20)$$

where $n'_j = \pm 1, \pm 2, \ldots m_j$ are the harmonic numbers, therein $|n'|_j = |n_j|$, however at least one of numbers n'_j *has opposite sign* with respect to the analogous number n_j. In this case the general resonant condition (1.2.18) can be written more elegantly

$$\left| \frac{d\theta}{dt} \middle/ \frac{d\psi}{dt} \right| \leqslant \varepsilon << 1, \qquad (1.2.21)$$

In the framework of the *hierarchical theory of oscillations and waves* [16] the condition (1.2.21) gives the criterion for classifying resonances according to the system's hierarchical levels. Namely, in the case of the FEL model the fast combination phase ψ is related to the *zeroth hierarchical level*, whereas the slow combination phase θ characterizes the dynamical variables of the *first hierarchical level* [15,16,18]. But let us discuss this problem in more detail.

In the general case of the multi-frequency model we can have, in principle, a few combination phases θ_l ($l = 1,2,\dots$) may be related with the first hierarchical level. These phases formally describe several resonances realized in parallel and simultaneously. Such systems are called *multi-resonance systems*. From the 'point of view of the zeroth hierarchical level' (that characterized by the fast combination phases ψ_k, $k = 1,2,\dots$) all slow combination phases θ_l are characterized by the same order in their velocities of varying. However, essentially different situations occur in the zeroth and first hierarchical levels, correspondingly. We take in view that the first level comprises only slow phases θ_l without any 'fast background' of fast phases ψ. Hence different combination phases θ_l begin 'compete' (with respect to their velocities) amongst themselves. Therefore amongst these slow combination phases, the more and less slow phases can be distinguished. Some of these phases can be regarded as new '*super-slow*' and '*quasi-fast*' ones, similarly with situation in he zeroth hierarchical level. In turn, some of such 'quasi-fast' phases can form new '*super-slow*' combination phases Θ in the same manner (see further Chapter 4 and 5 for more details). We can regard this as a realization of peculiar '*slow resonances*', i.e., the *resonances of the next* (i.e., second) *hierarchical level*, and so on. Further, the described calculational procedure can be repeated many times until no oscillation phases (resonant as well non-resonant) remain at the considered stage of the calculational.

Let us mention that, generally, we can distinguish two types of the multi-resonance systems. The first is characterized by *bound* (in particular, *coupled*) resonances. In this case one of oscillation phases can *take part in a few resonances simultaneously* [17,22,23]. The coupled resonances always are *resonances of the same hierarchy*. In contrast, the *non-bound resonances* (which are characteristic for systems of the second type) can belong to the same hierarchical level as well as to different hierarchical levels. In detail the concept of hierarchy in nonlinear multi-frequency oscillatory systems is discussed in Chapter 4. The corresponding examples are given in Chapter 5. The general ideology of arbitrary hierarchical dynamical systems is described in Chapter 2.

Then we turn again to discussion of resonant condition (1.2.21). At the first sight it could be seem that the essence of the latter essentially differs from the analogous condition (1.2.18). Nevertheless, it is not really so. The point is that both quantities in (1.2.21), (1.2.18): ψ and $(n_j p_j)$, respectively, are characterized by an *approximately equal scale of varying velocity*, $\sim \varepsilon << 1$. Hence, formal differences are not essential and they do not influence in strictness of corresponding calculational accomplished on such basis.

The sense of the latter affirmations can be obviously illustrated at the example discussed above of the FEL model (see Fig. 1.1.1 and corresponding comments). This particular case is described by following parameters:

$$j, i = 1, 2; \ p_s \equiv p_1; \ p_p \equiv p_2; \ n_1 = n_1' = +1; \ n_2 = -n_2' = -1;$$
$$\text{i.e., } \theta = p_1 - p_2; \ \psi = p_1 + p_2. \tag{1.2.22}$$

The changing velocity of the phase θ in this case is essentially lesser than changing velocity of the phase ψ if the condition

$$\frac{dp_1}{dt} \approx \frac{dp_2}{dt}, \tag{1.2.23}$$

is satisfied, in the case with resonance condition (1.2.13) held. Hence, velocities of changing $p_{1,2}$ and ψ are of the same order in the small parameter ε.

Then let us define the concepts of '*resonant point*' and '*vicinity of the resonant point*'. For this, it is convenient to introduce the concepts of '*resonant function*' and '*resonant curve*'. We assume nonlinear resonant system is *passing through a resonant* state during some interaction process. The function

$$\Phi(t) = \frac{d\psi}{dt} \Big/ \frac{d\theta}{dt}, \tag{1.2.24}$$

we consider as the resonant function. Further, we again use the model of *FEL*. Conventionally, the effectiveness of the energy exchange between of the electron and the electromagnetic signal wave for the time interval $\Delta t = t_2 - t_1$ in the theory of *electron devices with long-time interaction* is characterized by *single-particle efficiency of the interaction*

$$\eta = \frac{\mathcal{E}(t_2) - \mathcal{E}(t_2)}{\mathcal{E}(t_1)}, \tag{1.2.25}$$

where η is the efficiency, $\mathcal{E}(t_{1,2})$ is the energy of the electron as function of time at t_1 and t_2. Analysis shows that the qualitative characteristic of dependency of the efficiency (1.2.25) on time could be explicitly resonant. To illustrate the latter we change variables $t \rightarrow \Phi^{-1}$ in the function $\eta(t)$ (for instant, using the definition of the resonant function (1.2.24)) and pass to corresponding dependency $\eta(\Phi^{-1})$. The latter is described graphically by the *resonant curve*. A typical form of resonant curve is given in Fig. 1.2.2.

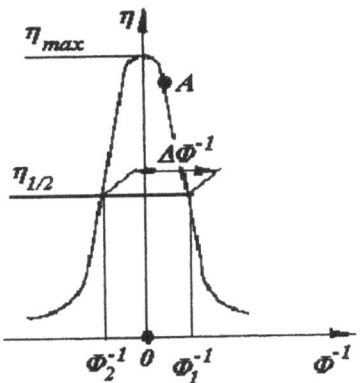

Figure 1.2.2. Qualitative dependency (resonant curve) of electron efficiency η on Φ^{-1}. Here η_{max} is the maximum of the efficiency η, as a function of Φ^{-1}; $\eta_{1/2} = (1/2)\eta_{\max}$, $\Delta\Phi^{-1} = \Phi_2^{-1} - \Phi_1^{-1}$ is the width of resonant vicinity (width of resonance band); the point $\Phi^{-1} = 0$ is called the *resonant point*; A is the current working point which characterizes the state of the system at the moment t.

The sense of the resonant point and resonant vicinity is obvious from this figure. Namely, the resonant point corresponds to the point of maximum $\eta_{\max}$. The vicinity of the resonant point is determined by the interval $\Delta\Phi^{-1}$ on which the efficiency $\eta(\Phi^{-1})$ varies from $\eta_{1/2}(\Phi_2^{-1}) = (1/2)\eta_{\max}$ through the point $\eta(0) = \eta_{\max}$ until $\eta_{1/2}(\Phi_1^{-1}) = (1/2)\eta_{\max}$.

We mention that the realization of slow braking of the electron motion is a characteristic peculiarity of the FEL interaction mechanism [2–4]. Therein the transformation of kinetic energy of the electron into the signal energy of the electromagnetic wave occurs. This means that the velocity of the longitudinal electron motion in the situation considered is a slowly varying function of time t. As follows from definitions (1.2.14), (1.2.15), (1.2.19), (1.2.20), (1.2.24) the resonant function $\Phi(t)$ in the general case is a slowly varying function of time t, too. The latter appears in Fig. 1.2.2 as a slow motion of the working point A along the resonant curve from the right side to the left side. In principle the passing of working point A through the *resonant point* $\Phi^{-1} = 0$ can occur immediately as well as through its nearest vicinity. The resonant condition (1.2.21) can violate in time when the point A moves too far from the resonant point (and the resonant vicinity, correspondingly). In such a case the electron interaction with the electromagnetic field ultimately loses its *resonance* character and, therefore, the effectiveness (1.2.25) becomes too low. In the specialist literature, the above physical mechanism described is called *saturation of amplification* [2–4].

Let us point out that the physical peculiarities described of resonant systems essentially influence the mathematics of the problem. Namely, as shown below, calculational algorithm for the given class of calculational problems essentially depends on the kind of interaction mechanism, i.e., the resonant or the non-resonant (see for details in Chapters 7–11). Therefore specific algorithm use is needed for sewing resonant and non-resonant parts of solutions [17].

2.4   Slowly Varying Amplitudes and Slowly Varying Initial Phases. Complex Amplitudes

Let us discuss some peculiarity of description methods of the weak resonant nonlinear systems. As mentioned above, the classical form of the linear solutions like (1.2.9) is, generally, illegal in the weak nonlinear case. It is also mentioned that such a solution representation, nonetheless, could be disseminated at the weak nonlinear models, if some addition specific suppositions are accepted. These 'addition specific suppositions' concern new 'extended' definitions for the oscillation amplitude and initial phase. The difference with the linear case like to (1.2.9) is that at that time the oscillation amplitude and initial phase are found to be slowly varying functions on time t. So, the corresponding weak nonlinear solutions can be written in the following 'modernized' general form

$$z_k = A_k(t) \sin[\omega t + \varphi_{0k}(t)], \qquad (1.2.26)$$

where z_k is the k-th component of some n-dimensional vector z; $A_k(t)$ is k-th component of the *slowly varying amplitude* vector $A(t)$; $\varphi_0(t)$ is k-th component of the *slowly varying initial oscillation phases* vector $\varphi(t)$, ω is the (permanent) cyclic frequency. It is easily be convinced that some other version of introducing the slowly varying values is possible, too:

$$z_k = A_k(t) \sin[\omega(t) t + \varphi_{0k}], \qquad (1.2.27)$$

where $\omega(t)$ is the *slowly varying cyclic frequency*, φ_{0k} is the permanent (!) initial phase. Both discussed representations (1.2.26) and (1.2.27), as experience shows are, in principle, equivalent. However, the first version is more widespread in practice, inasmuch as really more convenient.

Thus the basic *weak nonlinear mathematical problem* could be formulated as a determining the slow dependencies $A(t)$ and $\varphi_0(t)$ (or $A(t)$ and $\omega(t)$), respectively. One of main interest in this book is description of methods for the constructing such kind dependencies.

Apart from the real amplitude $A(t)$ and phase $\varphi_0(t)$ the so called *complex slowly varying amplitudes* are used for practical calculations. The definition of complex slowly varying amplitude can easily be obtained from (1.2.26) utilizing the well known de Moivre formula:

$$\sin\alpha = \frac{1}{2i}\left(e^{i\alpha} - e^{-i\alpha}\right) = \left(\frac{1}{2i}e^{i\alpha} + \text{c. c.}\right). \qquad (1.2.28)$$

Using (1.2.28) in (1.2.26), we obtain the representation for (1.2.26) via the slowly varying complex amplitudes:

$$z_k = \left[\frac{1}{2i}A_k(t)\exp\{i\varphi_{0k}\}\right]e^{ip} + \left[\frac{-1}{2i}A_k(t)\exp\{-i\varphi_{0k}\}\right]e^{-ip}$$
$$= \left[\frac{1}{2i}A_{ck}(t)e^{ip} + \text{c. c.}\right], \qquad (1.2.29)$$

where 'clear' *oscillation phase p* is defined as

$$p = \omega t, \qquad (1.2.30)$$

$A_{ck}(t) = A_k(t)\exp\{i\varphi_{0k}\}$ is the slowly varying complex amplitude; the designation c. c. $= A_k^*(t) = A_k(t)\exp\{-i\varphi_{0k}\}$, i.e., it is the *complex conjugated* term.

Thus the weak nonlinear oscillations formally could be described by means the concept of slowly varying amplitudes and phases. Therein the oscillations similar to (1.2.26), (1.2.27), and (1.2.29) are called in references *the quasi-harmonic oscillations*, because, as mentioned above, they are not 'true' harmonic oscillations. But let us discuss the concepts of harmonic and non-harmonic oscillations in more detail.

2.5 Harmonic and Non-Harmonic Oscillations

Thus we introduce further the concepts of *harmonic* and *non-harmonic oscillations*.

According to the classical definition an *oscillation* is a motion process in which the object studied *systematically* comes back to its initial state.

This definition includes a very wide spectrum of various types of oscillations, including the *periodic, non-periodic, nearly periodic, conditionally periodic; harmonic* and *non-harmonic oscillations*, etc.. In what follows we briefly discuss a few of these concepts which will be needed below.

One says about the *periodical oscillations* in the case only if the 'systematical' process in the system is strictly periodical. Earlier, in our illustration examples with the linear pendulum (see Fig. 1.2.1 and corresponding commentaries) we considered that the action of the external and the internal forces causes the oscillation processes, having pure periodical. In a simplest case they can be described by some harmonic (see definitions (1.2.9), (1.2.26), (1.2.27), (1.2.29)) functions. This particular case of the periodical oscillations is called the *harmonic oscillations*.

In what follows we use the model of free electron laser (FEL — Fig. 1.1.1 and corresponding explanations). As can easily be seen in Fig. 1.1.1, the electron in the working bulk of FEL moves under the action of two periodic fields. They are the pumping static (permanent) magnetic H-ubitron field with space period λ_p and the electromagnetic signal wave field, which is characterized by the space period λ_s and the temporary period $T_s = 2\pi/\omega_s$. Therefore the electron moves along some complex trajectory, having no fixed period (as it is mentioned in Subsection 1.1, we have at least two periods in the considered FEL case). Thus the *two-periodic oscillations* are realized in the discussed illustration example. In the general case of many oscillations we can speak about the n-periodic (multi-periodic) oscillations (here $n = 1, 2, \dots$).

The important peculiarity of nonlinear systems is their ability to transform any initially harmonic oscillations into *non-harmonic* (but, periodic) *oscillations*. For example, in the case of the nonlinear pendulum the linear oscillations take place for relatively small amplitudes only (see solutions (1.2.9)). A number of higher harmonics occurs with the growth of the oscillation amplitude. This characteristic process can be illustrated by nonlinear pendulum. Here the oscillation process is described by the equation (1.2.5). In the simplest case of linear approximation the solution of (1.2.5) can be given by a sine-like harmonic function (1.2.9). In the second, third, and so on orders of approximation, oscillation harmonics appear in the solution (see further Chapter 5 for more details). The nonlinear pendulum represents the evident example of so called *multi-harmonic periodic oscillations*.

Similarly, higher harmonics are generated really in FEL by pumping and signal oscillation electron phases, simultaneously (see Volume II, for instance). Besides that, by virtue of nonlinear nature of the oscillation process various *combination oscillations* (see definition of the combination phases (1.2.19), (1.2.20)) could be exited, too. In this case we get a complex case of the *multi-periodic multi-harmonic oscillations*.

3. BASIC CONCEPTS AND DEFINITIONS OF THE THEORY OF WAVES IN WEAK NONLINEAR SYSTEMS

3.1 Definition of the Wave

The definition of the wave is based on the concept discussed above of oscillation. Namely, *a wave is oscillations occurring in time and space simultaneously*. Let us show this by the simplest example of a sound transversal wave in a solid medium (see Fig. 1.3.1 and Fig. 1.3.2).

Under such an interaction neighboring material points oscillate too. Let us note that any perturbation caused by such interactions propagates in the medium with some finite velocities. Therefore, initial oscillation phases of different material points are found to be shifted in time (see curve 2 in Fig. 1.3.1).

Let us consider that we have to do with waves in some transversally unbounded homogeneous stationary non-dissipated elastic medium. In addition, we choose within this medium some small part of it which can be treated as the *material point* at coordinate $x = x_1$. In what follows, this material point starts to oscillate with the frequency $\omega = 2\pi/T$ (where T is the *temporal* oscillation period). These oscillations occur in the transverse plane under action of some external perturbation. The discussed state of system is illustrated by curve 1 in Fig. 1.3.1.

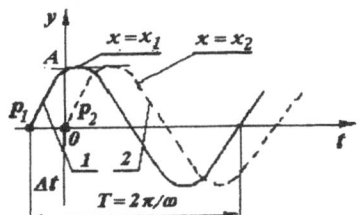

Figure 1.3.1. Illustration of oscillation process of two neighboring material points p_1 and p_2 (in some elastic medium) with spatial coordinates $x = x_1$ (curve 1) and $x = x_2$ (curve 2). $T = 2\pi/\omega$ is temporal wave period, ω is cyclic frequency, $\Delta t = t_1 - t_2$ is temporal lag of reproducibility wave processes at points x_1 and x_2.

Then we consider that all medium points always interact with one another. According to the above suppositions this interaction should be elastic.

We marked above already that initial phases of different point oscillations are shifted in time. Therein relative magnitude of this shift depends essentially on each point position in each instant of time. Hence all oscillations of other points occur with some '*time lag*'. This lag is larger the more remote the point is. The physical picture discussed is illustrated in Fig. 1.3.2. It shows explicitly the lag of oscillations of the second material point p_2, placed at $x = x_2$.

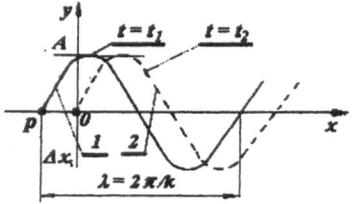

Figure 1.3.2. Instantaneous pictures' of the wave process at moments $t = t_1$ (curve 1) and $t = t_2$ (curve 2) Δx is spatial displacement of the given point p on wave front on time $\Delta t = t_1 - t_2$, $\lambda = 2\pi/k$ is spatial *wave period* (the length of the wave), k is wave number.

Thus oscillations of different medium points have different initial oscillation phases because of the *lag-effect*. Therefore, some spatial distribution of elementary oscillators (i.e., the medium points) over their initial oscillation phases occurs. Let us photograph mentally this distribution at two different time instant t_1 and t_2. We can see this spatial distribution has an explicit oscillatory nature (see Fig. 1.3.2). However, contrary to the situation depicted in Fig. 1.3.1, these oscillations have an explicitly expressed spatial nature. Let us remember that the time instants t_1 and t_2 are chosen arbitrary. This means the spatial oscillation process is realized simultaneously with the temporary oscillations. So the wave indeed is a superposition of two simultaneous oscillation processes, one of which occurs in time, and the other is realized in space.

The spatial oscillation period $\lambda = 2\pi/k$ is called the *wavelength*. The value k is named the *wave number*. Therefore the definition for the *wave amplitude* does not distinguish from the above given definition for the oscillation amplitude (see formulas (1.2.26)–(1.2.29) and corresponding commentaries).

3.2   The Phase and Group Wave Velocities

Thus between oscillations in two spatial points x_1 and x_2 these is a time lag $\Delta t = t_1 - t_2$. Consequently the propagation velocity of interactions between material points can be determined in the following manner

$$v = \Delta x / \Delta t \qquad (1.3.1)$$

with all notations given above. However, it should be mentioned that the notation v, which is determined by such a method, describes, in the general case, two different physical values. The first of them is the motion velocity of a fixed phase point on the wave front. And the second is the velocity of energy transfer by the wave. (Let us recall which the *wave front* is a surface in space that is formed by points with equal oscillation phases). Both these values in the simplest case considered coincide. However, these velocities can be different in the case of some disperse anisotropy arbitrary medium. They are called the *phase* and *group velocities*, respectively. We must distinguish these concepts in the general case.

The following rigorous definition for the phase velocity is accepted:

$$\vec{v}_{ph} = \omega \left(\frac{1}{k_x} \vec{e}_x + \frac{1}{k_y} \vec{e}_y + \frac{1}{k_z} \vec{e}_z \right), \qquad (1.3.2)$$

where $\vec{k} = (k_x, k_y, k_z) = \vec{k}_0 k = k_x \vec{e}_x + k_y \vec{e}_y + k_z \vec{e}_z$ is the *wave vector*, $\vec{e}_{x,y,z}$ are the unit vectors along the axes x, y, and z, respectively; $\left| \vec{k}_0 \right| = 1$, $k = \sqrt{k_x^2 + k_y^2 + k_z^2}$ is the wave number, $k_{x,y,z}$ are the corresponding modules of the vector $\vec{k}$ components. The group velocity $\vec{v}_{gr}$ could be defined as:

$$\vec{v}_{gr} = \frac{d\omega}{d\vec{k}}. \qquad (1.3.3)$$

Let us note that the directions of vectors of the phase and group velocities coincide only in the simplest case of isotropic medium. In the general case these vectors are directed arbitrary. Both these cases can occur in the above discussed FEL model. For instance, the phase and group velocities of the signal wave in traditional FEL H-ubitron arrangements, as a rule, are coincided [2–4]. However, some other rather exotic physical situations could be realized in some Dopplertron type of FELs. Their characteristic design feature is the use intense electromagnetic waves (including microwaves) as FEL pumping [1]. Therefore the situations with opposite directions of the phase and group velocities can occur really [16, 24, 25]. The physical models of such type are discussed in Chapters 10–13, Volume II.

3.3   The Phase of a Wave

Thus each j-th oscillation point in the considered model medium can be characterized by proper oscillation phase (see Fig. 1.3.1)

$$p_j = \omega t + \varphi_0, \tag{1.3.4}$$

Therefore each next point p_{j+1} oscillates owing to the *lag-effect* with the shifted phase:

$$p_{j+1} = \omega \left(t - \Delta x / v\right) + \varphi_0, \tag{1.3.5}$$

where all designations are given above. However, we should member that choice of the j-th observation point is arbitrary. Hence, it could be written for any oscillating point of the medium

$$p = \omega \left(t - x/v\right) + \varphi_0 = \omega t - kx + \varphi_0, \tag{1.3.6}$$

where $k = \omega/\nu = 2\pi/\lambda$, as before, is the *wave number*, λ is the *wavelength* (see Fig. 1.3.2).

Taking into consideration the definitions for phase velocity (1.3.2), we can generalize the definition for the wave phase (1.3.6) in the case of an arbitrary wave process:

$$p = \omega t - \vec{k}\vec{r} + \varphi_0, \tag{1.3.7}$$

where $\vec{r} = (x, y, z)$ is the radius-vector of observation point in the three-dimensional space.

3.4   Transverse and Longitudinal Waves

We discussed earlier the examples in which the plane of particle oscillations in a medium is normal to the direction of wave propagation. The waves of such a type are called *transversal waves*. In the opposite situation (i.e., when the material points oscillate along the direction of wave propagation) we have *longitudinal waves*. It should be mentioned that rather exotic wave types with mixed electrodynamic structure can exist in some complex electrodynamic systems. They can not be classified as either purely transversal waves or longitudinal ones. For example, the L *waves* in some type of retarding electrodynamic systems can be excited. The other example: the TM or TE *waves* in waveguides, etc.. The electromagnetic signal waves in FELs can, as a rule, be classified as the transverse type waves, whereas the pumping wave in an FEL Dopplertron can be purely transversal (laser radiation [26], for instance) as well as complex mixed type of waves [1–4]. *Space charged (plasma) waves* can be excited within the working bulk of relativistic electron beams. Amongst these, both transversal and longitudinal waves can be found. Corresponding examples of interaction of transverse and longitudinal waves are discussed in Chapters 12, 13, Volume II.

3.5   Surface and Volumetric Waves

Surface and *volumetric waves* are distinguished by their physical nature. Surface waves are characterized by surface physical mechanisms of excitation. The total wave energy in this case is localized within nearest to the *surface* layer of medium. The characteristic thickness of such a layer roughly equals the length of the wave. Usually waves on water are evidently surface waves. Volumetric waves propagate through volumes whose characteristic size can essentially exceed the wavelength. The light and sound waves serve as other obvious examples of volumetric waves.

3.6   The Concept of Dispersion

Dispersion is characterized by presence of a dependency of the phase wave velocity on frequency:

$$\vec{v}_{ph} = \vec{v}_{ph}\left(\omega\right), \qquad (1.3.8)$$

Taking this into account we can rewrite the definition (1.3.2) in more general form

$$k_{x,y,z} = \frac{\omega}{v_{x,y,z}(\omega)}. \qquad (1.3.9)$$

Relationships like to (1.3.9) are called the *dispersion relations* or the *dispersion laws*. Solving the corresponding *dispersion equation*

$$D\left(\omega, \vec{k}\right) = 0, \qquad (1.3.10)$$

can obtain them. Here $D\left(\omega, \vec{k}\right)$ is the *dispersion function*. The waves whose dispersion law can be found from dispersion equations like to (1.3.9) are defined as the *proper waves* of the considered system. For instance the light waves are proper waves for vacuum, as an electrodynamic system, and so on. The opposite relation

$$D\left(\omega, \vec{k}\right) \neq 0 \qquad (1.3.11)$$

is known as the condition of *improper waves'* existence. The stimulated electron waves in plasmas of electronic beam or H-ubitron (magneto-undulated) pumping field in FELs are examples of improper waves (see numerous illustration examples in Chapters 12, 13, Volume II).

3.7   Waves with Negative, Zero and Positive Energy

Performing corresponding calculational for the considered model of sound wave in an elastic medium, we can obtain the well known expression for the *wave energy density* [27]:

$$u = \frac{dE}{dV} = \frac{\rho A^2 \omega^2}{2}, \qquad (1.3.12)$$

where dE is the wave energy in an elementary volume dV; ρ is the medium density; A and ω are the wave amplitude and the frequency, respectively. It seems at first sight that the wave energy u should be a quite positive-determined quantity, because ρ, $A^2, \omega^2 > 0$. As is widely known, the energy describes the ability of bodies to work. Therefore the existing the '*negative energy*' or the '*zeroth energy*' seems to be impossible. However, the concepts 'the *wave with negative energy*' and 'the *wave with zeroth energy*' are often used in the modern physics, too. What is the matter? Let us shortly discus this problem in more detail.

As a simplest analysis shown, the mentioned seeming contradiction carries purely terminological nature. The point is that, talking about the wave energy we really always take in view the *energy difference* (!). It is the difference between the medium energy with a wave and the medium energy without the wave. Hence, strictly speaking, we should write for the wave density in the general case the following more precise definition instead (1.3.12):

$$u = \frac{d\left(E_{\text{with}} - E_{\text{witout}}\right)}{dV}. \qquad (1.3.13)$$

It can easily be seen that the sign of the energy density u (and the wave energy, too) depends on the correlation of magnitudes of the two different energies E_{with} and E_{without} in (1.3.13). In particular, we have the 'usual case' of *positive wave energy* in the case

$$E_{\text{with}} > E_{\text{without}} \qquad (1.3.14)$$

The opposite case describes the system with the negative wave energy. If both energies are equal we have to do with the zeroth wave energy. All these three types of waves are used further in the book. For instance, the electromagnetic signal wave in FELs is characterized by the positive energy always. The slow space charged wave (SCW) in electron beam plasmas in parametric Raman's FELs is the wave with negative energy. In addition, at last, an example of a wave with zeroth energy is shown by increased and dissipated waves in the beam models with two-stream instability. In more detail, all these examples are discussed in Chapters 12, 13, Volume II.

4. OTHER BASIC DEFINITIONS AND EQUATIONS OF PHYSICS

A number of physical concepts and definitions have been used above. All of them will be utilized in the following Chapters. However the author speaks in view the situation, when a reader has not required knowledge in the field of general physics. So, let us add in what follows the portion of widely known physical information. Here a number of concepts, definitions, and most popular formulae necessary in the future are given (see also [27] or any suitable textbook fore more details).

4.1   Linear Velocity and Linear Acceleration

Let us start with the mechanics of material point. The *(linear) instantaneous velocity* is the value

$$\vec{v} = d\vec{r}/dt, \tag{1.4.1}$$

where $\vec{r}$ is the radius-vector of the material point. The *linear instantaneous acceleration* could be determined as

$$\vec{a} = d\vec{v}/dt \tag{1.4.2}$$

Both them describe the *kinematics* of material point in three-dimensional space.

4.2   Linear Momentum and Force

The concept of the *linear momentum*

$$\vec{p} = m\vec{v} \tag{1.4.3}$$

is related to the *dynamics*. In Newtonian dynamics the *force* is the only cause that determines the motion of a material point. The *second law of dynamics* describes this relation mathematically:

$$\vec{F} = d\vec{p}/dt, \tag{1.4.4}$$

where $\vec{F}$ is the *force*. In the nonrelativistic case the second law can be rewritten as

$$\vec{F} = m\vec{a}. \tag{1.4.5}$$

It should be mentioned that the discussed definitions for the *momen-tum* and the *force* are valid in the nonrelativistic only. Some other definitions should be given in the relativistic case.

4.3   Curvilinear Motion

In the case of curvilinear motion the concepts of the moment of force

$$\vec{M} = \left[\vec{r}\vec{F}\right],\qquad(1.4.6)$$

the moment of momentum

$$\vec{L} = [\vec{r}\vec{p}],\qquad(1.4.7)$$

and the moment of inertia of material point

$$I = mr^2\qquad(1.4.8)$$

are used to describe the dynamics of material point. Comparing expressions (1.4.6)–(1.4.8) with corresponding quantities for linear motion one can see that $\vec{M}$ is functional analog of the force $\vec{F}$, the moment of momentum $\vec{L}$ corresponds to the linear momentum $\vec{p}$, and the moment of inertia I is the 'curvilinear analog' of the mass of material point m. These concepts were used above to illustrate nonlinear pendulum (see (1.2.1)-(1.2.4)).

4.4   Rotation

The simplest case of curvilinear motion is rotation. In the rotation kinematics, besides the linear values, the angular values are introduced also. Including, the angular displacement is:

$$\vec{\varphi} = \varphi\vec{n},\qquad(1.4.9)$$

where φ is the angle of a turn of the radius-vector of material point $\vec{r}$, $\vec{n}$ is the unit vector normal to the plane of vector $\vec{r}$ rotation (see Fig. 1.2.1 and corresponding commentaries). The angular velocity and the angular acceleration (see also (1.2.1)–(1.2.4)) are defined for the linear motion as:

$$\vec{\omega} = \frac{d\vec{\varphi}}{dt};\quad \varepsilon = \frac{d\vec{\omega}}{dt}.\qquad(1.4.10)$$

The different scalar components of the vector of angular velocity $\vec{\omega} = \vec{\omega}\{\omega_x, \omega_y, \omega_z\}$ describe the different circular frequencies of rotations of the material point around the rotation axes. This displays the physical meaning some of algorithms, which are set forth further in Chapters 3, 4.

The correspondence between linear and angular quantities is determined by the relationships:

$$\vec{v} = \left[\vec{\omega}\vec{R}\right]; \ \vec{a} = a_n\vec{n} + a_\tau\vec{\tau}; \ a_n = v^2/R = \omega^2 R; \ a_\tau = dv/dt = \varepsilon R,$$

$$(1.4.11)$$

where $\vec{R} = -R\vec{n}$ is the vector along the rotation directed to the rotation center, $\vec{v} = v\vec{\tau}$, $\vec{n} \perp \vec{\tau}$, $\vec{n}$ and $\vec{\tau}$ are corresponding unit vectors. The generalization of the second law of dynamics onto rotation can be given in the form (1.2.1).

4.5 Energy. Field of Forces

The ability of the body to work

$$A = \int_{r_1}^{r_2} \vec{F} d\vec{r}$$

$$(1.4.12)$$

is called the *energy*. One distinguishes the *kinetic* and the *potential energy*. The kinetic energy characterizes the motion dynamics of a body. In general relativistic case it is

$$\mathcal{E}_k = \frac{mc^2}{\sqrt{1 - v^2/c^2}} - mc^2,$$

$$(1.4.13)$$

where all designations are given above. The value mc^2 is called the rest energy, m is the rest mass. In the case of nonrelativistic motion the expression (1.4.13) can be reduced to the following well known form:

$$\mathcal{E}_k = mv^2/2.$$

$$(1.4.14)$$

The definition of the potential energy is related to the definition of *field of forces*. The part of space for each point of which some vector $\vec{A}$ can be corresponded, is called the field of the vector $\vec{A}$.

The *second law of dynamics* (1.4.4) can be expressed in the form describing the motion of quasi-continuous flow of material points in the field of forces:

$$\frac{\partial \vec{p}}{\partial t} + \vec{v}\,\mathrm{grad}\,\vec{p} = \vec{F}\left(\vec{r}, \vec{p}, t\right).$$

$$(1.4.15)$$

You must distinguish the *potential* and *vortex* fields. The condition of field potentiality can be written as

$$\oint_L \vec{A}d\vec{l} = 0, \qquad (1.4.16)$$

where L is the integration contour, $d\vec{l}$ is the differential element of this contour. The potential field can be described by the *potential function* $U(\vec{r}, t)$. The force $\vec{F}$ in each point of the field can be expressed as the negative gradient of the function $U(\vec{r}, t)$:

$$\vec{F} = -\operatorname{grad} U(r, t). \qquad (1.4.17)$$

Below we will use both above mentioned field types. Therein the electric field can exist in the potential form (it can be generated by charged particles) as well as the vortex field (that is excited by a time-varying magnetic field — see Volume II for more details).

Thus the kinetic energy is related to the body motion, whereas the potential energy corresponds to the position of the body in a force field.

4.6   Motion Integrals

Characteristic mathematical peculiarity of any dynamical system is the possibility of constructing the so called *motion integrals*. These are such combination of dynamical parameters, which are conserved during the system motion process. It is known that $3N - 1$ motion integrals can be constructed for any closed system consisting of N material points. It is remarkable that only three from them are characterized by the *property of additivity*. These integrals are connected with the corresponding *conservation laws*. Including, the energy conservation law

$$\mathcal{E}_k + U(\vec{r}, t) = \sum_{n=1}^{N} \mathcal{E}_n = \mathcal{E} = \text{const}, \qquad (1.4.18)$$

the momentum conservation law

$$\vec{p} = \sum_{n=1}^{N} \vec{p}_n = \text{const}, \qquad (1.4.19)$$

and the moment of momentum conservation law

$$\vec{L} = \sum_{n=1}^{N} \vec{L}_n = \text{const}. \qquad (1.4.20)$$

All these motion integrals are related to the inherent properties of the space-time and they are common for all dynamical systems. Apart from the other particular motion integrals discussed, which could be specific for a studied system, can be obtained, too. In Chapter 5 and in Volume II we will widely use various motion integrals in analysis.

4.7   Canonical Variables. Hamiltonian Equations

Mechanics knows two versions of the problem of material point motion — Newtonian and Hamiltonian ones. Above we had to do with the first of them only. As was mentioned, the Newtonian formalism treats the force as the main motion cause that is represented by the second law of dynamics (1.4.4). In contrast to from the *Newtonian formalism*, the *Hamiltonian formalism* does not treat the force as a necessary basic concept. The motion of the material point in this case is completely described by the *Hamilton function*. The latter is the total energy written in terms of momentum and coordinates. The description by means of *canonical variables* is used here. Let us use the three-dimensional coordinate $\vec{r}$ and the canonical momentum $\vec{P}$ as the canonical variables. Hamilton function in this case can be written as

$$\mathcal{H} = \mathcal{E}\left(\vec{r}, \vec{P}, t\right).$$
(1.4.21)

The dynamics of a material point in an external field of force is governed by the *Hamiltonian equations*

$$\frac{d\mathcal{H}}{dt} = \frac{\partial \mathcal{H}}{\partial t}; \quad \frac{d\vec{P}}{dt} = -\frac{\partial \mathcal{H}}{\partial \vec{r}}; \quad \frac{d\vec{r}}{dt} = \frac{\partial \mathcal{H}}{\partial \vec{P}}.$$
(1.4.22)

As mentioned above, the dynamics of a system considered can be completely described by the equations (1.4.22). These equations are widely used further in the book (see, for instance, Chapter 5, and Chapters 9, 11–13 in Volume II.

4.8   Electromagnetic Field

The electric and magnetic fields are specific particular cases of the *electromagnetic field*. Dynamics of the latter is described by the Maxwell equations:

$$\left[\vec{\nabla}\vec{E}\right] = -\frac{1}{c}\frac{\partial \vec{B}}{\partial t}; \qquad \left[\vec{\nabla}\vec{H}\right] = \frac{1}{c}\frac{\partial \vec{D}}{\partial t} + \frac{4\pi}{c}\left(\vec{j}_0 + \vec{j}\right);$$

$$\left(\vec{\nabla}\vec{D}\right) = 4\pi\left(\rho_0 + \rho\right); \qquad \left(\vec{\nabla}\vec{B}\right) = 0,$$
(1.4.23)

where $\vec{E}$ and $\vec{D}$ are the vectors of intensity (strength) and displacement (induction) of the electric field, $\vec{H}$ and $\vec{B}$ are the vectors of strength and induction of the magnetic field; $\vec{j}_0$ and $\vec{j}$ are the current density vectors resulting from external and intrinsic sources of fields, ρ_0 and ρ are the space charge densities caused by similar sources, $\vec{\nabla}$ is the nabla operator.

The *intensity vector* of electric field is force that acts on some positive unit charge $(+q)$ in a given point of the electric field:

$$\vec{E} = \vec{F}/(+q). \qquad (1.4.24)$$

As mentioned earlier, there are two different types of the electric fields. They are the *scalar (potential)* and the *vortex* electric fields. As follows from equations (1.4.23), electric charges create the scalar electric fields, whereas a time-varying magnetic field creates the vortex electric field. The vector of electric displacement $\vec{D}$ is connected with the intensity vector $\vec{E}$ by the following relationship:

$$\vec{D} = \varepsilon\vec{E} \qquad (1.4.25)$$

where ε is the *dielectric permittivity* of medium. It is the auxiliary value that facilitates description of the electric field within *dielectrics* (*magnetodielectrics*).

Electric current (i.e., directed motion of charges) creates magnetic field. Analogously to situation with the vector $\vec{E}$, the vector of magnetic induction $\vec{B}$ is the force characteristic of the magnetic field. The intensity of magnetic field $\vec{H}$ is an auxiliary quantity to describe magnetic field in substance

$$\vec{H} = \mu^{-1}\vec{B}, \qquad (1.4.26)$$

where μ is the *magnetic permeability* of the medium.

4.9   Lagrange and Euler Variables

Electrodynamic plasma-like systems are, as a rule, described in terms that are usually accepted in the hydrodynamics and aerodynamics. For instance, intensive electron beams are treated as flows of charged fluids or gases. The *Lagrange* or *Euler variables* are used for such method of description.

In the first case, a researcher considers an *individual particle* (electron or compact aggregation of electrons — *large particle*) and fixes its position $\vec{r}(t)$ in any time instant t. Correspondingly, phases of the electron oscillations $\psi(\vec{r}(t), t) = \psi(t)$ (see above Subsection 1.2.2) expressed via the Lagrange coordinates $\vec{r}(t)$ are referred to as *Lagrange oscillation phases*. Analogously, we can determine the linear Lagrange velocity $\vec{v}(t)$, angular velocity $\vec{\omega}(t)$, etc..

In the second approach the researcher considers a spatial point with the radius-vector $\vec{r}$ and it registers the velocities $\vec{v}$ of electrons passing through this point in each time instant t. Therefore, each point of the beam volume is put in relation to a certain point of the field of velocities $\vec{v}\,(\vec{r}, t)$. The velocity function $\vec{v}\,(\vec{r}, t)$ is an *Euler variable*, as well as the phase $\psi\,(\vec{r}, t)$, the angular velocity $\vec{\omega}\,(\vec{r}, t)$, etc., could be regarded also as the Euler variables.

Thus the Lagrange formalism is related with the *one-particle description* of the system, whilst Euler formalism is associated with its *many-particle description*. This distinction will help us to avoid misunderstanding in the analysis of the nature of quantities to be considered in what follows. In the account below, we employ widely both: the Lagrange as well as Euler approaches.

5.   CONCEPT OF SYSTEM

Most of the objects investigated in physics and engineering are various types of systems. Therefore, the knowledge of general features of the systems allows to better understand the problems studied. It concerns both: the mathematical description of the problem, as well as the analysis of results obtained. Apart from that, peculiarities of some systems (for instance, the *hierarchical dynamical systems*) allow us to construct corresponding highly effective asymptotic calculational methods. In other words, the systems properties can be useful in solving the mathematical part of the problem, too. Lastly, the set of concepts and ideas considered in the general system theory is very interesting in itself. So let us discuss some general features of dynamical systems.

5.1   What is a System?

It seems incredible that until today there is no precise and clear the *system* definition. To illustrate this we use one of the best definitions known in the literature: " We call *the system* an object of any nature (or an aggregate of interacting objects of any nature, including objects of different nature) that has explicit *'system' property (properties)*, that does not belong to any individual part of the system for any method of its decomposition, and that does not follow from properties of system parts" [28].

Thus a specific feature of a system is that it has principally new properties what do not reduce to those of any of its parts. For instance, air has properties (wind, tornado, etc.) what its molecules do not have at all. Another example, two young people decide to form their own family. The problem arises inevitably: which of them should be the head of the family? It is obvious that before the wedding this problem does not appear. This example describes also the peculiar property of the so called *hierarchical systems* to be discussed in the next Chapter. Here in this Section we confine ourselves by the most general properties of systems (including hierarchical ones). Therein we discuss the *dynamical systems* most, keeping in mind that dynamical systems include *static systems* as a particular case.

Let us note that separate parts of the system could have proper system properties differing from those of the system as a whole. These parts are called the *subsystems*. The association of systems is called the *supersystem*. Hierarchical systems are a specific variety of supersystems where each hierarchical level consists of some subsystems.

Any system, supersystem, or subsystem has *inputs* and *outputs*. Besides that, any object of such a type is in some *surrounding medium*. The system communicates with the surrounding medium by its inputs and outputs. In the general case the communicating with the medium is driven by energy, matter, information, etc..

Hence, the concept of system is relative. Each subsystem can be regarded as a system (with respect to any its 'subsubsystem') or, equivalently, each supersystem can be considered as a subsystem of corresponding 'supersupersystem'. As is shown in the following Chapters, this system property is characteristic for the hierarchical systems.

What is the *medium* from the 'system point of view'? The medium in the system theory is some external surrounding the system interacts with. Therefore, other parts of the system serve as corresponding medium for each subsystem.

The *open* and *closed systems* are distinguished. The open system has an exchange with the surrounding medium, whereas in the closed system there isn't any exchange.

5.2   State of the System

The ordered aggregate of system parameters that determine the evolving in the system is called the *state of the system*. For instance, the state of some thermodynamically equilibrium gas is completely determined by three parameters: gas pressure p, temperature T, and volume V.

5.3   General Classification of the Systems

All systems can be divided into four large classes:

a) the *determined systems*, where all possible processes are determined within the interval $[0, \tau]$ (where τ is the time of the system life);

b) the *stochastic systems*, where the behavior of its elements has essentially stochastic character);

c) the *chaotic systems*, in which there is dynamical chaos (for instance, the motion of a dense plasma beam in the case of essential collisions);

d) the *complex systems*, i.e. the systems characterized by the essential complexity (the complexity concept is given below in this Chapter). The distinguishing feature of the complex system is that it consists of a large number of structural elements.

The nature of the first three types of systems is rather obvious. However, it is not so evident essence of the complex systems. The latter are different in principle from others and they have a number of interesting and unusual peculiarities. Our Universe, human society, the human organism, plasmas and plasma-like systems, and many other similar objects are the most vivid examples of the complex systems.

5.4 Some General Properties of Complex Systems

The most important properties of complex systems for understanding their nature are:

1. The *uniqueness*. Each complex system exists in one copy only or else it is a very rare object. For instance, we know only one Universe, only one human society, etc.. Any doctor can say that every human organism is unique and exists in Nature only in one copy. Each husband can confirm that his wife and his mother in law, as complex systems, are unique objects and each of them exists in Nature in one copy only, too.

2. The *weak predictability*. One can have as much as he like detail information about the system elements or he may know its behavior at some interval $(-T, 0]$. However, this does not allow him to recognize the *exact* behavior of the complex system, as a whole, in the later interval $(0, \tau]$. The examples of the Universe, human society, a human organism, a wife and a mother in law illustrate this property rather evidently.

3. The *negentropyness* or the *purposefulness*. A complex system can control (in some range) its own entropy (or negentropy) for any random influences of the external medium. As is known (see below this subsection) the entropy characterizes the chaos within the system. Correspondingly, the negentropy is the measure of order within the same system. In the dynamical sense, the negentropy dynamics characterizes an aspiration of the system to achieve some purpose. The purposefulness expresses the analogous thing; namely, the ability of the complex system to conserve and to amplify some main dynamical process leading to this purpose. Therefore both these concepts are very close.

As is demonstrated later, there are such types of processes in vacuum electronics. For instance, the cooling effect of particle beams within the EH-accelerators (see Chapter 9, Volume II) [16, 29–31]. A corpuscular beam, as a complex system, behaves in this case in such a manner that the beams 'warmth' is taken away from the system. Here, it seems that we have a violation of the second law of thermodynamics. However, it is not so, because any accelerator (including EH-ubitron one) is thermodynamically an open system. It means that the required energy for the 'cooling' procedure is taken from the surrounding medium (in the present case from external source).

Other similar examples are the effects of phase and polarization discrimination (see Chapter 12, Volume II) [31–33]. In the first case the initial slowly varying oscillation phases of electromagnetic waves always strive for the same magnitudes during the nonlinear resonance interaction, irrespective of their initial magnitudes. In the case of polarization discrimination the wave polarization behavior is characterized by the analogous trend.

6.   BASIC POSTULATES AND PRINCIPLES OF THE GENERAL SYSTEM THEORY

Systems in physics and engineering are rather specific objects of the surrounding world and this specificity is reflected by some general principles. We point out three such principles, namely:

a) the *principle of physicality*;

b) the *ability to model principle*;

c) the *purposefulness principle*.

6.1   Principle of Physicality

According to this principle, processes in systems are governed by the principles of physics. It means that in the system theory any observed effect always has its proper cause (that, however, can be unknown). Hence, application of the known physical principles is quite sufficient to explain any phenomena in systems. In other words, the idea of the existence in Nature of some unknown nonphysical 'mystical power' does not find any place in this field of human activity. It should be mentioned, however, that the latter affirmation does not deny the existence of God. It says only that God should be considered as a real physical object, whose activity does not contradict the basic physical principle. Hence, the concept of God might be formulated in terms of physics. Such an attempt is undertaken in the next Chapter, in particular.

In turn, the principle of physicality includes a few basic postulates [28]. Let us consider the most important of them.

6.2   Integrity postulate

Integrity postulate: any *complex system* should be regarded as a single system entire. This postulate in the case of complex systems is based on the existence of the above mentioned 'system property'. Hence, any decomposition of the system violates the system integrity, inasmuch as such a decomposition leads to the vanishing of this property. Let us illustrate this thought by the following example. Let there be some volume of a gas with a number of molecules $N \gg 1$, so this object could be classified as a complex system. We divide this volume into N separate parts. It is supposed that there is only one molecule within each such partial volume. It is obvious, that the integrity of the system, as a whole, is violated as the result of decomposition. Indeed, any separate molecule has no collective properties of the gas. In principle we can decompose a given gas volume into j parts by another method, but the condition $N_j \gg 1$, therein, is satisfied for each j-th partial separate volume (N_j is the number of molecules in a j-th part). In this case each partial j-th gas volume can be represented as some subsystem of many molecules. However, the specific 'whole system' property of each such subsystem differs from the system collective properties, as a whole.

Thus the system as a whole can never be considered as an exact equivalent of any of its partial parts (with any of its subsystems). This postulate can also be illustrated by the so called hierarchical systems. In this case each level of the hierarchical system (i.e. each partial hierarchical subsystem), on the one hand, has a resemblance to the whole system (the hierarchical resemblance principle — see below in the next Chapter for details). On the other hand, this resemblance does not have the total scale system nature and has a particular specific character.

6.3   Autonomy Postulate

As a rule, every type of physical phenomenon corresponds to a certain transformation group. Such a group generates a proper geometry for the space problem. Therefore the general theory of systems also allows a geometrical interpretation. One postulates that there is some autonomy (with respect to a given system) type of geometry describing a complex system. This geometry should be an invariant for any system decomposition. Each part of the system is described by the same geometry which the system has as a whole. The latter phrase expresses the *autonomy postulate*.

6.4   Principle of Ability to Model

A finite number of various models can describe any given complex system. Each model can reflect certain peculiarity of the system only. In other words, each complex system can be modeled by various methods. In the general theory of systems this statement is called the *principle of ability to model*. A number of consequences follow from it. The most important is that each particular group of properties of some complex system can be studied by specific theoretical model.

Let us mention that two types of such models can be constructed. They are particular models and the total one. Particular models can describe a part of the system properties only. However, the particular models only have a practical significance. Turing's well known theorem explains this paradox. The point is that any total model is as complex for studying as the system itself. Therefore only partial simplified models can be used for practical purposes really. However, let us recall that simplifying the model we, at the same time, simplify the system analysis by loosing part of information about true system properties. So the choice sing of the theoretical model is always connected with a search for some compromise between its completeness, on the one hand, and quantity of the lost information concerning its real properties.

The other system properties can be formulated as the following postulates.

6.5   Complementarity Postulate

Niels Bohr formulated the *complementarity postulate* in quantum mechanics. However, its application in the system theory shows generality of its nature. The essence of the complementary postulate is the following: a *complex system* reacts differently revealing its various properties under the action of different external influences. In the general case, these reactions demonstrate the system properties which can even exclude each other. For instance, an electron has corpuscular properties (photo-effect, Compton effect, etc.) in certain cases and it behaves like a wave (diffraction of electron on a crystal lattice, tunnel effect, etc.) in other situations. Hence the electron is a particle and a wave simultaneously. Such a situation seems impossible from the point of view of 'ordinary common sense'. However, physical experiments prove this is true. The unusualness of the discussed situation becomes clear if we take into account the property of ability to model the complex systems.

We can illustrate such a physical picture by the following virtual experiment. Let us suppose that the object studied (for instance, the marble statue of your supervisor) is in a box. Therefore we have a possibility of observing the system behavior through only a few small apertures which are placed on different sides of the box. It is obviously that the observed pictures which we can see through different apertures can be essentially different. Each such picture (i.e., the supervisor projections) demonstrates some separate group of the object properties only. Therefore each such group can be described by some 'proper' theoretical (i.e., particular) model. Thus in spite of the initial single object we obtain, eventually, a number of theoretical models describing different properties of the same model. It is obviously that some of these properties can, in principle, exclude each other.

6.6   Action Postulate

All known complex systems of people's everyday life and different fields of science demonstrate one interesting peculiarity which can be formulated in the form of an *action postulate*. Accordingly with this postulate, any complex system reacts to an external action in the *threshold* way. There are a number of examples in physics and engineering that illustrate this system property. For instance, an electron transition within an atom occurs only when the energy of perturbed photon exceeds some threshold value (experiments by Frank and Hertz, the short-wavelength limit of breaking radiation, the red limit of photoeffect, etc.). In this book the efficacy of this postulate is demonstrated by examples of the threshold effects realized in FELs and EH-accelerators (see Chapters 9–13, Volume II).

6.7   Uncertainty Postulate

This postulate was originally introduced in quantum mechanics (*Heisenberg's uncertainty principle*). Practice shows that this principle holds for macroscopic complex systems too. It is known here as the *uncertainty postulate*. Let us point out once more that it is applicable to both: quantum and classical complex systems. Its physical nature can be clarified by the known 'device problem'. The point is that we have a possibility to obtain any information about a system only through a corresponding measuring. However, any measuring process changes the state of the treated system. Therefore, the error of measuring is always finite. Hence, the measured state always differs from the 'true' one, i.e., the strictly 'true' behavior is uncertain, in principle. In the case of a strong enough (with respect to corresponding characteristic scale) external 'device' influence, its true dynamics can be essentially different from the measured one.

6.8   Purposefulness Principle

In the general system theory the purposefulness is a functional trend to achieve corresponding state of the complex system or to conserve (or to amplify) some system process. The sense of the *purposefulness principle* is that complex system has specific purposefulness behavior. It is the most incomprehensible and intriguing feature of real complex systems, because neither modern physics nor engineering contains any grounds for it. However, in practice, the purposefulness is always a result of evolution of the whole system properties. Obvious examples are the Earth biosphere, human society, the Universe, etc.. Realization of the discussed principle in electrodynamic systems is presented by the cooling effect in EH-accelerators (Chapter 9, Volume II), and the effects of phase and polarization discrimination in FELs (Chapter 12, Volume II).

6.9   Entropy and Negentropy

As has already been mentioned , entropy determines the degree of chaos in the system. Negentropy (negative entropy) is a measure of order in the system. Three types of entropy are distinguished: physical entropy, thermodynamic entropy, and informational entropy [10].

Let us define the thermodynamic and physical entropies in the usual way accepted in general physics. We assume simplest system consisting of two identical particles moving within the volume V (see Fig. 1.6.1). The state of volume V with both particles characterizes the *microstate* of discussed system. The system shown in Fig. 1.6.1 is characterized by only one microstate.

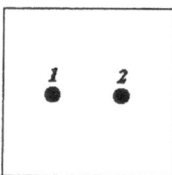

Figure 1.6.1. The example of the simplest system, which consists of only two identical particles with number 1 and 2.

Then let us assume the volume V is divided into two parts (see Fig. 1.6.2). The particles can take their positions within identical volumes V_1 and V_2 in four different ways. However, only two of them really are different. Inasmuch as both particles (and the volumes $V_{1,2}$) are identical, the states of the system, as a whole, described by drawings a) and d), should be considered equivalent. The same situation takes place with the states shown in drawings b) and c). Therefore the system shown in Fig. 1.6.2 can have two different microstates. Let us continue this imaginary experiment with the increasing number of particles. Doing this, we can obtain some generalization for the case of system consisting of N interacting particles and Λ partial volumes. Doing this we obtain the expression for calculational of the total numbers of microstates W:

$$W = \frac{N!}{\prod\limits_{i=1}^{\Lambda} N_i!}. \tag{1.6.1}$$

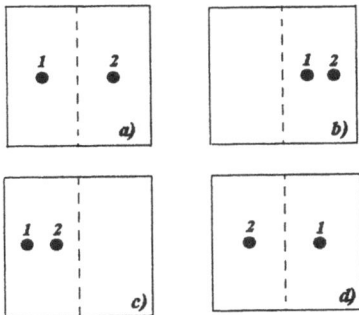

Figure 1.6.2. The illustration of the concept of 'number of microstates of the system'. It can easily be seen that the system can basically have only two different microstates (it is considered that all particles have the same properties, i.e. in this case only a number of particles in each subvolume $V_{1,2}$, is important).

The number of system microstates W (1.6.1) is called the *thermodynamic probability*. It, in contrast to from the 'usual' one, is characterized by the following inequality: $W > 1$. As follows from (1.6.1), for the system consisting of K subsystems, the total thermodynamic probability of the system as a whole can be given as

$$W = \prod\limits_{j=1}^{K} W_j, \tag{1.6.2}$$

where W_j is the partial thermodynamic probability characterizing j-th subsystem. The multiplication in (1.6.2) of partial thermodynamic probabilities W_j is very inconvenient for real calculations. However, it is widely known that the logarithmic function transforms the multiplication procedure into addition. Hence using the logarithm we can introduce instead the 'multiplied' thermodynamic probability W the logarithmic (thermodynamic) entropy S:

$$S = k \ln W, \tag{1.6.3}$$

where k is the Boltzmann constant. It is evident that, in contrast to the thermodynamic probability W, the thermodynamic entropy S is more convenient for practical use:

$$k \ln \prod_{j=1}^{K} W_j = k \sum_{j=1}^{K} \ln W_j = \sum_{j=1}^{K} S_j. \tag{1.6.4}$$

In what follows we consider the concept of *information entropy*. Let there be some system behaving as a source of information. We assume this source is discrete with respect to the variety of Λ possible steady states. Each state has (usual!) probability of realization P_j. The uncertainty in prediction of the system state, as a whole, in a certain time can be given as [10]:

$$S_{\text{inf}} = - \sum_{i=1}^{\Lambda} P_j \log_2 P_i, \tag{1.6.5}$$

where S_{inf} is the information entropy. It attains the maximum:

$$(S_{\text{inf}})_{\max} = \log_2 \Lambda. \tag{1.6.6}$$

It can be shown (see, for instance, [10]) that both entropies are related by some constant factor. This means that, in principle, both these concepts could be considered to be equivalent in the system theory. Let us note, however, that the information entropy characterizes more general property of the surrounding world than the thermodynamic or physical ones. The concept of information entropy will be used further in Chapter 2 to discuss peculiarities of hierarchical systems.

Three fundamental laws form the basis of thermodynamics. The first is the generalization of the energy conservation law for thermodynamic systems (the *first law of thermodynamics*). The two other laws are specific characteristics of thermodynamic system property. There are known seven classical equivalent formulations of the *second law of thermodynamics*. For instance, the formulation of Clausius is: the entropy of a closed system in thermodynamic equilibrium can only increase or remain unchanged. According to the *third law of thermodynamics* (*Nernst's theorem*), the entropy tends to a finite limit, vanishing for zero system temperature.

It should be mentioned that the applicability of entropy to describe complex systems is an open question in modern literature (see, for instance [28]). This is explained by the discussed above peculiarities of complex systems. The author's position coincides with the position given in [10].

6.10   Complexity

Complexity is rather clear intuitively, however it is relatively difficult for the formal description. We distinguish *structural and algorithmic complexity*. The first one characterizes the number of structural elements of a system. The second one is related to the degree of development of reciprocal connections between them. However, both concepts are closely connected.

The structural complexity can be illustrated by examples of the theory of electrical circuits. For instance, let us assume an electrical circuit characterized by m inputs and m' outputs. The dynamics of the system can be given by some multi-terminal network (see Fig. 1.6.3).

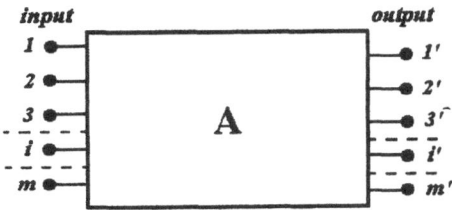

Figure 1.6.3. Illustration of the concept of 'multi-terminal network'. In general, the number of input poles m is not equal to the number of output poles m.

We assume that the currents I_i flow into the system through the system inputs, and the currents $I_{i'}$ flow through the outputs. There are some interactions between currents of both types within the system volume. Therefore each current I_i can correlate (in the general case) with any current $I_{i'}$:

$$I = \|A\| \, I', \qquad (1.6.7)$$

where I and I' are the vectors with currents I_i and $I_{i'}$ as components; $\|A\|$ is the *matrix of transformation*.

It is obvious that all information about the structural complexity of the system considered is contained in elements of the transformation matrix $a_{ii'}$. In the simple case when all elements $a_{ii'}$ ($a_{ii'} = 0$, $i \neq i'$) are vanishing, besides the diagonal elements $a_{ii'}$ ($i = i'$), each system input is connected with one corresponding output only. In the limit case $a_{ii'} = 1$ the system transforms into a system of noninteracting ideal conductors. So the structural complexity is minimum in this situation.

As can be shown, the algorithm describing such system turns out to be the shortest one. If all matrix elements are random and all $a_{ii'} \neq 0$ the structural complexity is maximum. It is obvious that the corresponding algorithm describing this model should be much longer. The *minimum length of corresponding algorithm, describing the system, characterizes its algorithmic complexity*. The algorithmic complexity is used most often in informatics and cybernetics, [10]. Below, in Chapter 2, talking about the complexity, we will have in the mind the algorithmic complexity.

References

[1] R.A. Silin, V.V. Kulish, Ju.I. Klymenko. Electronic device. Inventions Bulletin, 15 may 1991. Soviet Inventors Certificate, SU No. 705914, priority of 18.05.1972, Published in non-secret USSR press after removal of relevant stamp of secrecy:.

[2] T.C. Marshall. *Free electron laser*. MacMillan, New York, London, 1985.

[3] C. Brau. *Free electron laser*. Academic Press, Boston, 1990.

[4] P. Luchini, U. Motz. *Undulators and free electron lasers*. Clarendon Press, Oxford, 1990.

[5] F.B. Hildebrand. *Introduction to numerical analysis*. Dover Publishers, New York, 1987.

[6] D. Kahaner, C. Moler, S. Nash. *Numerical Methods and Software*. Prentice-Hall International, Inc, 1989.

[7] E.A. Grebennikov. *Averaging method in applied problems*. Nauka, Moscow, 1986.

[8] N.N. Bogolubov, Mitropolskii. *Methods of averaging in the theory of nonlinear oscillations*. Publising House Academy of Science of USSR, Moscow, 1963.

[9] N.N. Moiseev. *Asymptotic methods of nonlinear mechanics*. Nauka, Moscow, 1981.

[10] J.S. Nicolis. *Dynamics of Hierarchical Systems. An Evolutionary Approach*. Springer-Verlag, Berlin, Heidelberg, New York, Tokyo, 1986.

[11] R.K. Dodd, J.C. Eilbeck, J.D. Gibbon, H.C. Morris. *Solutions and nonlinear wave equations*. Academic Press, London, 1982.

[12] M.M. Khapaev. *Asymptotic methods and equilibrium in theory of nonlinear oscillations*. Vysshaja shkola, Moscow, 1988.

[13] L.D. Landau, E.M. Lifshitz. *Hydrodynamics*, volume 6 of *Theoretical Physics*. Nauka, Moscow, 1986.

[14] A.N. Kondratenko, V.M. Kuklin. *Principles of plasma electronics*. Energoatomizdat, Moscow, 1988.

[15] V.V. Kulish, S.A. Kuleshov, A.V. Lysenko. Nonlinear self-consistent theory of superheterodyne and parametrical free electron lasers. *The International Journal of Infrared and Millimeter Waves*, 14(3):451–560, 1993.

[16] V.V. Kulish. Hierarchical oscillations and averaging methods in nonlinear problems of relativistic electronics. *The International Journal of Infrared and Millimeter Waves*, 18(5):1053–1117, 1997.

[17] V.V. Kulish. *Methods of averaging in nonlinear problems of relativisticelectrodynamics*. World Federation Publishers, Atlanta, 1998.

[18] V.V. Kulish, A.V. Lysenko. Method of averaged kinetic equation and its use in the nonlinear problems of plasma electrodynamics. *Fizika Plazmy*, 19(2):216–227, 1993.

[19] A.P. Sukhorukov. *Nonlinear wave-interactions in optics and radiophysics*. Nauka, Moscow, 1988.

[20] N. Bloembergen. *Nonlinear optics*. Benjamin, New York, 1965.

[21] J. Weiland, H. Wilhelmsson. *Coherent nonlinear interactions of waves in plasmas*. Pergamon Press, Oxford, 1977.

[22] S.S. Kohmanski, V.V. Kulish. To the nonlinear theory of free electron lasers with multi-frequency pumping. *Acta Phys. Polonica*, A68(5):740, 1985.

[23] S.S. Kohmanski, V.V. Kulish. Parametric resonance interaction of electron in the field of electromagnetic waves and longitudinal magnetic field. *Acta Phys. Polonica*, A68:725–736, 1985.

[24] V.V. Butuzov, V.P. Zakharov, V.V. Kulish . Parametric instability of high current relativistic electron flux in the field of dispersed electromagnetic waves. Deposited in Ukrainian Scientific Research Institute of Technical Information, Kiev, feb 1983.

[25] V.P. Zakharov, V.V. Kulish. Explosive instability of electron flux in the field of dispersing electromagnetic waves. *Ukrainian Physical Journal*, 6:878–881, 1985.

[26] A.M. Kalmykov, N.Ja. Kotsarenko, V.V. Kulish. Possibility of transformation of the frequency of the laser radiation in electron flux. *Pisma v Zhurnal Technicheskoj Fiziki*, 4(14):820–822, 1978. (Soviet: Letters in the Journ. of Technical Physics).

[27] D.H. Menzel. *Fundamental Formulas of Physics*. Dover Publications, Inc, 1960.

[28] V.V. Druzshynin, D.S. Kontorov. *System-techniques*. Radio i Sviaz, Moscow, 1985.

[29] V.V. Kulish, P.B. Kosel, O.B. Krutko, I.V. Gubanov. Effect of cooling of relativistic beams of charged particles during of their acceleration in the crossed

eh-ubitron fields. *Pisma v Zhurnal Tekhnicheskoi Fiziki*, 22(17), 1996. (Letters in Russ. Journ. Techn. Phys.).

[30] V.V. Kulish, P.B. Kosel, A.G. Kailyuk. New acceleration principle of charged particles for electronic applications. *The International Journal of Infrared and Millimeter Waves*, 19(1):33–93, 1998.

[31] V.V. Kulish, P.B. Kosel, A.G. Kailyuk, I.V. Gubanov. New acceleration principle of charged particles for electronic applications. examples. *The International Journal of Infrared and Millimeter Waves*, 19(2):251–329, 1998.

[32] V.P. Zakharov, V.V. Kulish. Polarization effects in parametric interaction between transverse electromagnetic waves and the high current electron flow. *Zurnal Tekhnicheskoj Fiziki*, 53(10):1904–1908, 1983. (Sov. Journal of Technical Physics).

[33] V.P. Zakharov, V.V. Kulish, S.S. Kohmanski. Effect of phase discrimination of electromagnetic signal in modulated relativistic electron flow. *Radiotekhnika i elektronika*, 34(6):1162–1172, 1984. (Sov. Journal of Radioengineering and Electronics).

Chapter 2

HIERARCHY AND
HIERARCHICAL SYSTEMS

1. BASIC CONCEPTS OF THE THEORY OF HIERARCHICAL DYNAMICAL SYSTEMS

What is *hierarchy* in itself? Why does Nature need a hierarchy? What are specific features of *hierarchical systems*? We will try in this Chapter to give the simplest answers to these (and some other similar) questions.

1.1 Why Does Nature Need Hierarchy?

We start with the questions: *what is sense of the hierarchy* in the Nature and what it is originally? It could seem strange but the modern science has not strict and clear answers to these questions. Taking this into consideration we discuss the following such explanations, which seem most attractive amongst various others [1–4].

The complex systems, containing a large number of elements and connections, should be essentially unstable. Therein this instability increases with growth of the system *complexity*. However, it is well known that there are many examples of complex systems in Nature which are obviously stable. The explanation of this paradox is the following. As mentioned above, all complex natural systems have the property of *purposefulness*. Hence theoretically some special mechanisms (purposefulness) providing and supporting the stable state of such systems should exist. Experience shows that indeed, similar mechanisms exist really. For instance, astrophysical analysis of the cosmic catastrophes on the Earth during the last three billions years [5] shows that our biosphere has the obvious property of purposefulness. Every time after a next global catastrophe the Earth reacts in a specific way such that the influ-

ence of any negative results of this catastrophe on the biosphere turns
out to be minimal [5]. In other words, owing to purposefulness the stable
state is reached every time by the shortest path.

As analysis has shown, the concept of purposefulness is closely connected with *self-organization* [3]. Therefore in speaking about purposefulness we, at the same time, should take in view the self-organization.
But the question arises: what is the eventual result of realization of the
purposefulness and self-organization in complex systems from the structural point of view? The answer is very simple: it is the hierarchical
systems. The point is that amongst various known complex systems
only hierarchical ones are found to be stable [1, 6]. This means that existing natural complex systems self-organize by such manner that their
hierarchical arrangement always is realized eventually. I.e., we can say
that *the 'purpose' of an evolution process in any natural complex system
is forming its hierarchical structure* because only such structures can be
stable really.

1.2   Two Approaches to the Theory of Hierarchical Systems

As we noted above, conventional version of hierarchical theory (see
[1–6]) is most popular now. The most essential feature of this version
is that all hierarchical levels of a considered dynamical system are described by *the same dynamical variables*. I.e., these variables are common to each hierarchical level, as well as for the system, as a whole. As
a result, mathematical structures of relevant dynamical equations for
each hierarchical level in this case, as a rule, are found to be *essentially
different*.

Another (new) version of hierarchical approach differs from the conventional one because *a proper set of dynamical variables* for each hierarchical level is introduced [7–16]. Therein the proper variables are
chosen in such way that the *mathematical structure of the dynamical
equations for every hierarchical level* (in these new variables) *is roughly
the same*. We call this feature the *self-modeling* (self-resemblance) *principle* [12, 13, 15, 16]. The self-modeling principle is completed by some
set of other hierarchical principles.

Hence in our version of hierarchical theory essentially new basic concepts are used. The self-modeling principle is a key point of this version.
Taking this into consideration we begin discussion of the new hierarchical
concepts with the self-modeling principle.

1.3   Self-Modeling Principle

First, it should be noted that really the hierarchical idea, in itself, is very ancient. One can find relevant explicit examples in ancient Indian, Greek, Roman, Egyptian, American mythologies [17,18], etc.. However, a more vivid and perfect abstraction of its original sense was given by cosmogonic ideas of the Kabbalah [19–23]. According to the latter our hierarchical world is organized within the above mentioned 'self-modeling *(self-resemblance) scheme'*. This means that every hierarchical level of the world system and the world, as a whole, has the same formal hierarchical structure. This structure is called the *tree of life* (hierarchical *tree* of life). The general idea of this concept is illustrated in Fig. 2.1.1. According to established tradition [19–23] the tree of life depicts complex topological combination of specific force centers (referred as to *sefirot* (sefirs)) and the relevant net of dynamical connections, which are called the *ways* (see Fig. 2.1.1).

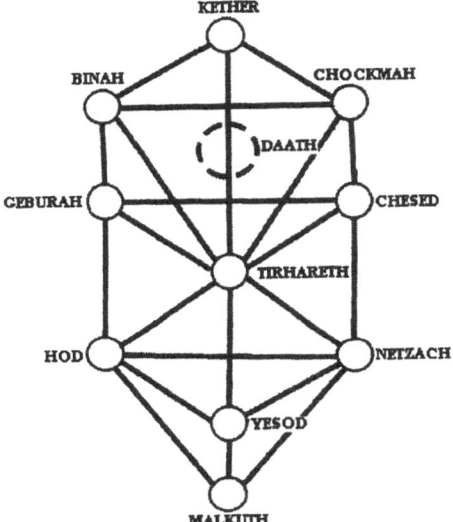

Figure 2.1.1.   Simplest version of the tree of life according to the Kabbalah. Here the sefirot (sefirs) are pictured by the circles, the straight lines illustrate the concept of ways. It is readily seen that the tree of life represents a hierarchical system consisting the seven hierarchical levels.

It is considered that the tree of life is an *elementary universal scheme* of everything existing in our world. I.e., the Universe, as a whole, a people, an atom, an electronic device, and many other objects of the surrounded world have *the same scheme of arrangement* (!). Using modern scientific language we can say that *all natural hierarchical systems* in the surrounding world satisfy the self-modeling principle. It means that each higher hierarchical level should model each lower one and, at the same time, the system as a whole, too. (In modern philosophy, this idea is also known as 'the holographic *principle*').

It indicates that a single fundamental principle, which should be described by a certain fundamental equation (or a set of equations), really determines the Universe, as a whole, as well as its all hierarchical levels. However, the 'little problem' remains here: *how to find this equation(s) really?* Today we are not ready to answer this question. But we hope that the discussed version of hierarchical theory of dynamical systems will give for us a suitable answer in future. Unfortunately, now the hierarchical tree can not be factorized completely (i.e., mathematically described) in the terms of modern physics. However, we can use today simplified versions of the self-modeling principle for practical purposes. Namely this approach is used in the present book. Here we confine ourselves to studying some particular hierarchical dynamical problems only. Including that we will try to use the general self-modeling idea for studying some oscillatative and wave resonant electrodynamic objects. Here the ideology and methods of constructing the specific hierarchical calculational technology, which is based on the use of the self-modeling principle, are the main objects of our attention.

1.4   Main Idea of the Hierarchical Method

The topic discussed is interesting mainly owing to the possibility of treating any natural nonlinear dynamical system as some kind of hierarchical systems. So some specific mathematical method can be used for description of such systems. We bear in mind the set of analytical–numerical methods which are based on the use of the hierarchical concept discussed above, generally, and the self-modeling principle, in particular.

The main idea of such *hierarchical methods* can be expressed as *using some hierarchical properties of the treated real dynamical system* for elaboration of relevant algorithms for asymptotic integration of equations governing its dynamics. Therefore in what follows we will discuss the hierarchical properties of real dynamical systems, which are used further in the hierarchical calculational algorithms. Formally to discuss relevant hierarchical algorithms, we again come back to the sacramental question: *what is hierarchy* originally? At that time, however, we are interested in the quantitative aspect of this problem.

1.5   Again: What is Hierarchy Originally? The Structural Hierarchy

The concept of hierarchy seems so obvious that there is no difficulty in defining it. However, the original nature of the hierarchical concept is much deeper and more complex than it looks at first sight.

We define *hierarchy* as *specific system of preferences with respect to certain hierarchical parameters.* However, what is 'specific system of preferences'? Looking for the answer to this question, let us primarily concern the simplest example of hierarchical system referred to as the '*Russian matryoshka*'. The traditional its version is shown in Fig. 2.1.1.

The matryoshka consists of the series of figures distinguished from each other by their size only. Therefore these figures can be included one within other; i.e., the largest matryoshka (Fig. 2.1.2) contains all smaller matryoshkas.

Figure 2.1.2. Usual Russian matryoshka in the assembled state. We have obvious illustration of the structural hierarchy. Here the height of each matryoshka can be regards as the relevant hierarchical scale parameters a_κ.

The characteristic sizes of the matryoshkas can be arranged as a set of *hierarchical scale* parameters. We form the above mentioned 'system of preferences' with respect to these scale parameters Owing to the obviously expressed resemblance of matryoshkas, we can use only one characteristic size as the hierarchical scale parameter. This can be explained by that all matryoshkas possess the same spatial proportion. For example, the height of the matryoshka a_κ can be regard as an univalent (and rather convenient) hierarchical scale characteristic.

When let us open the assembled matryoshka in Fig. 2.1.2. As a result we may discover that it is a *hierarchical series* of resembling separate matryoshkas (Fig. 2.1.3). Thus we can consider the height of any κ-th matryoshka as a peculiar *structural hierarchical scale parameter* a_κ. We can construct the *hierarchical series* for all these scale parameters (see Fig. 2.1.3):

$$a_1^{-1} < a_2^{-1} < \cdots < a_\kappa^{-1} < \cdots < a_m^{-1} < a_{m+1}^{-1}, \qquad (2.1.1)$$

here m is the total number of structural hierarchical levels of the considered system, a_{m+1} is the normalization characteristic constant (for instance, the height one of experimentalists). But in the discussed illustration example we have to do with the *static hierarchical system* only, i.e., the system *without any motion*. Hence, it can be said that the system (matryoshka) possesses *structural hierarchy*.

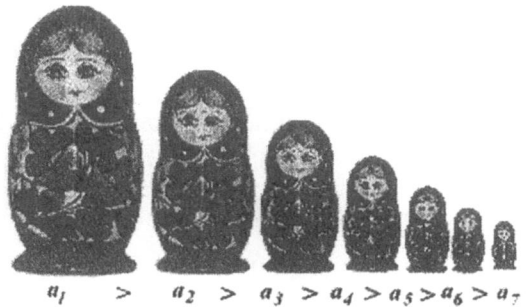

$$a_1 \quad > \quad a_2 \quad > \quad a_3 \quad > \quad a_4 > a_5 > a_6 > a_7$$

Figure 2.1.3. The illustration of the 'hierarchical series' concept. We can be convinced that each following matryoshka (or, in other words, the matryoshka of each higher hierarchical level) has less (!) height. Besides that, each following matryoshka should contain a lesser number of structural elements (atoms and molecules in the present particular case).

In what follows we can disclose one important peculiarity of the hierarchical system considered. Namely, every subsequent matryoshka contains a fewer number of structural elements (atoms and molecules of the matryoshka's material). It means that every higher hierarchical level-matryoshka is less complex than the lower one (the concept of complexity is discussed above in Subsection 6.10). Generalizing, one can be convinced that this property is common for all natural hierarchical systems with structural hierarchy. We especially fix our attention to this feature because it will become necessary later.

1.6   Dynamical Hierarchy

All hierarchical systems known in nature and society, besides the structural hierarchy, have *dynamical (functional) hierarchy*, too. This type of hierarchy is characteristic for systems *with motion*. Yearly, in the case of structural hierarchy, we had to do with hierarchy of *structural* scale parameters. In this (second) case relevant *dynamical* parameters play a similar role. To characterize dynamical hierarchy we use the *hierarchy of velocities of varying system dynamical parameters* on time. The

Universe is a most convenient illustration of the concept of dynamical hierarchy.

The Table 2.1.1 illustrates some dynamical features of the Universe. Information pictured is interesting and somewhat unexpected. There are both *structural and dynamical hierarchies* in given natural system *simultaneously*. Besides that, a number of hierarchical levels in both these cases are equal, similarly to the tree of life (see Fig. 2.1.1). It should be noted especially that this fact is characteristic for many natural hierarchical systems such as human society, national and world economic systems, military systems, various electronic systems, etc..

Table 2.1.1. Characteristics of the Universe as a Hierarchical System

Hierarchical level numbers	Hierarchical level name	Average characteristic size	Characteristic velocity of varying dynamical parameters
[numbers]	—	[cm]	$[c^{-1}]$
1	elementary particles	10^{-17}	10^{25}
2	nucleus	10^{-10}	10^{16}
3	atoms and molecules	10^{-5}	10^{13}
4	planet systems	10^{8}	10^{-5}
5	star systems	10^{15}	10^{-8}
6	galaxy systems and molecular clouds	10^{23}	10^{-15}
7	metagalaxy	$> 10^{30}$	$< 10^{-37}$

In principle, the *hierarchical systems* with different numbers of structural hierarchical levels and dynamical ones can be realized. However, it should be pointed out especially that the systems with coinciding number of structural and dynamical hierarchical level are most widespread in the Nature. Therefore, later we shall take an interest in systems of such a type only.

Similarly to the matryoshka, a *lesser quantity of structural elements characterizes every higher hierarchical level* of any natural dynamical hierarchical system. For instance, the total number of elementary particles in the Universe are substantially more than the number of nuclei, the number of star systems are more than the number of galaxies, and so on (see Table 2.1.1). However, the number of structural elements in a system in thermodynamics is connected with degree of chaos in the system. It means that *the lower is the occupied hierarchical level the higher is the degree of chaos* in the Universe, as a hierarchical system. Similarly, in every human social structure the chaos increases with a decrease of number of a social level, etc..

Then we pay attention to the important fact that *different dynamical variables* (and, consequently, corresponding dynamical equations) *describe the dynamical processes at different hierarchical levels*. Or, in other words, a proper set of the dynamical equations exists for each hierarchical level of the hierarchical system. In the methodological plane this can be treated as an inevitable sequence of appearance of the above discussed self-modeling principle, really acting in the Nature. Indeed, in the general case the same form of dynamical equations for different hierarchical levels can take place really in the case only, when the proper dynamical variables for these levels are different.

From the physical point of view this can be explained by the fact that different types of fundamental interactions play the main role on the different levels of hierarchy. Table 2.1.2 gives evident illustration of this natural phenomenon.

Table 2.1.2.   Fundamental Interactions

Interaction	Field quantum	Quantum mass	Interaction radius	Source	Connection constant
—	—	[GeV]	[cm]	—	—
Strong	gluon	0	$\leqslant 10^{-15}$	color charge	$\alpha_s \sim 1$ (for large r) $\alpha_s < 1$ (for small r)
Weak	intermediate bosons $W\pm, Z^0$	82, 93	10^{-18}	weak charge	$(4\pi)^2 (Mc/h)^2 G/hc$ $= 1,02 \cdot 10^{-5}$
Electro-magnetic	photon	0	∞	electric charge	$\alpha = 2\pi e^2/hc = 1/137$
Gravitation	graviton	0	∞	mass	$2\pi G_N M^2 2/hc$ $= 0,53 \cdot 10^{-38}$

Here M is the nucleon mass, G_N is the gravitation constant, G is the Fermi constant, h is the Plank constant, c is the light velocity in vacuum.

But let us discuss this phenomenon in more detail. As mentioned above, every hierarchical level has its proper set of the fundamental interactions that determine characteristic dynamical processes here. Therefore one or two types of interactions prevail on each hierarchical level. In particular, dynamics of processes at the first and second hierarchical level of the Universe, as a hierarchical system (elementary particles and nucleus), is determined by the weak and strong interactions (see Table 2.1.2). At the same time the electromagnetic interaction plays the main role at the third level, and so on. It should be mentioned that other interactions also play some role here. However, the proper mechanism (or mechanisms) of fundamental interactions determines a general physical picture of the given level. For example, gravitation interaction practically does not exert any influence at the processes at first three levels. But it completely determines the dynamics of main processes at higher hierarchical levels.

Later we turn to the above mentioned affirmation that simultaneous presence of the structural and dynamical hierarchies, as a rule, are characteristic for natural dynamical systems. We note that both: the structural as well as dynamical characteristic scale parameters turn out to be essentially different (in magnitude) for different levels of the system (see Table 2.1.1). The higher the hierarchical level the slower the velocity of characteristic processes, and the fewer the number of structural elements. This observation opens the possibility for *using the ratios of neighboring scale parameters as relevant expansion parameters of the dynamical problem considered. This idea is put in the basis of the hierarchical method.* It turns out to be very effective for construction of different versions of the hierarchical calculational algorithms.

Thus each hierarchical system level can be characterized by dynamical scalar scale parameter $b_\lambda (\lambda = 1, 2, 3, \ldots, k;$ k is a total number of dynamical hierarchical levels). This parameter characterizes the varying velocities of the proper (to the given level) dynamical variables. Correspondingly, analogously to the case of structural hierarchy (see, for example (2.1.1)), the relevant *dynamical hierarchical series* can be constructed in the case of dynamical (functional) hierarchy, too:

$$b_1 < b_2 < \cdots < b_\lambda < \cdots < b_k < b_{k+1}. \qquad (2.1.2)$$

In general, as mentioned already, the numbers m and k are different. However, as in the hierarchical model of Universe, their equality is more typical in practice. Therefore, further we consider everywhere: $m = k$ for simplicity, i.e., *the total number of terms of both series is equal* and that each term a_κ^{-1} of the series (2.1.1) has corresponding term b κ in the series (2.1.2). Hence, one of two hierarchical series (2.1.1) and (2.1.2) is enough to describe the hierarchy in a dynamical system.

1.7   Hierarchical Series in Dimensionless Form

We consider that the hierarchy in the system is evidently expressed (like to the Universe; for instance — see Table 2.1.1), i.e., $a_\kappa^{-1} <<$ $a_{\kappa+1}^{-1}$, $b_\lambda << b_{\lambda+1}$. Moreover, we suppose $\kappa = \lambda$ and $a_\kappa^{-1} \sim b_\lambda$. Then normalizing series (2.1.1) and (2.1.2) with respect to scale parameters a_{m+1}^{-1} or b_{k+1}, we obtain relevant *hierarchical series* in the normalized dimensionless form. Bearing in mind these assumptions, we rewrite the series (2.1.1) in the so called 'strong' normalized dimensionless form

$$\varepsilon_1 << \varepsilon_2 << \cdots << \varepsilon_\kappa << \cdots << \varepsilon_m << 1 \qquad (2.1.3)$$

where $\varepsilon_\kappa = a_\kappa^{-1}/a_{m+1}^{-1} \sim b_\lambda/b_{\lambda+1}$ is a normalized scale parameter of κ-th hierarchical level. In the framework of the hierarchical method described below, these parameters play the role of relevant expansion parameters for the corresponding dynamical functions at κ-th level.

2.   HIERARCHICAL PRINCIPLES. HIERARCHICAL DESCRIPTION

2.1   Hierarchical Principles

The system of *hierarchical principles* is put in the basis of the discussed version of hierarchical method. These principles [10–16] are generalization of the above discussed self-modeling idea, and they summarize known today experimental facts and observations. The five following hierarchical principles can be formulated in this way, e.g. one *general hierarchical principle* and four particular fundamental ones.

General hierarchical principle.   *Everything in the Universe* is *of hierarchical nature.* Indeed, it is difficult to find in nature or society any object that, on the one hand, not belonging to relevant hierarchical level of some hierarchical system and having no intrinsic hierarchical structure, on the other hand. The other four principles concern the structure, general dynamical and thermodynamic features of the hierarchical systems.

The principle of information compression.   The principle of information compression: *each higher hierarchical level is always simpler than the preceding one.* The concept of *system complexity* is defined above (see subsection 6.10). The Universe (see Table 2.1.1), human society, and so on are evident examples of realization of this principle. Here the smaller a number of structural elements (and corresponding connections between the elements), the higher a number of hierarchical level of the system. Indeed, we always have one President (or other Head) of a state, a few heads of other power branches, a few tens of ministers, and so on. The millions of citizens are on the opposite end of the hierarchical stairs. Similar examples can be found in the astrophysics, the theory of long-time interacting electron devices [7–16] (see Chapter 4, Subsection 1.3, and Volume II), etc..

The principle of hierarchical resemblance. The principle of hierarchical resemblance (self-modeling principle or holographic principle): *each hierarchical level in its general properties represents the system as a whole.* In other words, the same type of dynamical equations describes as well every hierarchical level of the system and the system, as a whole. As mentioned already, *each hierarchical level is described by proper set of dynamical variables.* We regard this principle also as a peculiar *hierarchical invariance.* The Boltzmann kinetic equation or the quasi-hydrodynamic equation (or the equation for current density) can serve as evident illustrations of appearance of the principle of hierarchical resemblance. These equations have the same mathematical structure at every hierarchical level, but every time with respect to another (proper) set of dynamical variables (see relevant Chapters 6–8 and Volume II).

The hierarchical analogue of the second thermodynamic principle. The hierarchical analogue of the second thermodynamic principle: *each higher hierarchical level has less information entropy* than the preceding one (the concept of *information entropy* is discussed above in item 1.6.9). Numerous examples can be taken from our everyday life. Indeed, one can see each lower hierarchical society level has greater chaos. Because information from a lower level is compressed at each higher hierarchical level that this principle has interesting consequence: *the information influence in dynamical system* is directed from lower hierarchical levels to higher ones, whereas the managing action is directed oppositely. I.e., the managing influence is directed from up to down (correlate these affirmations with the diagram in Fig. 1.1.2).

Hierarchical analogue of the third thermodynamic principle. Hierarchical analogue of the third thermodynamic principle: *the highest level of a closed hierarchical system* is characterized by vanishing information entropy. As before, we can find the most vivid illustration in everyday life. For instance: only one person (i.e. only one structural element of a present system) occupies the top place in any social pyramid (i.e. only one person can be, say, the President of the Ukraine at the same time). It is obvious that the entropy of the subsystem (highest

hierarchical level as a subsystem) consisting of one element only must
be zero. So, the hierarchical principle here holds strictly. The principle
is applicable to any other natural and social hierarchical systems. However, one can readily be convinced that it holds especially ardently in
everyday social practice.

All above discussed hierarchical principles are applicable in their en-
tirety only to the *closed natural* hierarchical systems. There are many
another theoretical hierarchical constructions (which always turn out to
be *purely artificial* systems) those are *open systems*. These principles
can be satisfied only partially for such systems.

It may seem interesting and even incredible that all these principles
hold in multi-frequency nonlinear hierarchical electrodynamic systems
[7–16, 24], too (Chapters 6–8, and Volume II).

2.2   Dynamical Equation of the Zeroth Hierarchical Level

Let us assume that some dynamical system can be described by the
following exact differential vector equation

$$\frac{dz}{dt} = Z\left(z, t\right), \qquad (2.2.1)$$

where $z = \{z_1, z_2, ..., z_n\}$ is some vector and $Z = \{Z_1, Z_2, ..., Z_n\}$ is the
relevant vector-function in *n-dimensional* Euclidean space R_n, $t \in [0, \infty]$
(the *laboratory time*, for example). The initial conditions are standard
ones

$$z\left(t = 0\right) = z_0. \qquad (2.2.2)$$

Concerning properties of vector-function Z and initial conditions, we
separate three different situations:
determined system, function Z satisfies the uniqueness of the solution to
equation (1.2.1) and initial conditions are given uniquely;
stochastic systems characterized by random initial conditions;
chaotic systems with properties caused by nonlinear nature of the system and other causes. For instance, chaos can be generated by both:
nonlinear interactions of structural elements at a given hierarchical level
and a stochastic (or chaotic) component in the forming of hierarchical
structure of the system, as a whole.

Below (in this and the following Volumes) we shall discuss most of
these situations.

Now we take into account the general hierarchical principle. Let us
assume that our dynamical system possesses a hierarchical structure.

According to the principle of hierarchical resemblance each hierarchical level of the system is described by some equation with resemble (with respect to the expression (1.2.1)) mathematical structure. Therefore for every hierarchical level we have proper (specific only for this level) set of dynamical variables.

Further, we consider the hierarchy strongly expressed and the numbers of structural and dynamical hierarchical levels coincide. As above, in the case of strong hierarchy (definition (2.1.3)) the set of the hierarchical structural and functional (dynamical) parameters is a set of small parameters for corresponding asymptotic expansions.

Thus solving the hierarchical problem we first give an adequate formal definition for the problem small parameters. To do this, the hierarchical analogy of the second and third thermodynamic principles can be used. According to the first, each higher hierarchical level has less information entropy than the preceding one. One can prove the latter affirmation equivalent to the assertion that *each higher hierarchical level has less characteristic velocities of the varying of dynamical variables than the preceding one.* The affirmation that characteristic velocities of dynamical variables varying of the highest level of the closed hierarchical system are zero is the consequence of the fourth hierarchical principle. In particular, the example of the Universe as a natural hierarchical dynamical system (see Table 2.1.2) obviously illustrates both these consequences in practice. Using these suppositions we formulate the *general definition of the hierarchical scale parameter* of a problem for the κ-th level:

$$\varepsilon_\kappa \sim \left| \frac{dz^{(\kappa+1)}}{dt} \right| \bigg/ \left| \frac{dz^{(\kappa)}}{dt} \right| \ll 1. \qquad (2.2.3)$$

where $z(\kappa)$ is the vector of the proper variables of a κ-th hierarchical level. In this way we can classify the initial set of variables according to hierarchical levels. Possibility occurs to construct corresponding hierarchical series of the type (2.1.3) (with the scale parameters (2.2.3)). The initial dynamical equation (2.2.1) (the *zero-level equation*) generally depends on all hierarchical small parameters (2.2.3) simultaneously:

$$\frac{dz^{(0)}}{dt} = Z^{(0)} \left(z^{(0)}, t, \varepsilon_1, \dots, \varepsilon_m \right), \qquad (2.2.4)$$

Later we expand the right part of the equation (2.2.1) in a power series with respect to small parameters ε_κ. Because hierarchy here is expressed strongly (2.2.3), we can temporarily neglect the influence of all other higher hierarchical levels but the nearest lower hierarchical level:

$$\frac{dz^{(0)}}{dt} = Z^{(0)}\left(z^{(0)}, t, \varepsilon_1\right), \quad z^{(0)}\left(t_0\right) = z^{(0)}, \tag{2.2.5}$$

where $\varepsilon_1 \ll 1$ is the scale parameter for the first hierarchical level.

The physical sense of the approximation is that influences of neighboring hierarchical levels are accounted only. The model of the Universe above can be used as illustration supporting this assumption. Indeed, the Sun exerts the main influence upon the motion, for instance, of the Earth. At the same time its influence upon the Moon's motion is inessential, whereas the motion of the latter is mainly determined by influence of the Earth (see Fig. 2.1.2). However, other higher hierarchical levels also exert some small influence upon the given level (the other planets and Sun weakly perturb the motion of the Moon in the gravitational field of the Earth). This calculational situation is illustrated in Fig. 2.2.1. The illustration of the scheme of interactions between the zeroth and other hierarchical levels of the dynamical system is given. Taking into consideration the above formulated supposition about accounting only neighboring interactions only, we obtain the system any hierarchical level interacts with neighboring levels only. It means that in such a case we can 'travel' through the hierarchical system only 'step by step', from one level to neighboring next level and in back direction.

The calculational technology taking into account the influence of other (higher) hierarchical interactions (see illustration in Fig. 2.2.1) can be elaborated on the way of further developing of the approach discussed here (for instance, by perturbations). In this book, however, we are not interested in such calculational situations. We propose this problem for the reader, as an teaching tusk.

2.3 Structural and Functional (Dynamical) Operators

Now we introduce the concepts of the *structural* and *functional (dynamical) operators* regulating the structural and functional (dynamical) hierarchies. To describe structural hierarchy one introduces *the structural operator M*

$$Z^{(\kappa+1)}\left(z^{(\kappa+1)}, t\right) = M^{(\kappa)} Z^{(\kappa)}\left(z^{(\kappa)}, t\right), \tag{2.2.6}$$

where $Z(\kappa)$ is the vector-function defined as the right part of relevant standard dynamical equation (of the type (2.2.5)) for κ-th hierarchical level. In other words, every such function characterizes nonlinear dynamical properties of κ-th hierarchical level of a system.

Thus the *structural operator M* is *responsible* for the hierarchical resemblance in the discussed dynamical system. To obtain the functions

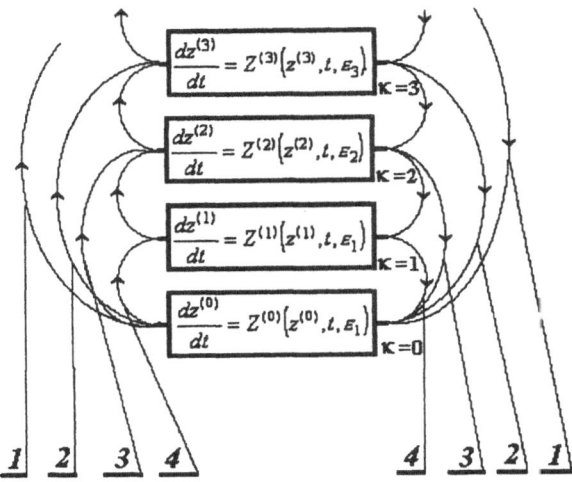

Figure 2.2.1. Illustration of the scheme of interactions between the zeroth and other hierarchical levels of a dynamical system. Here 1 are the connections between the zeroth and a κ-th hierarchical levels, 2 are the connections between the zeroth and third hierarchical levels, 3 are the connections between the zeroth and the second hierarchical levels, 4 are the connections between the zeroth and the first (neighboring) hierarchical levels. We suppose that interactions between neighboring hierarchical levels only are important, and they are taken into account only in our calculational algorithms.

$Z^{(\kappa+1)}\left(z^{(\kappa+1)},t\right)$ (for a known function $Z^{(\kappa)}\left(z^{(\kappa)},t\right)$) various integral transformations are used. For electrodynamic problems most promising are the averaging operator, the Fourier or Laplace transformations, the convolution type transformations, the conformal mapping, and some others. The differential or mixed differential–integral structural operators can be used for problems of another type. But in this book we will discuss only integral operators (the averaging and Fourier transformations, correspondingly).

The *transformation functional (dynamical) operator* $\hat{U}$ (a function in the case of integral structural operators) describes the dynamics of connections between proper variables of two neighboring hierarchical levels. In other words, it *provides* the realizing of the hierarchical resemblance (self-modeling) principle in the given system. We define the transformation functional (dynamical) operator as

$$z^{(\kappa)} = \hat{U}^{(\kappa+1)}\left(z^{(\kappa+1)},t\right). \tag{2.2.7}$$

As one can be convinced, the introduced operators (structural (2.2.6) and functional (2.2.7)) also satisfy other hierarchical principles (the principle of information compression, the hierarchical analogies of second and third laws of thermodynamics, and the general hierarchical principle, respectively).

Then we recall once more that according to the assumption $\varepsilon_\kappa \ll 1$, only interactions of neighboring hierarchical levels are taken into account everywhere. Correspondingly, other types of functional operators $\hat{U}^{(\kappa+1)}$, describing different non-neighboring interactions in the system, can be introduced in principle, too. However, as already mentioned, in this book we study the simplest dynamical models with neighboring interactions only.

2.4   Classification of the Hierarchical Problems

In the framework of the hierarchical theory presented in the general case, the operators of hierarchy $M^{(\kappa)}$ and $\hat{U}^{(\kappa)}$ are not independent mathematical objects. In fact, one of them should always be considered given characteristic of the hierarchical system. In this connection we distinguish two types of hierarchical problems:

1) The analysis problem — having structure of the operator $M^{(\kappa)}$ one must construct the functional operator $\hat{U}^{(\kappa)}$, i.e. to determine the character of functional connections between hierarchical levels.

2) The synthesis problem — having the operator $\hat{U}^{(\kappa)}$ one must construct the operator $M^{(\kappa)}$, i.e., to synthesize the hierarchical structure of the system.

Up to now the main attention in literature was paid only to the analysis problem. In should be noted that the theory and calculational technology of the synthesis problem is not being developed today. Therefore in this book we also have to do with the analysis problem only.

In turn each mentioned problem can be divided into a direct and its *inverse problem*. The straight problem (see Fig. 2.2.2) is characterized by the formulating of a problem for the highest (m-th) hierarchical level. Further, the relevant functional dependencies for other lower hierarchical levels are constructed by special calculational.

In the case of the inverse problem (see Fig. 2.2.3) the initial problem is formulated for a certain hierarchical level conditionally called the zeroth level. Then we 'lift' along the 'vertical hierarchical axis up' toward the highest hierarchical level and, therefore transform this problem into some higher (highest m-th, for example) hierarchical level.

Specific feature of the hierarchical methods which are discussed in this book is that here both the above mentioned problems are used at the same time. At the first stage of the solution algorithm the initial problem is transformed into some higher (highest, for instance) hierarchical level (the straight problem — see Fig. 2.2.2). Then solving the problem

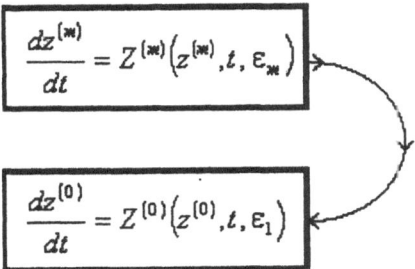

Figure 2.2.2. Illustration of the main idea of the direct hierarchical problem. Here the problem formulated for some higher (m-th highest, for example) hierarchical level transforms into the problem for the zeroth hierarchical level.

at this level we further 'slip down' the found solutions (the inverse problem — Fig. 2.2.3). In what follows we find required solutions for the zeroth hierarchical level (see also Fig. 1.1.2 and relevant commentaries). Namely this version of hierarchical calculational technology is the main object of our interest in this book.

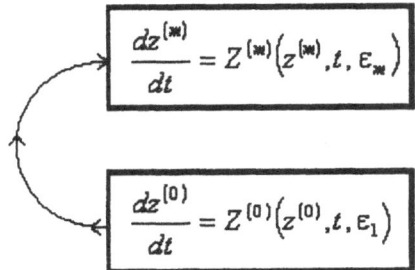

Figure 2.2.3. Illustration of the main idea of the inverse hierarchical problem. Here the problem formulated for the zeroth transforms into the problem for some higher (the highest m-th, for example) hierarchical level.

Then we turn our attention to the problem of numbering of hierarchical levels in a dynamical system. There are a few methods of the numbering. In this book we used a peculiar method smaller numbers correspond to levels that are more complex. Therein we conveniently consider some level as the zeroth one. Numbering in such a way, we can realize, in principle, negative numbers of hierarchical levels also. It should be mentioned that the discussed case $\kappa \in [-\infty, m]$ can not be realized within the framework of other versions of numbering methods [2].

2.5   Hierarchical Tree

To illustrate hierarchical structures rather obvious graphic method is used. It is called the *hierarchical tree method*. Its graphic figures are called the *hierarchical trees* [2] (it should not mistaken the concepts of (hierarchical) tree of life (see Fig. 2.1.1) and hierarchical tree, correspondingly). The simplest version of the hierarchical tree is shown in Fig. 2.2.4. Here the specific hierarchical tree is given, where each two structural elements of some hierarchical level form one element of the level one step higher. We illustrate this graphical situation by the following example.

Let it be some abstract military sub-unit each chief has two followers only. On the top of hierarchical structure a general stays, who has two colonels as direct subordinates. It is obvious (that is important) that without these subordinates, the general 'is not a general' functionally. Thus colonels are responsible to forming one step higher level — the *level of the general* (unfortunately, generals in a real life very often forget this rather important circumstance).

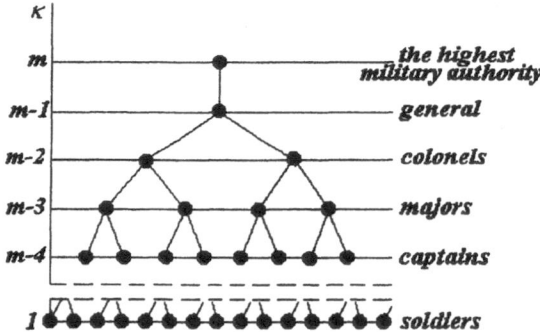

Figure 2.2.4. The simplest illustration of the concept of 'hierarchical tree'. It is a military sub-unit, where each chief has only two followers. Each two structural elements of a κ-th hierarchical level forms relevant element of the $(\kappa + 1)$-th hierarchical level. The general occupies the top of the hierarchical pyramide, related with military unit. Any other higher really existed hierarchical levels are represented by only one point. It is the point 'the highest military authority'.

Further, we suppose that each colonel has lieutenant colonels representing the *level of lieutenant colonels* and governing two majors, and so on. It is obviously that any real military chief has, as a rule, more than two followers. However, in the given illustration example this is not important. By this we only try to illustrate one of basic features of all known hierarchical structures. Namely, *some group* (that is specific for the given system) of structural elements at any κ-th hierarchical level forms *one structural element* at the $(\kappa + 1)$-th hierarchical level, and so on. This idea is illustrated in this example by the following manner. The general without his subordinates 'is not a general' functionally, because in such case he could be lose any possibilities to be a 'general leading activity'. Analogously, without major-subordinates colonels 'are not colonels', and so forth.

Then we turn our attention to the highest point of the hierarchical tree in Fig. 2.2.4. This point plays very important role for understanding of physical nature of real dynamical hierarchical structures. It is well known that any general has his own military authority. In turn, this authority also has some higher authority, and so on. On the top of the structure in Fig. 2.2.4 the point 'the highest military authority' depicts all possible (including, unknown) higher authority. It is important that this point also represent some hierarchical system. However, the structure of the latter can not be disclosed in the framework of lower its part. Indeed, all structural element of the system pictured in Fig. 2.2.5 should known all down structure of managed 'branch' of the hierarchical tree. But any such element should know only one higher chief. In turn, this chief knows 'all down structure of managed 'branch' of the hierarchical tree', and so on. Or in other words, *any information in the considered hierarchical system is directed from lower hierarchical levels to higher ones, whereas the managing action is directed oppositely* (it is the sequence from the hierarchical analogue of the second thermodynamic principle).

Thus we can affirm that highest separate point in the discussed hierarchical system embodies the 'highest military power' in this military structure.

The analogous, in principle, situation is realized in our Universe, too. The hierarchical tree of the Universe, given in Fig. 2.2.5, illustrates this observation. It is obvious that contrary to the previous thought model a number of structural elements at each hierarchical level can be essentially much more than two. However, as mentioned above, this circumstance is not important: both hierarchical trees really are similar.

Further we again turn to discussion of the problem concerned to the highest point on the top of hierarchical tree. Analyzing all known today natural complex hierarchical dynamical systems, one can be convinced that all relevant hierarchical threes (without any exceptions!), analogously to Fig. 2.2.4, Fig. 2.2.5, and Fig. 4.1.1, always contain a corresponding separate point at the top. Hence we have weighty grounds to affirm that this property is universal for our world and it is characteristic for all hierarchical dynamical systems existing in the Nature. On the

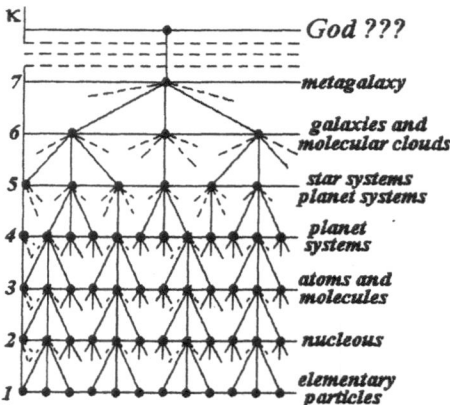

Figure 2.2.5. Our Universe as a hierarchical dynamical system. Here κ is the number of a hierarchical level. An aggregate of structural elements on some κ-th hierarchical level forms relevant element on the $(\kappa + 1)$-th hierarchical level. Comparing the general arrangement of the hierarchical trees in Figs. 2.2.4 and 2.2.5 we could ask: what the physical sense of the highest point in the diagram in Fig. 2.2.5 in view of the treatment of the analogous point in Fig. 2.2.4?

other hand, our Universe also can be considered as a natural dynamical complex system. Consequently, it, in principle, should also contain such an element, in spite of the circumstance that it strictly is not yet fixed in official astrophysics. On the other hand, the question arises: what is the physical sense of the highest point in the diagram in Fig. 2.2.5 in view of the treating of the analogous point in Fig. 2.2.4?

Thus following to this logic we inevitably come to the fundamental problem: *what hierarchical level (and the separate point on its) really occupied higher the 'metagalaxy' level?* What is the real physical sense of these point and level? The author considers that this point (and level, correspondingly) represents all unknown for us highest (with respect to our world-hierarchic system) hierarchical levels. This can be treated as a *physical definition of the concept of God.* Thus as a result of our reasoning we obtain a possibility for formulating of the concept of God in traditional terms of the physics.

It could seem incredible, but, as shown below in Chapter 4, analyzing wave resonance (oscillation–resonance) processes in electronic devices with long-time interactions [11, 12, 16] (see Fig. 4.1.1 and corresponding comments) the analogous *hierarchical trees* can be obtained, too. In this case the discussed highest separate point represents some source of energy for acceleration of the electron beam that comes in input of the system. Taking into consideration our above reasoning we can say that functionally this source could be treated as a peculiar 'god' for this device-hierarchic system. In this connection it is interesting to note that general hierarchical approach, which is used here (the theory of hierarchical waves and oscillations [7–16, 24]), really have essentially wider framework of utilization than it seems at first sight. As corresponding analysis shown, the most natural hierarchical systems (including, Universe, human society, etc.) can be successfully described in the framework of this wave-like (oscillatative-like) approach. Moreover, the hierarchical asymptotic calculational methods, some of them are described in this book below (see also [7–16, 24]), can be used for studying of these objects also.

Thus the concept of hierarchical tree indeed turns out to be universal and very effective means for description of various dynamical hierarchical systems of different nature.

References

[1] J.S. Nicolis. *Dynamics of Hierarchical Systems. An Evolutionary Approach.* Springer-Verlag, Berlin-Heidelberg-New York, Tokyo, 1986.

[2] R. Rammal, G. Toulouse, M.A. Virasoro. Ultrametricity for physicists. *Reviews of Modern Physics*, 58(3):765–788, 1986.

[3] H. Haken. *Instability Hierarchies of Self-Organizing Systems and Devices.* Advanced Synergetic. Springer-Verlag, Berlin-Heidelberg-New York- Tokyo, 1983.

[4] Kaivarainen. *Hierarchical concept of matter and field.* Earthpuls Press, 1997.

[5] B.M. Vladimirskij, L.D. Kislovskij. *The outer space influences and the biosphere evolution*, volume 1 of *Astronautics, Astronomy*. Znanije, Moscow, 1986.

[6] V.V. Druzshynin, D.S. Kontorov. *System-techniques.* Radio i Sviaz, Moscow, 1985.

[7] V.V. Kulish. Nonlinear self-consistent theory of free electron lasers. method of investigation. *Ukrainian Physical Journal*, 36(9):1318–1325, 1991.

[8] V.V. Kulish, A.V. Lysenko. Method of averaged kinetic equation and its use in the nonlinear problems of plasma electrodynamics. *Fizika Plazmy*, 19(2):216–227, 1993. (Sov. Plasma Physics).

[9] V.V. Kulish, S.A. Kuleshov, A.V. Lysenko. Nonlinear self-consistent theory of superheterodyne and free electron lasers. *The International journal of infrared and millimeter waves*, 14(3):451–568, 1993.

[10] V.V. Kulish. Hierarchical oscillations and averaging methods in nonlinear problems of relativistic electronics. *The International Journal of Infrared and Millimeter Waves*, 18(5):1053–1117, 1997.

[11] V.V. Kulish. Hierarchical approach to nonlinear problems of electrodynamics. *Visnyk Sumskogo Derzshavnogo Universytetu*, 1(7):3–11, 1997.

[12] V.V. Kulish, P.B. Kosel, A.G. Kailyuk. New acceleration principle of charged particles for electronic applications. hierarchical description. *The International Journal of Infrared and Millimeter waves*, 19(1):33–93, 1998.

[13] V.V. Kulish. Hierarchical method and its application peculiarities in nonlinear problems of relativistic electrodynamics. general theory. *Ukrainian Physical Journal*, 43(4):83–499, 1998.

[14] V.V. Kulish, P.B. Kosel, O.B. Krutko, I.V. Gubanov. Hierarchical method and its application peculiarities in nonlinear problems of relativistic electrodynamics. theory of eh-ubitron accelerator of charged particles. *Ukrainian Physical Journal*, 43(2):33–138, 1998.

[15] V.V. Kulish. Hierarchical method and its application peculiarities in nonlinear problems of relativistic electrodynamics. single-particle model of cyclotron-resonant maser. *Ukrainian Physical Journal*, 43(4):98–402, 1998.

[16] V.V. Kulish. *Hierarchical theory of oscillations and waves and its application to nonlinear problems of relativistic electrodynamics.* Causality and locality in modern physics. Kluwer Academic Publishers, Dordrecht/Boston/London, 1998.

[17] Andre Nataf. *Dictionary of the Occult.* Wordsworth Editions Ltd, Hartfordshire, 1988.

[18] F. Kapra. *Dao of physics.* ORIS, St. Petersburg, 1994.

[19] M. Laitman. Kabbalah. Israel, 1991.

[20] D. Fortune. *The mystical Qabalah.* Alta Gaia Books, New York, 1989.

[21] N. Garbo. Cabal. New York-London, 1978.

[22] C. Ponce. *Quest books.* Kabbalah. The Thesophical Publishing House, Wheaton, Illinois, USA Adyar, Madras, India, 1995.

[23] W. Parfitt. *The New Living Qabalah.* Element, Shaftesbury, Dorset Rockport, Massachusetts, Queesland, 1995.

[24] V.V. Kulish. *Methods of averaging in non-linear problems of relativistic electrodynamics.* World scientific Publishers, Atlanta, 1998.

Chapter 3

HIERARCHICAL ASYMPTOTIC METHODS. GENERAL IDEAS

A specific feature of this Chapter is that it is recommended to be read two times. The first time should help to 'find one's bearing on the ground'. The second might be helpful for the obtaining entire picture of the considered scientific direction. A reader (we fear) after acquaintance with the book might be too overloaded by the variety of proposed calculational schemes and peculiarities of their practical application. So we hope that this Chapter can help to generalize main impressions and obtained knowledge in the field of the hierarchical theory and the hierarchical methods, in particular.

We shall discuss below the main ideas and peculiarities of the hierarchical calculational method. It will be shown how specific calculational ideology can be developed immediately from the above discussed characteristic properties of hierarchical dynamical systems. In particular, the general idea of hierarchical approach by *direct* and *inverse problems* of analysis (see Figs. 2.2.1–2.2.5 and relevant commentaries) will be illustrated quantitatively (i.e., by means relevant mathematical description). As mentioned above a specific combination of these problems allow formulating the general calculational hierarchical problem. But let us make a few remarks on this topic.

All hierarchical problems (and, correspondingly, hierarchical systems) could be divided into oscillatative and wave ones. The first of them are characteristic for so called *lumped systems* and they could be described by relevant exact differential equations. Their typical examples are the problems about motion of a charged particle in some external electromagnetic field (see example set forth below in Chapter 5). The equations like to (2.2.4), (2.2.5) may be proposed as a general formulation of such type problems.

Characteristic feature of the problems of second type is description of the studied object by some system with partial derivatives. Such objects usually is classified as the *distributed systems*. The problems about propagation of different nature waves in various mediums (including nonlinear) are described, as a rule, by the system with partial derivatives. Examples of such type will be given in Chapter 7 and Chapter 9 (Chapter 8, Section 7).

The problems of both type are the object of attention in this book. They are deeply correlated one with other. Therein mathematicaly specific features of methods for their solution, as well as physical nature of the considered objects determine this. That is why, in spite of their unlikeness, the mathematical structure of methods for their solution roughly is found to be the same. Therein some of methods for solution of the second type problems (like to the method of averaged characteristics — see Chapter 6), include immediately the methods for asymptotic integration of ordinary differential equations as a key point of the calculational algorithm. Other methods (such as the slowly varying amplitude method — see Chapter 8) use also the general ideology developed for the problem of the first type, but for the system with partial derivatives only.

Thus the main objects of our future discussion are the mathematical description of the straight and inverse problems in framework of the hierarchical theory of lumped and distributed oscillatative wave systems.

1. DETERMINED HIERARCHICAL SYSTEMS
1.1   Determined Hierarchical Systems

Let us begin our general discussion by the studying some hierarchical system described by initial dynamical equation like (2.2.4). It is considered for simplicity that we deal with the *determined hierarchical systems* (see the system classification given in the preceding Chapter). Then, accepting the supposition discussed above about interacting only neighboring hierarchical levels, we come to the 'hierarchicised' equation for the zeroth level of type (2.2.5). We use the definition (2.2.7) (formal solutions for equation (2.2.5)) for the connection between variables of the zeroth and first hierarchical levels:

$$z^{(0)} = \hat{U}^{(1)}\left(z^{(1)}, t, \varepsilon_1\right). \qquad (3.1.1)$$

Differentiating expression (3.1.1) by the time t we consider the dynamical equation for the zeroth hierarchical level (2.2.5). After transformations, we yield the so called *generalized hierarchical equation of the first order* for the unknown functional operator (transformation function) $\hat{U}^{(1)}$:

$$\left(\frac{\partial \hat{U}^{(1)}}{\partial z^{(1)}}, Z^{(1)}(z^{(1)}, t, \varepsilon_1)\right) + \frac{\partial \hat{U}^{(1)}}{\partial t} = Z^{(0)}(\hat{U}^{(1)}(z^{(1)}, t, \varepsilon_1), t, \varepsilon_1). \quad (3.1.2)$$

The first term of equation (3.1.2) is product of Jacobi matrix $\partial D^{(1)}/\partial z^{(1)}$ of the order $n \times n$ and the column vector $Z^{(1)}$. It is considered that the structural operator $M^{(0)}$ (see definition (2.2.6) and corresponding comments) is given. It means that, in view of definition (2.2.6), the new dynamical function of the first hierarchical level $Z^{(1)}$ can be considered as a known.

In the coordinate form the equation (3.1.2) can be written as

$$\sum_{i=1}^{n} \frac{\partial \hat{U}_k^{(1)}}{\partial z_i} Z_i^{(1)}\left(z_1^{(1)}, \ldots, z_n^{(1)}, t, \varepsilon_1\right) + \frac{\partial \hat{U}_k^{(1)}}{\partial t} = Z_k^{(0)}\left(\hat{U}_1^{(1)}, \ldots, \hat{U}_n^{(1)}, t, \varepsilon_1\right).$$
$$(3.1.3)$$

Accordingly to the resemblance hierarchical principle, relevant dynamical equation for the next hierarchical level should has mathematical structure analogous to (2.2.5). But at that time it takes place with respect to new (proper) dynamical variable set of the first level only:

$$\frac{d z^{(1)}}{d t} = Z^{(1)}(z^{(1)}, t, \varepsilon_1), \quad z^{(1)}(t_0) = z_0^{(1)}. \quad (3.1.4)$$

The new dynamical equation for the first hierarchical level (3.1.4), in correspondence with the above discussed hierarchical principles, should be *simpler* that the initial equation for the zeroth hierarchical level (2.2.5). Therefore, the equation of this type usually referred to as the *truncated equations.* Relevant solutions of equation (3.1.4) allow to find the dynamical (functional) operator $\hat{U}^{(1)}$ (see equations (3.1.2), (3.1.3)), and as the result, to construct the solutions for variables of the zeroth hierarchical level (see definition (3.1.1)). Owing to application of the hierarchical method, this calculational algorithm is found to be actually *simpler* than in the case of solving the initial problem (2.2.4) immediately.

In the case the discussed dynamical system, besides the zeroth, has two higher levels of hierarchy else (the first and the second, correspondingly), we could follow in the analogous way. Namely, we consider problem (3.1.4) as a new initial one (analogous to problem (2.2.4)). This means that it could be solved by the analogous method (see also illustrations in Fig. 1.1.2 and corresponding commentaries):

$$z^{(1)} = \hat{U}^{(2)}(z^{(2)}, t, \varepsilon_1, \varepsilon_2); \qquad (3.1.5)$$

$$\frac{d z^{(2)}}{d t} = Z^{(2)}(z^{(2)}, t, \varepsilon_1, \varepsilon_2), \quad z^{(2)}(t_0) = z_0^{(2)}; \qquad (3.1.6)$$

$$\left(\frac{\partial \hat{U}^{(2)}}{\partial z^{(2)}}, Z^{(2)}(z^{(2)}, t, \varepsilon_1, \varepsilon_2) \right) + \frac{\partial \hat{U}^{(2)}}{\partial t}$$
$$= Z^{(1)}(\hat{U}^{(2)}(z^{(2)}, \varepsilon_1, \varepsilon_2, t), t, \varepsilon_1, \varepsilon_2); \quad (3.1.7)$$

$$\cdots \cdots \cdots \cdots \cdots \cdots \cdots \cdots \cdots \cdots \cdots \cdots$$

$$z^{(m-1)} = \hat{U}^{(m)} \left(z^{(m)}, t, \varepsilon_1, \ldots, \varepsilon_m \right); \qquad (3.1.8)$$

$$\frac{d z^{(m)}}{dt} = Z^{(m)} \left(z^{(m)}, t, \varepsilon_1, \ldots, \varepsilon_\kappa, \ldots, \varepsilon_m \right); \quad z^{(m)}(t_0) = z_0^{(0)}; \qquad (3.1.9)$$

$$\left(\frac{\partial \hat{U}^{(m)}}{\partial z^{(m)}}, Z^{(m-1)}(z^{(m)}, t, \varepsilon_1, \ldots, \varepsilon_m) \right) + \frac{\partial \hat{U}^{(m)}}{\partial t}$$
$$= Z^{(m-1)}(\hat{U}^{(m)}, t, \varepsilon, \ldots, \varepsilon_m). \quad (3.1.10)$$

Thus summarizing, we can formulate the general idea of the hierarchical method. This idea consists in the use of the fact that each new truncated dynamical equation for a higher κ-th hierarchical level is *simpler* than the preceding one. At the same time, it possesses the *analogous* mathematical structure with the initial (and preceding) equation. (This is a result of practical realization of the compression and resemblance hierarchical principles — see the preceding Chapter). This means that the solution procedure for the dynamical equation of the last hierarchical level should be essentially *simpler* than the analogous procedure for initial equation of the zero level.

Further we propose some unexpected analogy of the discussed calculational scheme with one like situation in biology. In this case the truncated equation of the highest hierarchical level (3.1.9) can be treated as some peculiar 'gene' of the treated dynamical system. Whereas the complete set of equations (3.1.8)–(3.1.10) we can treated as some 'embryo'. Within this 'gene' complete set of properties of the original system (2.2.4) are 'enciphered'. This situation is like that, as in the 'true' gene the information about some organism is enciphered. Providing some 'culture *medium'* for our 'mathematical gene' one can evolve it into the original system (2.2.4) (similarly, for instance, to situation with developing of the human embryo). In our formal case the functional operators $\hat{U}^{(\kappa)}$ and the generalized hierarchical equations of the type (3.1.2), (3.1.7), (3.1.10), etc. play the role of such peculiar 'culture medium' for developing our 'equations embryo' (3.1.10). It is clearly that in this case we indeed can hope for simplifying of general solution procedure because any embryo in nature always is simpler the whole organism.

Thus this analogy once more illustrates that all natural hierarchical dynamical systems apparently have a common nature. Moreover we might hope that these 'common nature' could be described in framework of the proposed new hierarchical approach.

At last, we should mention that this method, generally, has some difficulties. Most of them originate from peculiarities of integration of generalized hierarchical equations (3.1.2), (3.1.3), (3.1.7), (3.1.10). These difficulties relate to the special form of a chosen structural operator. It is obvious that such calculational schemes have of practical interest in the one case only. Namely, in the case when the total labor-intensity of the joint solution procedure (for hierarchical dynamical equations (2.2.4), (3.1.4), (3.1.6), (3.1.9) and corresponding generalized ones (3.1.2), (3.1.3), (3.1.7), (3.1.10)) is less than the procedure of solving initial equation (2.2.1) 'immediately'. This result can be attained really if some *analytical* algorithms could be constructed for the solving generalized hierarchical equations like to (3.1.2), (3.1.3). This holds at a practice for special structural operators M only. General situation is simplified in the case of periodic (oscillatative) systems. A number of structural operators of such kind can be constructed for such situations. It should be mentioned, however, that most widely used such operators are based on the well known Riemann theorem of integral averaged value [1]:

$$\lim_{p \to \infty} \left[\int_a^b f\left(x\right) \sin\left(px\right) dx \right] \to 0. \qquad (3.1.11)$$

Let us discuss some of structural (averaging) operators M, which can be constructed on the basis of Riemann theorem (3.1.1).

1.2   Averaging Operators

Averaging (smoothing) *operators* satisfy the above mentioned labor-intensity test in optimal manner. It is explained by the wide prevalence of these operators in the problems of electrodynamics and mechanics. Most often, averaging operators are used for some explicit function of time t, space coordinate z_i or of the t and z_i simultaneously. Formally, these operators can be constructed as

$$M_z\left[Z\left(z,\varepsilon\right)\right] = \bar{Z}\left(\varepsilon\right) = \frac{1}{(2\pi)^n}\int\limits_0^{2\pi}\cdots\int\limits_0^{2\pi} Z\left(z,\varepsilon\right)dz_1\ldots dz_n, \quad (3.1.12)$$

the *averaging operator* on spatial coordinates, and

$$M_t\left[Z\left(z,t,\varepsilon\right)\right] = \bar{Z}\left(z,t_0,\varepsilon\right) = \lim_{T\to\infty}\frac{1}{T}\int\limits_t^{t_0+T} Z\left(z,t,\varepsilon\right)dt \quad (3.1.13)$$

is the *averaging operator* on time. The condition of the function periodicity in the latter case

$$Z\left(z,t+(2\pi),\varepsilon\right) = Z\left(z,t,\varepsilon\right)$$

is satisfied. In the first case (3.1.12) the function $Z\left(z,\varepsilon\right)$ is supposed 2π-periodical on all components of n-dimensional vector z within the domain of definition $z \in G^n$, i.e.,

$$Z\left(z+(2\pi),\varepsilon\right) = Z\left(z,\varepsilon\right), \quad (3.1.14)$$

where by $(z+(2\pi),\varepsilon)$ the vector $(z_1+2\pi,\ldots,z_n+2\pi)$ is denoted. If the function Z is periodical only on a part of coordinates $z_1,\ldots,z_n$ the operator M_z (3.1.12) can be determined as

$$M_{zs}\left[Z\left(z,\varepsilon\right)\right] = \bar{Z}\left(z_{s+1},\ldots,z_n,\varepsilon\right)dz_1\ldots dz_s, \quad (3.1.15)$$

if

$$Z\left(z_1+(2\pi),\ldots,z_s+(2\pi),z_{s+1},\ldots,z_n,\varepsilon\right)$$
$$= Z\left(z_1,\ldots,z_s,z_{s+1},\ldots,z_n,\varepsilon\right);\ \left(0\leqslant s\leqslant n\right). \quad (3.1.16)$$

It means that the averaging procedure is performed only on part of the variables $z_1,\ldots,z_s$, and that the function Z is periodical in these variables only. While integrating in (3.1.13), the variables $z_1, \ldots, z_s$ are considered as relevant constants. Here and below in this subsection we neglect hierarchical index κ.

In spite of formal resemblance of both versions of the averaging operator, their nature is essentially different. Namely, the averaging in (3.1.13) is performed over the independent scalar variable t(the laboratory time, for instance). Whereas, in similar procedures (3.1.12) we have dealing with the components of some vector z. But in the general case, the latter s relevant function of variable t. (This functional dependence is determined by dynamical equation (2.2.1)). Nevertheless, the coordinates z_s are considered as constants at performing relevant averaging procedures in (3.1.12). It should be noted that procedures of constructing of corresponding averaging operator with accounting of dependencies of the type $z_s(t)$ is possible also (see, for instance, [2–4]). But below in this book we will not have to do with such situations.

In practice *mixed* version of function $Z(z, t, \varepsilon)$ periodicity (in time t and coordinates of vector z simultaneously) is realized. In view of definitions discussed, the construction of *averaging operator* is obvious. However, let us mention one special case of the mixed averaging operators. It realizes in situations, when acting forces are of wavy form. The discussed in Chapter 1 *free electron laser (FEL)* (analyzed also in Chapters 8–11, Volume II) is evident case of mixed problem. Characteristic peculiarity of the latter is linear combination of time t and space z_j $(j = 1, 2, \ldots, s)$ coordinates. Such special combinations are called *scalar phases* p_i $Z(z, t, p, \varepsilon)$ $(i = 1, 2, \ldots, q)$ (see, for instance, definition (1.3.7) in Chapter I). The function $Z(z, t, p, \varepsilon)$ (written in *parametric form* with respect to p) is periodical in components of vector p. By p here the following vector is denoted

$$p = \{p_1, \ldots, p_i, \ldots, p_q\}; \quad p_i = \omega_i t - \sum_j \left(k_j^{(i)} z_j \right), \qquad (3.1.17)$$

where ω_i are components of the vector of cyclic frequencies ω and $k_j^{(i)}$ are components of the wave vectors $k^{(i)}$ in n-dimensional space:

$$\omega = \{\omega_1, \ldots, \omega_i, \ldots, \omega_q\}; \quad k^{(i)} = \left\{ k_1^{(i)}, \ldots, k_j^{(i)}, \ldots, k_n^{(i)} \right\} = k(\omega_i). \tag{3.1.18}$$

In problems of electrodynamics and mechanics the case of function $Z(z, t, \varepsilon)$ expandable in a Fourier series takes a special place. It means that the *Dirichlet–Jordan theorem* [5] about development in a Fourier series holds here. Then the n-multiplied Fourier series express the function Z in the domain G_n. For the last case the relevant Fourier expansion can be written as

$$Z\left(p,\varepsilon\right) = \sum_{\|m\|\geqslant 0} Z\left(\varepsilon\right) \exp\left\{i\left(mp\right)\right\}, \qquad (3.1.19)$$

where $\|m\| = |m_1| + \cdots + |m_q|$ is the norm of the q-*dimensional* harmonic vector $m = \{m_1, \ldots, m_q\}$, the components of which are integer numbers $m_i = 0, \pm 1, \pm 2, \ldots;\ i = 1, \ldots, q;\ (mp)$ is a scalar product of two q-dimensional vectors:

$$(mp) = \sum_{i=1}^{q} m_i p_i.$$

According to these definitions, numbers m_i play the role of *harmonic numbers* and $Z_m\left(\varepsilon\right)$ are *Fourier amplitudes*.

2. STOCHASTIC HIERARCHICAL SYSTEMS

The general characteristics of the oscillation-like problems have discussed above in the previous Section. In what follows let us dwell brief on the hierarchical wave-like problems. Therefore the stochastic *distributed systemss* are chosen as a convenient illustration example of such kind.

2.1   Factors of Stochasticity in Dynamical Systems

Let us discuss mathematical peculiarities of the description of stochastic hierarchical systems. Within the framework of the presented approach, the stochastic properties of a model can be determined by three different causes:

a) random character of initial conditions for scalar components of vector z for corresponding equations (2.2.1), (3.1.4), (3.1.6), (3.1.9) (for instance, a spread in thermal particle velocities in plasma-like system input);

b) stochastic dynamics in physical mechanisms described by functions $Z^{(\kappa)}$ (collision processes in kinetic systems);

c) stochasticity of the constructing the hierarchical tree, i.e. by the non-synonymy of relevant solutions for the transformation functions $\hat{U}^{(\kappa)}$.

Due to stochastic mechanisms, we have the transformation of initially determined hierarchical tree into a *stochastic ensemble of hierarchical trees* (about the concept of hierarchical tree see in Chapter 2, Subsection 2.5). Here we discuss only general ideas of possible stochastic versions of the presented hierarchical theory. Later (see, for example, Chapters 12, 13, Volume II) we will consider and illustrate various particular features of these systems in the framework of the method of averaged kinetic equation (see Chapter 6).

2.2 Example of Hierarchical Model of Stochastic System

For simplicity we consider that only the first two of above mentioned processes determine stochasticity of a system. To describe this system we construct *distribution function*

$$f = f\left(z_1, \ldots, z_j, \ldots, z_n, t, \varepsilon_1, \ldots, \varepsilon_\kappa, \ldots, \varepsilon_m\right), \qquad (3.2.1)$$

where z_j are the components of z-vector, and $\varepsilon_\kappa = \varepsilon_1, \ldots, \varepsilon_\kappa, \ldots,$ $\varepsilon_m \ll 1$ are scale parameters. So the dynamics of phase volume is determined by the single-particle dynamics of components z_j. In fact, amongst these components there are space coordinates of particles and their velocities.

The motion of each separate particle is described by single-particle equation of the type (3.1.4), (3.1.6). In the stochastic model discussed here , the fact stochasticity of motion is determined by only the first two causes, i.e., random character of initial conditions for scalar components of vector z and stochastic dynamics in physical mechanisms described by functions $Z^{(\kappa)}$, respectively. Besides that we assume 'determined part' Z' can be separated from the function $Z(z, t, \varepsilon)$ (see (3.1.4), (3.1.6)) in additive way:

$$Z = Z' + Z'', \qquad (3.2.2)$$

where the function Z'' describes the stochastic component of the particle dynamics (for instance, stochastic addendum in the Langevin equation). So, in the case $Z' = 0$ the pure stochastic model realizes. In the opposite case $Z'' = 0$ the stochasticity of the system is determined by stochasticity of initial (boundary) conditions only. Further, we accept for simplicity that discussed stochastic system has properties of a quasi-continuous medium. We differentiate the definition (3.2.2) in time t and change the momentum differential into particular derivatives traditionally

$$\frac{d}{dt} \rightarrow \frac{\partial}{\partial t} + \sum_{j=1}^{n} \frac{dz_j}{dt} \frac{\partial}{\partial z_j} \qquad (3.2.3)$$

In what follows we consider single-particle equation (2.2.5) and assumption (3.1.2). As a result we obtain the so called *kinetic equation*:

$$\left(\frac{\partial}{\partial t} + \sum_{j=1}^{n} Z_j' \frac{\partial}{\partial z_j} \right) f(z_1, \ldots, z_n, t, \varepsilon_\kappa)$$

$$= \left(\sum_{j=1}^{n} Z'' \frac{\partial}{\partial z_j} \right) f(z_1, \ldots, z_n, t, \varepsilon_\kappa). \quad (3.2.4)$$

Quite often the function Z'' describes the collision processes in stochastic systems. Therefore, we call the right side of the equation (3.2.4) *the collision integral* ςI_{st}. Then the equation (3.2.4) can be rewritten in the following more compact form:

$$\left(\sum_{j=1}^{n} Z''(z_1, \ldots, z_n, t, \varepsilon_\kappa) \frac{\partial}{\partial z_j} \right) f(z_1, \ldots, z_n, t, \varepsilon_\kappa) = \zeta I_{st}, \quad (3.2.5)$$

where $\zeta \ll 1$ or $\zeta \gg 1$ is the *collision scale parameter*. In the case of the 6-dimensional space (three space coordinates plus three components of canonical momentum), we formally reduce (3.2.4), (3.2.5) to traditional form of the *Boltzmann equation*.

Thus we see that stochastic dynamics of the system, as a whole, really is determined by single-particle dynamics of its elementary parts. This circumstance allows us to use the version discussed above of the hierarchical approach to investigating kinetic systems, too. Indeed, using the formula (3.3.1) and definition (3.2.5) we reduce kinetic equation (3.2.4) to the form with the total differential

$$\frac{df}{dt} = Z_f(f, t, \varepsilon_1, \ldots, \varepsilon_m), \quad (3.2.6)$$

where all notations are obvious in view of context of the about said. It is seen that mathematical structure of equations (3.2.6) and, for example, (2.2.5) is the same. Hence, hierarchical calculational scheme set forth above can be applied for solving the stochastic problem, too. However, there are here some methodical difficulties. The latter are connected with some peculiarities of calculational procedures for back transformations from the system with total derivatives to the system with particular derivatives. The essence of idea of these transformations and the technical ways of overcoming of the mentioned difficulties will be discussed below in other Chapters. In particular, one example of such hierarchical technology is described in this book in a form of the method of averaging characteristics, method of averaged quasi-hydrodynamic equation, and the method of averaged kinetic equation, respectively (see, for example, Chapters 6–8 in this Volume and in Chapters 12 and 13, Volume II). Here we only point out that generally problems of such a type are particular variety of the wider class of problems called the *hierarchical wave* ones. Let us shortly discuss some characteristic properties of systems described by such kind equations.

3. WAVE RESONANT HIERARCHICAL SYSTEMS

As we mentioned above, some peculiar aspects of analyzing distributed (wave) systems require relevant modifications of hierarchical methods. The mathematical specificity here is reflected by partial derivatives in equations governing the processes in the system. Therefore all problems of this kind can be divided into two separate groups. Namely, those problems which can be reduced to systems of exact differential equations, and those that cannot be reduced to such systems. Some class of wave systems described by so called *quasilinear equations* (with respect to the partial derivatives — see Chapter 6 for more details), including hyperbolic equations [6], quasi-hydrodynamic equation, Boltzmann kinetic equation (see Chapters 6–8, and Chapters 12, 13 in Volume) [7], etc., serve as explicit examples of the first kind. The list of the second kind is much wider. We mention here the nonlinear wave equations discussed by theoreticians, such as Korteveg–de Vries, Boussinesq, modified Korteveg–de Vries, nonlinear Schrodinger, sine-Gordon, and other equations [8]. Unfortunately, the averaging method in its classical form is not applicable in most cases of this type. The only exception is the class of weak nonlinear problems. In particular, the problems like to the problem about wave propagation in the media with weak non-linearity dispersion (about these concepts see in Chapter 1, Sections 2 and 3). Other numerous examples of such type could be found in the physics of condensed matter, spectroscopy of atoms and molecules, etc.. In electrodynamics problems of this kind were arose at first in studying the radio wave propagation in layers of the Earth atmosphere. M.A. Leontovich [9] proposed and accomplished an idea similar to the Van der Pol [10] and averaging (Krylov–Bogolyubov) [6,11–20] methods. The amplitudes of interacting waves are taken as the slowly varying variables (see Chapter 1, Subsection 2.4). Since the non-linearity is weak, the solution formally reproduces the solution of the equivalent linear system. The distinction is that wave amplitudes became slowly varying quantities and their dynamics is described by smoothed (truncated) equations (equations of the first hierarchical level — see Chapter 8).

Similarly to the Van der Pol and Krylov–Bogolyubov approaches, the evolution of *the slowly varying amplitude* method undergoes necessary stages of generalization and mathematical substantiation. A.V. Gaponov, M.I. Rabinovich, and L.A. Ostrovsky [21–23] have done this most completely. However, practice shows the application of their version in electrodynamic problems requires modification of the calculational procedure [2, 24–26]. The latter should account the specific features of physical conditions in plasma-like electrodynamic systems. For instance, it concerns various effects of nonlinear generation of quasistationary electromagnetic fields, realization of degenerate mechanisms of wave resonant interactions, etc.. The *modified version* of such a type is described and illustrated below in Chapters 8, 12 and 13.

3.1   Standard Hierarchical Equations in the Case of Hierarchical Wave Problem

N.M. Krylov and N. N. Bogolyubov [11, 12] elaborated (in the form of the averaging method [6, 11–19]) the first calculational approach for asymptotic solution of nonlinear oscillation problems that we now regard as hierarchical one. Bogolyubov's algorithm, in itself, is the simplest version of the two-level hierarchical calculational scheme (see Section 5). So, hierarchical technologies developed here (see Chapters 4–8) can be considered as relevant generalization of Bogolyubov's ideas [6, 7, 12–15, 19].

As mentioned previously, the essential peculiarity of the methods of such class (as well as their hierarchical variety) is that they have to do with *exact differential equation*, which are written in some specific forms. In references these forms are called as *standard systems* (see, for example, equations (1.1.4), (2.2.1), relevant materials of Chapter 4, and Section 5). Such calculational situations are typical, as mentioned above, for lumped systems. Apart from the lumped systems the *distributed systems* are rather widespread in electrodynamics. As mentioned above, the use of *particular derivatives* is most characteristic point for formal methods of description of the distributed systems, from the mathematical point of view. Therefore, at the first step of the constructing of relevant hierarchical calculational procedures for the case of distributed systems (i.e., the systems with particular derivatives) we should formulate also the analogous concept of a standard system.

In what follows let us remind the above said that general ideology of the hierarchical calculational approaches in both these cases are rather close. This circumstance in some specific cases allows to consider the calculational technologies developed for systems with exact differential equations and systems with particular derivatives from some common point of view. In other cases we have not such possibility. However, different types of equations possess their different specific features that can be expressed in the form of corresponded calculational procedures. Therefore below we undertake the attempt to classify the wave resonant problems with respect to this criteria. The concept of the *general standard form* is put in the basis of such classification.

Thus we consider dynamical wave system described by the following differential matrix equation, which we below consider *general standard form*:

$$A\frac{\partial U}{\partial t} + (BP)U + CU = R(U, \partial U/\partial t, (PU), z, t), \qquad (3.3.1)$$

where A, B, C are square matrices of size $n \times n$, $U = U(z, t)$ is some vector-function in Euclidean n-dimensional space R^n with coordinates $\{z_1, z_2, \ldots, z_n\}$, i.e. $\forall z \in R^n \ z = (z_1, z_2, \ldots, z_n)^T$, $\forall z_I \in (-\infty, +\infty)$, $I \in (1, 2, \ldots, n)$, P is some linear differential operator in the space R^n, $R(\ldots)$ is a given weak nonlinear vector-function, t is some scalar variable, for instance, the laboratory time. The equation (3.3.1) in the coordinate form can be written as

$$\sum_{l=1}^{n} \left(a_{pl}\frac{\partial}{\partial t} + b_{pl}P_{pl} + c_{pl} \right) U_l = R_p (U, \partial U/\partial t, (PU), z, t), \qquad (3.3.2)$$

where a_{pl}, b_{pl}, c_{pl} are the elements of the matrices A, B, C, P_{pl} are elements of the operator P; U_l, R_p are the components of the vectors U and R.

One may be convinced that *Maxwell's equations, the Boltzmann kinetic equation*, the quasi-hydrodynamic equation, and many others can be reduced to general standard form (3.3.1), (3.3.2) [6, 13–18]. The relevant examples are given below in Chapter 8 and Chapters 12, 13 in Volume II). Unfortunately, we should state that up to today there is no general algorithm for integration of equations (3.3.1), (3.3.2) immediately. Therefore researchers usually consider particular cases only. We also follow this way.

At first we use hierarchical principles (see Chapter 2 for more details). We begin with general hierarchical principle that the considered dynamical system has a hierarchical structure. Relevant dynamical equations for each higher level are simpler than for preceding one (the principle of information compression). Moreover, each hierarchical level of the system is described by equations of the same mathematical structure (the principle of hierarchical resemblance). Therefore for each hierarchical level we have set of another dynamical variables proper to it (the principle of information compression), and so on. Then we introduce structural and functional operators.

The structural hierarchical operator $M^{(\kappa)}$ can be introduced similar to (2.2.6):

$$R^{(\kappa+1)} = R^{(\kappa+1)}\left(U^{(\kappa+1)},\ldots\right) = M^{(\kappa)} R^{(\kappa)}\left(U^{(\kappa)},\ldots\right), \qquad (3.3.3)$$

where $R^{(\kappa)}$ is the given vector-function defined before (the right part of (3.3.1)). This function characterizes nonlinear dynamical properties of κ-th hierarchical level. It is obvious that the operator (3.3.3) satisfies all hierarchical principles, too.

The *transformation functional operator* (some function in the given particular case) describes the dynamics of connections between proper variables of two neighboring hierarchical levels. We can determine it analogously to (2.2.7):

$$U^{(\kappa)}\left(z^{(\kappa)},t\right) = \hat{U}^{(\kappa+1)}\left(U^{(\kappa+1)}\left(z^{(\kappa+1)},t\right)\right). \qquad (3.3.4)$$

Both operators of hierarchy are not independent mathematical objects. We consider some integral structural operator $M^{(\kappa)}$ is given. Then we act on equation (3.3.4) by the differential operator defined by the left side of equation (3.3.1). Further, we use the second hierarchical principle with equation (3.3.1) and definition (3.3.4). The general hierarchical transformation equation can be obtained on this way

$$\left\{ A\frac{\partial}{\partial t} + \left[B\left(\frac{\partial z^{(\kappa+1)}\left(z^{(\kappa)}\right)}{\partial z^{(\kappa)}} \Big|_{z^{(\kappa)}=z^{(\kappa)}\left(z^{(\kappa+1)}\right)} \frac{\partial}{\partial z^{(\kappa+1)}} \right) \right] + C \right\}$$

$$\times \hat{U}^{(\kappa+1)}\left[U^{(\kappa+1)}\left(z^{(\kappa+1)},t\right) \right]$$

$$= \left[M^{(\kappa+1)} \right]^{-1} R^{(\kappa+1)}\left(U^{(\kappa+1)}, \frac{\partial U^{(\kappa+1)}}{\partial t}, \frac{\partial U^{(\kappa+1)}}{\partial z^{(\kappa+1)}}, z^{(\kappa+1)}, t \right), \qquad (3.3.5)$$

where required vector dependences $z^{(\kappa+1)} = z^{(\kappa+1)}\left(z^{(\kappa)}\right)$ and $z^{(\kappa)} = z^{(\kappa)}\left(z^{(\kappa+1)}\right)$ can be found by definition (3.3.4); $\left[M^{(\kappa+1)} \right]^{-1}$ is inverted operator $M^{(\kappa+1)} = M^{(\kappa)}|_{\kappa \to \kappa+1}$: these operators satisfy to the following condition: $M^{(\kappa+1)} \left[M^{(\kappa+1)} \right]^{-1} = I$, where I is the unit vector. Here also the corresponding operator $P = P^{(\kappa)}$ after relevant transformations is written in the variables of $(\kappa + 1)$-th hierarchical level. In this case we choose it in the simplest differential form: $P^{(\kappa+1)} = \partial/\partial z^{(\kappa+1)}$.

Therefore, with given structural operator $M^{(\kappa+1)}$ (or the inverse operator $\left[M^{(\kappa+1)}\right]^{-1}$) the equation (3.3.5) gives us relevant hierarchical transformation function $U^{(\kappa+1)}$ (3.3.4). Further, we assume the function $R^{(\kappa+1)}$ weakly nonlinear and expandable in Fourier series.

It should be mentioned that these assumptions hold for wide class of electrodynamic problems. One type of such problems formally correlates with the problem of a charged particle motion (beam) in predetermined electromagnetic fields [3, 4, 7, 27–35]. The self-consistent dynamics of plasma-like systems is the second example [2, 7, 24–26].

3.2 Classification of Problems

As mentioned above, the hierarchical algorithms can be conditionally divided into two groups:

a) the algorithms reducible to integration of standard 'single-particle' equation (1.3.4);

b) the algorithms reducible to the so called *Rabinovich standard form* [21–23].

Let us in what follows shortly to comment specific features of both these cases.

Case a).   This case realizes whether if some calculational procedure for constructing of some equivalent set of exact differential equations can be accomplished really. A number of procedures of such type are known in references. They are the method of characteristics (see Chapter 6, Section 3), the method of lines, the method of total differential (see above expression (3.2.3) and corresponding commentaries), and some others. Let us illustrate this idea in the simplest example of the method of total differential. We suppose for this that two first terms in left side of can be rolled up to the total derivatives because (3.2.3):

$$A = B = I;$$

$$(BP) = \left(\frac{dz^{(\kappa)}}{dt} \times \frac{\partial}{\partial z^{(\kappa)}}\right);$$

$$R = R^{(\kappa)}\left(U^{(\kappa)}, z^{(\kappa)}, t\right);$$

$$\frac{dz^{(\kappa)}}{dt} = Z^{(\kappa)}\left(z^{(\kappa)}, t\right), \tag{3.3.6}$$

where I is unit matrix, the expression in the right side of the second equation is the product of the diagonal Jacobian matrix of order $n \times n$

and column operator $\partial/\partial z^{(\kappa)}$; $Z^{(\kappa)}$ is a nonlinear dynamical vector function defining single-particle motion in the system. One may be confident that the stochastic hierarchical problem discussed above can be reduced to the form (3.3.1), (3.3.6). Using (3.3.6) after relevant transformation equation (3.3.1) can be rewritten as

$$\frac{dU^{(\kappa)}}{dt} = R'^{(\kappa)} \left(U^{(\kappa)}, z^{(\kappa)}, t \right); \qquad (3.3.7)$$

where $R'^{(\kappa)} = R^{(\kappa)} - CU^{(\kappa)}$ is given nonlinear wave vector-function of κ-th hierarchical level. Further, we introduce a new *extended coordinate vector* for κ-th hierarchical level:

$$z'^{(\kappa)} = \left\{ z^{(\kappa)}, U^{(\kappa)} \right\}, \qquad (3.3.8)$$

and a new *extended nonlinear dynamical vector-function*

$$Z'^{(\kappa)} = \left\{ Z^{(\kappa)}, R'^{(\kappa)} \right\} = Z'^{(\kappa)} \left(z'^{(\kappa)}, t \right). \qquad (3.3.9)$$

Therefore the initial system (3.3.1) obtains the form of relevant 'true' single-particle equation like (3.3.1) or (2.2.1):

$$\frac{dz'^{(\kappa)}}{dt} = Z'^{(\kappa)} \left(z'^{(\kappa)}, t \right). \qquad (3.3.10)$$

In the particular case $R'^{(\kappa)} = 0$ we obtain 'pure single-particle' version of the discussed hierarchical problem.

Thus a spatially distributed wave resonant dynamical problem described by equation (3.3.1) in the case (3.3.6) can be reduced to quasi-single-particle problem (3.3.10). This means that all calculational technologies, developed for asymptotic integration of the 'single-particle' exact equations (see Sections 1, 4–6, and Chapters 4, 5), can be successfully used for solution of this type distributed problems. However, there are some essential differences between the calculational algorithms in the 'purely lumped' and 'distributed' cases, respectively. Namely, it is readily seen that equation (3.3.10) allows to find solutions for the function $U^{(\kappa)}$ in the form $U^{(\kappa)} = U^{(\kappa)}(t)$. Whereas the initial form of initial 'distributed' equation (3.3.1) supposes the eventual solutions' dependency of the form $U^{(\kappa)} = U^{(\kappa)} \left(z^{(\kappa)}, t \right)$. This problem is known as the *back transformation* one. Specific calculational technology is developed for its solution. Let us shortly discuss the main ideas of such calculational technology.

Constructing of the truncated equation for the m-th hierarchical level

$$\frac{dz'^{(m)}}{dt} = Z'^{(m)}\left(z'^{(m)}, t, \varepsilon_1, \ldots, \varepsilon_m\right), \qquad (3.3.11)$$

and solving it according to the general hierarchical scheme, we obtain formal solutions for $(m-1)$-th level in the form (see definition (2.2.7) and corresponding commentaries):

$$z'^{(m-1)} = \hat{U}'^{(m)}\left(z'^{(m)}, t, \varepsilon_1, \ldots, \varepsilon_m\right). \qquad (3.3.12)$$

Taking into account definitions (3.3.8) and (3.3.9), we rewrite solution (3.3.12) as

$$U^{(m-1)}\left(z^{(m)}\right) = \hat{U}_U^{(m)}\left(z^{(m)}, U^{(m)}, t, \varepsilon_1, \ldots, \varepsilon_m\right) z^{(m-1)}\left(z^{(m)}\right)$$
$$= \hat{U}_z^{(m)}\left(z^{(m)}, U^{(m)}, t, \varepsilon_1, \ldots, \varepsilon_m\right), \qquad (3.3.13)$$

where all notations are evident. Further, we expand the function $U^{(m-1)}\left(z^{(m)}\right)$ in a series in small vicinity of the point $z(m-l)$:

$$U^{(m-1)}\left(z^{(m-1)}, t\right) \cong U_U^{(m)}\left(z^{(m-1)}, U^{(m)}\left(z^{(m-1)}\right), t, \varepsilon_1, \ldots, \varepsilon_m\right)$$
$$+ \frac{1}{2!}\frac{\partial \hat{U}_U^{(m)}}{\partial z^{(m)}}\Delta z^{(m-1)}\bigg|_{z^{(m)}=z^{(m-1)}} + \cdots, \qquad (3.3.14)$$

where Δz_{m-1} is the difference between $z^{(m)}$ and $z^{(m-1)}$, which can be found from the corresponded expansion of the function $z^{(m)}$ in small vicinity of the point $z^{(m-1)}$.

In what follows, we perform the analogous procedure with hierarchical levels $(m-1)$ and $(m-2)$, and so on. Moving by this method 'backward', we eventually can reach the 'last' zeroth level

$$U(z, t) \equiv U^{(0)}\left(z^{(0)}, t\right) = \hat{U}^{(1)}\left(z^{(1)}, U^{(1)}, t, \varepsilon_1, \ldots, \varepsilon_m\right). \qquad (3.3.15)$$

After performing expansion in a series (3.3.15) in small vicinity of the point $z^{(1)} = z$ we obtain the looked for dependency $U(z, t)$.

3.3 Krylov–Bogolyubov Substitution

As shown below, the physical situation with the function $Z^{(\kappa)}(z^{(\kappa)}, t)$ in the right part of equation (3.3.10) *periodical* (or *nearly periodical*, or *conditionally periodical*) is most interesting. Therein experience shows that the most promising versions of hierarchical method can be constructed within averaging methods (in more detail such type calculational algorithm is described below in Section 5). Let us say a few words about such calculational scheme based on *Krylov–Bogolyubov substitution*. This is interesting for another reasons, too. The point is that historically N.M. Krylov and N.N. Bogolyubov (and many followers after) [6,12–19] namely began studying the averaging methods. The latter can be regarded as the first practically effective realization of the hierarchical idea in practical calculational algorithm.

We assume that the function $Z'^{(\kappa)}$ in (3.3.10) is periodical and slowly varying

$$Z^{(\kappa)}\left(z^{(\kappa)}, t, \varepsilon_\kappa\right) = \varepsilon_\kappa \hat{Z}^{(\kappa)}\left(z^{(\kappa)}, t, \varepsilon_\kappa\right), \qquad (3.3.16)$$

where the 'prime' is ignored further for simplicity. In this case, the functional operator $\hat{U}^{(\kappa)}$ can be found from the *Krylov–Bogolyubov substitution*:

$$z^{(\kappa-1)} = \hat{U}^{(\kappa)}\left(z^{(\kappa)}, t, \varepsilon_\kappa\right) = z^{(\kappa)} + \varepsilon_\kappa \hat{U}'^{(\kappa)}\left(z^{(\kappa)}, t, \varepsilon_\kappa\right). \qquad (3.3.17)$$

Relevant general hierarchical equation for the vector-function $\hat{U}'^{(\kappa)}$ (3.3.2) can be written in some another (with respect to (3.3.2)) form

$$\varepsilon_\kappa \left(\frac{\partial \hat{U}'^{(\kappa)}}{\partial z^{(\kappa)}}, Z^{(\kappa-1)}\left(z^{(\kappa)}, t, \varepsilon_\kappa\right)\right) + \frac{\partial \hat{U}'^{(\kappa)}}{\partial t}$$

$$= Z^{(\kappa-1)}\left(z^{(\kappa)} + \varepsilon_\kappa \hat{U}'^{(\kappa)}, t, \varepsilon_\kappa\right) - Z^{(\kappa)}\left(z^{(\kappa)}, t, \varepsilon_\kappa\right) \quad (3.3.18)$$

Here, we would like especially to stress that equation (3.3.18), as it will be shown below in Section 5, can be solved analytically [11, 12]. Due to this, the application of hierarchical scheme based on substitution (3.3.17) turns out to be very effective.

3.4   Case b)

But let us turn to the problem discussed above in Subsection 3.2. Assume that function R in standard form (3.3.1) can be developed in a Fourier series. Separate out the zeroth Fourier harmonics

$$R_0 = R - \tilde{R}, \qquad (3.3.19)$$

where $\tilde{R}$ is the oscillatory part of the function R. Then we consider that function R can be represented as a series in powers of the highest hierarchical small parameter ε_1:

$$R = \sum_{n=1}^{\infty} \varepsilon_1^n R_U^{(n)} \left(U, \partial U / \partial t, (PU), z, t, \varepsilon_1, \ldots, \varepsilon_m \right). \qquad (3.3.20)$$

With (3.3.19), (3.3.20) we can write the standard equation (3.3.1) as a set of two equations

$$A \frac{\partial U_0}{\partial t} + (BP) U_0 + C U_0 = \sum_{n=1}^{\infty} \varepsilon_1^n R_{U0}^{(n)} \left(U, \frac{\partial U}{\partial t}, (PU), z, t, \varepsilon_1, \ldots, \varepsilon_m \right);$$

$$A \frac{\partial \tilde{U}}{\partial t} + (BP) \tilde{U} + C \tilde{U} = \sum_{n=1}^{\infty} \varepsilon_1^n \tilde{R}_U^{(n)} \left(U, \frac{\partial U}{\partial t}, (PU), z, t, \varepsilon_1, \ldots, \varepsilon_m \right),$$

$$(3.3.21)$$

where $R_{U0}^{(n)}, \tilde{R}_U^{(n)}$ are relevant expansion coordinates, $U = U_0 + \tilde{U}$.

Each of equations (3.3.21) describes essential different class of the objects of studying. In particular, the first of the latter governs a 'smooth' (quasi-stationary) dynamics of the system. For example, it might concern a generation of some 'smooth' quasi-static electric and magnetic fields within a plasma-like relativistic system (see, for instance, Chapters 12, 13 in Volume II). Standard numerical methods or exact or approximate calculational methods can be used for solving this part of the problem.

The second of equations (3.3.21) describes some nonlinear dynamics of waves in the system, including the wave resonant interactions of different kind. Characteristic feature of this problem is presence a number of fast oscillations. At that, the general situation is complicated by the non-linear and resonant nature (quite often) of these wave interactions. The problems of such types can be solved by the use of hierarchical analytical–numerical methods.

There are three versions of solution to the discussed problem are described further in the book:

1) Application of the slowly varying amplitudes method (see Chapter 8, Sections 1–3).

2) Application of the method of averaged characteristics (see Chapter 6, Section 2).

3) Application of the hierarchical transformations of coordinates method (see Chapter 8, Section 4).

Therefore accordingly with the above said, we can distinguish two versions of *the slowly varying amplitude* method:

a) traditional version (see Chapter 8, Sections 1, 2) that is rather popular in acoustics, nonlinear optics, radiophysics, etc.;

b) modernized version (see Chapter 8, Section 3), which takes into account specific peculiarities of non-linear wave resonant electrodynamic problems of plasma-like systems.

Following any of these paths we can obtain some truncated (shortened) equations of the first hierarchy and relevant set of transformation relationships. Accordingly to the general hierarchical idea further we again use the described calculational scheme, separating the following (in a hierarchical series) small parameter ε_2. But at that time this doing with respect to the truncated equations of the first hierarchy, and so on. Ultimately, after exhausting terms in the relevant hierarchical series, we obtain some truncated equations for the highest hierarchical level. Solving these equations and fulfilling corresponding back transformations we eventually obtain complete solution of the initial problem.

4.   VAN DER POL'S METHOD

4.1   A Few Introductory Words

The mathematical averaging procedure has been employed in celestial mechanics as early as in Gauss' time [19]. It became with time a basic element of the algorithm for the approximate solution of the systems of nonlinear differential equations. But this happened only after the Flemish radio engineer Van der Pol had applied the averaging idea to propose a simple refined method of solving the nonlinear problems of electrical and radio engineering [10].

In spite of being insufficiently substantiated, the method has been widely employed by practical engineers, while mathematicians thought it was not worth their attention, in the same manner as they had treated Heaviside's papers on operational calculus. N.N. Bogolyubov and N.M. Krylov initiated the crucial changes in 1930s. They employed Van der Pol's ideas to develop a rigorous mathematical approach to the analysis of nonlinear ordinary differential equations of some specific type. Having separated the slow and fast variables, they presented the solution in the asymptotic series form, with the first term reproducing the relevant Van der Pol result [12,13]. Later on, the approach has been developed and substantiated by Ukrainian mathematicians of Bogolyubov's school (Yu.A. Mitropolsky, B.I. Moiseenkov, A.M. Samoilenko and others) [6,14,15].

4.2   Van der Pol's Method's Variables

Let us suppose that some lumped oscillatative system is described by the following second order equation:

$$\ddot{z} + \omega^2 z = \varepsilon \varphi(z, \dot{z}); \quad z(t_0) = z_0, \tag{3.4.1}$$

where z is the spatial coordinate, ω is some constant (say, linear cyclic frequency — see its concept above in Chapter 1, Section 2), $\varepsilon << 1$ is *the small parameter of the problem*, $\varphi(z, \dot{z})$ is a given (in the general case, nonlinear) function of coordinate z and velocity $\dot{z}$. Here the symbol 'dot' is total (material) derivative on time t (i.e., $\dot{z} \equiv dz/dt$). It is importantly that the requirements to the function φ are not too strict — it can be, e.g., discontinuous. The condition $\varepsilon << 1$ implies that the system (3.4.1) must be weakly nonlinear (about this concept see in Chapter 1, Section 2). It means that it describes nearly linear oscillations.

Then we introduce the *generating equation*:

$$\ddot{z} + \omega^2 z = 0, \tag{3.4.2}$$

which immediately follows from (3.4.1) for $\varepsilon = 0$. As readily seen, the generation equation (3.4.2) is an ordinary linear differential equation (compare with (1.2.8)) whose solution is well known to be (see, for instance, (1.2.9))

$$z = x \cos \psi, \tag{3.4.3}$$

where x is the amplitude, and

$$\psi = \omega(t - t_0) \tag{3.4.4}$$

is the linear oscillatative phase (see Chapter 1, Section 2). The variables x and ψ are referred to as Van der Pol *variables*. Van der Pol's main idea is to assume that the solution of the nonlinear equation (3.4.1) is formally described by the same expression as the solution of the linear equation (3.4.2). The difference is that *the amplitude x is slowly varying function of t*. The physical meaning of the term *slowly varying function* is that the amplitude x weakly changes during one oscillation period $T = 2\pi/\omega$ (for more details about this concept see in Chapter 1, Section 2). The result becomes appreciable only after a long time $t >> T$, which includes many oscillation periods $N = \psi/2\pi >> 1$. If the amplitude x is slow, then the phase ψ, for reasons given above, is the *fast variable*. It is reasonable to assume that the weak nonlinearity does not modify the general behavior of the solutions though gives rise to small nonlinear corrections in the relevant weakly nonlinear equation. In what follows we derive such equations from the slow amplitude and the fast phase.

We require that the new functions $x(t)$ and $\psi(t)$ satisfy, along (3.4.3), the condition

$$\dot{z} = -\omega x \sin\psi. \tag{3.4.5}$$

We differentiate (3.4.5) with respect to t and substitute the result in (3.4.1), to obtain:

$$-\dot{x}\omega\sin\psi - \omega x\dot{\psi}\cos\psi + x\omega^2\cos\psi = \varepsilon\varphi\left(x\cos\psi, -\omega x\sin\psi\right). \tag{3.4.6}$$

Moreover, we differentiate (3.4.3) with respect to t and make use of the condition (3.4.5), to obtain:

$$\dot{x}\cos\psi - x\dot{\psi}\sin\psi + \omega x\sin\psi = 0. \tag{3.4.7}$$

Thus we have derived a set of two differential equations for $x(t)$ and $\psi(t)$. Solving the equations with respect to $\dot{x}$ and $\dot{\psi}$ reduces the system (3.4.6)-(3.4.7) to the form:

$$\dot{x} = -\frac{\varepsilon}{\omega}\varphi\left(x\cos\psi, -x\omega\sin\psi\right)\sin\psi \equiv \frac{\varepsilon}{\omega}\varphi_1\left(x,\psi\right); \tag{3.4.8}$$

$$\dot{\psi} = \omega - \frac{\varepsilon}{\omega x}\varphi\left(x\cos\psi, -x\omega\sin\psi\right)\cos\psi \equiv \omega - \frac{\varepsilon}{\omega x}\varphi_2\left(x,\psi\right). \tag{3.4.9}$$

Expressions (3.4.8), (3.4.9) evidently illustrate concepts discussed above of slowly varying and fast variables. Indeed, as readily seen the ration of velocities of changing variables x and ψ is proportional to the small parameter ε:

$$\left|\dot{x}\right|\Big/\left|\dot{\psi}\right| \sim \varepsilon \ll 1. \tag{3.4.10}$$

Taking into consideration the definition for the hierarchical small parameters (2.2.3) we can consequently conclude that these variables are related to different hierarchical levels of the system. Besides that, we can detect that the Van der Pol's description discussed above could be treated as simplest and earliest version of the hierarchical approach in the theory of oscillations and waves.

4.3   Truncated Equations and Their Hierarchical Sense

The set of equations (3.4.8), (3.4.9) is nothing but an alternative version of the initial equation (3.4.1) and, in principle, it has no evident advantages over (3.4.1). However, the form (3.4.8), 3.4.9 suggests the way of how to find the approximate solutions. Following Van der Pol, we replace the slow varying amplitude x by its time averaged value, using averaging operator (3.1.13), i.e.,

$$x \cong \bar{x};$$
$$\dot{x} \cong \dot{\bar{x}}. \tag{3.4.11}$$

For the phase we substitute:

$$\psi \cong \bar{\psi};$$
$$\dot{\psi} \cong \dot{\bar{\psi}}. \tag{3.4.12}$$

Here the symbol 'overline' is averaging symbol. Then we carry out the time averaging in (3.4.8) and (3.4.9) with regard for (3.4.11) and thus obtain an approximate system of equations:

$$\dot{x} = \frac{\varepsilon}{\omega} \bar{\varphi}_1 \left(x \right); \tag{3.4.13}$$

$$\dot{\psi} = \omega - \frac{\varepsilon}{\omega x} \bar{\varphi}_2 \left(x \right), \tag{3.4.14}$$

where

$$\bar{\varphi}_{1,2} = \frac{1}{2\pi} \int_{0}^{2\pi} \varphi_{1,2} \left(x, \psi \right) d\psi \equiv M_{\psi} \varphi_{1,2} \left(x, \psi \right). \tag{3.4.15}$$

Equations (3.4.13), (3.4.14) are referred to as *truncated*. In terms of the hierarchical ideology they are the *equations of the first hierarchical level*. Small parameter ε in this case, as mentioned already, can be treated as the hierarchical scale parameter of the first hierarchical level (compare definitions (3.4.10) and (2.2.3), respectively). Thus we have the possibility of being convinced that *the Van der Pol method* is indeed the simplest realization of the hierarchical methods. Schematically idea of the Van der Pol method is illustrated in Fig. 3.4.1.

Deriving the system of truncated equations (3.4.13), (3.4.14) (i.e., the equations of the first hierarchical level) is the most important point of the Van der Pol method. This system possesses simpler mathematical structure (principle of information compression) than both the initial equation (3.4.1) and the system (3.4.8), (3.4.9), since equations (3.4.13) and (3.4.14) can be disconnected. Indeed, here one can solve the first equation independently and write the solution for the phase ψ in quadratures. As a result, solving the truncated equations actually is simpler than any procedure of integrating (3.4.1) directly. Moreover, in many problems of electrodynamics the researcher is interested in obtaining the rate of change of the phase $\dot{\psi} = \omega^*$, rather than the phase ψ in itself.

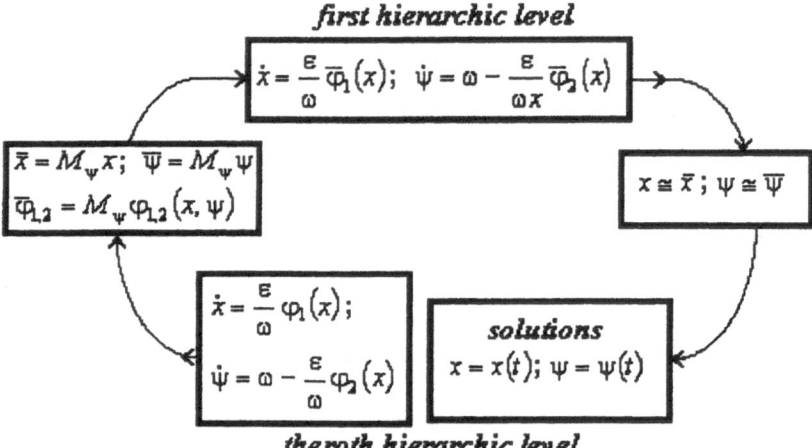

Figure 3.4.1. Illustration of hierarchical nature of the Van der Pol method. It is obviously seen that the averaged variables $\bar{x}$, $\bar{\psi}$ can be considered as proper variables of the first hierarchical level. It shown also, how resemblance hierarchical principle works in this version of the hierarchical method. Namely, here the mathematical structure of the dynamical equations for the zeroth hierarchical level are found to be similar to the corresponding equations of the first hierarchical level.

The quantity ω^* has physical meaning of the nonlinear cyclic oscillation frequency of the system. In particular, as will be shown in Chapter 8, Volume II, the dependency $\omega^*(t)$ (to be more exact, $\omega^*(x(t))$) is basic for realization of the elementary physical mechanism, which provide the operation of many electronic devices. Including, Free Electron Lasers (FEL), Cyclotron Resonance Maser (CRM), etc.. With $x(t)$ being given, the slow dependence $\omega^*(t)$ is immediately determined by truncated equation (3.4.14), i.e.,

$$\omega^*(x(t)) = \omega - \frac{\varepsilon}{\omega x(t)} \bar{\varphi}_2(x(t)), \qquad (3.4.16)$$

which makes this approach even more attractive for the researcher.

The idyllic situation is violated, however, by some difficulties. In particular, it had been unclear for a long time whether the solutions obtained by the Van der Pol method are sufficiently accurate, i.e., to what extent the exact solutions of the initial equation (3.4.1) are equivalent to those of the truncated systems (3.4.13), (3.4.14). This made the essence of the mathematicians' main object against the Van der Pol method [36]. This aspect has disturbed the practical engineers far less. They have been quite satisfied with the observation that in most cases the results obtained by this method had been in fairly good agreement with the experimental data. Few rare cases of obvious discrepancies did not influence the favorable attitude. However, the negative experience accumulated as the range of application of the method expanded. The problems were solved by N.M. Krylov and N.N. Bogolyubov, then Yu.A. Mitropolsky, D.I. Zubarev, A.A. Andronov, L.S. Pontriagin, A.N. Tichonov, V.N. Volosov, E.A. Grebennikov and other [6, 12–19]. Van der Pol's ideas obtained a new mathematically rigorous form and made a foundation for an extended mathematical theory, Bogolyubov's approach being its part. The latter made it possible to develop efficient asymptotic integration algorithms of the hierarchical type for a wide spectrum of differential equations.

5. METHODS OF AVERAGING. STANDARD VERSION

The main object of application of the hierarchical versions of the averaging methods is the mathematical model, which can be described by the following vector equation:

$$\frac{dz}{dt} = Z\left(z, t, \varepsilon_1, \ldots, \varepsilon_m\right), \tag{3.5.1}$$

where $z = \{z_1, z_2, \ldots, z_n\}$ is some vector and $Z = \{Z_1, Z_2, \ldots, Z_n\}$ is the relevant vector-function in n-dimensional Euclidean space R^n, $t \in [0, \infty]$ (for instance, the laboratory time), ε_κ are the hierarchical small parameters. In the present Section and below in Chapter 4 we, predominantly, will have to do with particular varieties of this model, where only influence of neighboring hierarchical levels really is important. In this case, as shown above in Chapter 2, Subsection 2.4, the general hierarchical problem (3.5.1) can be reduced to a successively catching hold of two-level hierarchical systems (see illustration in Fig. 2.2.1). As a result we eventually obtain that the *basic element* of any hierarchical problem of the type (3.5.1) is the algorithms of asymptotic integration of the *simplest two-level hierarchical system*. Therefore, here and below in next Chapter we will pay the main attention to investigation of two-level hierarchical systems of various types. Our studying we begin with the so called *Bogolyubov's standard system* [19].

5.1 Bogolyubov's Standard System

Traditionally the particular variety of the general systems (3.5.1)

$$\frac{dz}{dt} = \varepsilon Z\left(z, t, \varepsilon\right), \quad z\left(0\right) = z_0 \tag{3.5.2}$$

is referred to as *Bogolyubov's standard system*. Here we accept for simplicity: $\varepsilon \equiv \varepsilon_1$. Let us construct the asymptotic algorithm for its in-

tegration, basing on averaging operator (3.1.13). This means that the
function $Z(z, t, \varepsilon)$ for the first hierarchical level can be written in the
form:

$$\bar{Z}(z, \varepsilon) = M_t Z(z, t, \varepsilon) = \lim_{T \to \infty} \frac{1}{T} \int_0^T Z(z, t, \varepsilon) dt. \qquad (3.5.3)$$

We assume that the function $Z(z, t, \varepsilon)$ is 2π-periodical on t, i.e.,

$$Z(z, t + 2\pi) \equiv Z(z, t) \qquad (3.5.4)$$

for all $z \in G_n$.

Accordingly with the general hierarchical scheme (see above Section 1)
the dynamical equation for the first hierarchical level should have similar
to (3.5.2) mathematical structure:

$$\frac{d\bar{z}}{dt} = \varepsilon \bar{Z}(\bar{z}, t, \varepsilon), \quad \bar{z}(0) = z_0. \qquad (3.5.5)$$

The connection of the proper variables of the zeroth and first hierarchical level we find in form of the *Krylov–Bogolyubov substitution*
(transformation — see also (3.3.17) and commentaries)

$$z = \bar{z} + \varepsilon u(\bar{z}, t, \varepsilon). \qquad (3.5.6)$$

The generalized hierarchical equation of the first order (3.1.2) for the
unknown functional operator (transformation function) u in this case
can be obtained in the form:

$$\varepsilon\left(\frac{\partial u}{\partial \bar{z}}, Z_0(\bar{z})\right) + \frac{\partial u}{\partial t} = Z_0(\bar{z} + \varepsilon u, t, \varepsilon) - \bar{Z}(\bar{z}, \varepsilon). \qquad (3.5.7)$$

Then we expanse the function Z in a Fourier series (see (3.1.19) and
relevant comments):

$$Z(z, t) = \sum_{|k| \geqslant 0} Z_k(z) \exp\{ikt\}, \qquad (3.5.8)$$

$$\bar{Z}(\bar{z}) = Z_0(\bar{z}). \qquad (3.5.9)$$

Using the expansion (3.5.8), (3.5.9) we rewrite general hierarchical
equation (3.5.7)

$$\varepsilon \left(\frac{\partial u}{\partial t}, Z_0 \left(\bar{z} \right) \right) + \frac{\partial u}{\partial t}$$

$$= Z_0 \left(\bar{z} + \varepsilon u \right) - Z_0 \left(\bar{z} \right) + \sum_{|k| \geqslant 1} Z_k \left(\bar{z} + \varepsilon u \right) \exp \left\{ ikt \right\}. \quad (3.5.10)$$

Further we accept the following supposition: the function $Z \left(z, t \right)$ is considered analytical with respect to z within domain G_n. In this case we have a possibility of constructing a formal solution for $u \left(\bar{z}, t \right)$ [19].
We will look for change of variables (3.5.6) in the form

$$z = \bar{z} + \sum_{n=1}^{\infty} \varepsilon^n u_n \left(\bar{z}, t \right). \quad (3.5.11)$$

Then we substitute the change (3.5.11) in general hierarchical equation (3.5.10) equalizing here terms of equal order (with respect to the small parameter ε) in left and right sides, respectively. As a result after relevant calculations we obtain the following catching hold of system of equations

$$\frac{\partial u_1}{\partial t} = Z \left(\bar{z}, t \right) - Z_0 \left(\bar{z} \right) = \sum_{\|k\| \geqslant 1} Z_k \left(\bar{z} \right) \exp \left\{ ikt \right\};$$

$$\frac{\partial u_2}{\partial t} = \left(\frac{\partial Z \left(\bar{z}, t \right)}{\partial \bar{z}}, u_1 \right) - \left(\frac{\partial u_1}{\partial \bar{z}}, Z_0 \right);$$

$$\frac{\partial u_3}{\partial t} = \left(\frac{\partial Z \left(\bar{z}, t \right)}{\partial \bar{z}}, u_2 \right) + \frac{1}{2!} \left(\left(\frac{\partial^2 Z \left(\bar{z}, t \right)}{\partial \bar{z}^2}, u_1 \right), u_1 \right) - \left(\frac{\partial u_2}{\partial \bar{z}}, Z_0 \right);$$

$$\dotsfill \quad (3.5.12)$$

Very important feature of system (3.5.12) is a possibility to obtain its approximate analytical solutions:

$$u_1 \left(\bar{z}, t \right) = \sum_{\|k\| \geqslant 1} \frac{1}{ik} Z \left(\bar{z} \right) \exp \left\{ ikt \right\} + \varphi_1 \left(\bar{z} \right);$$

$$u_2\left(\bar{z},t\right) = \sum_{\|k\|\geqslant 1} \frac{1}{ik}\left(\frac{\partial Z_{-k}\left(\bar{z}\right)}{\partial \bar{z}}, Z_k\left(\bar{z}\right)\right) t$$

$$+ \sum_{\|k\|\geqslant 1}\sum_{\|s\|\geqslant 1} \frac{1}{i^2\left(k+s\right)}\left(\frac{\partial Z_k\left(\bar{z}\right)}{\partial \bar{z}}, Z_s\left(\bar{z}\right)\right)\exp\left\{i\left(k+s\right)t\right\}$$

$$- \left(\sum_{\|k\|\geqslant 1} \frac{1}{i^2 k^2}\frac{\partial Z_k\left(\bar{z}\right)}{\partial \bar{z}}\exp\left\{ikt\right\}, Z_0\left(\bar{z}\right) + \varphi_1\left(\bar{z}\right)t\right) + \varphi_2\left(\bar{z}\right);$$

$$\cdots\cdots\cdots\cdots\cdots\cdots\cdots\cdots\cdots\cdots\cdots\cdots\cdots\cdots \quad (3.5.13)$$

For obtaining of total solution of initial problem (3.5.2) (zeroth hierarchical level) we must:

a) to substitute solutions (3.5.13) into (3.5.11);

b) to find relevant solutions of the dynamical equation of the first hierarchical level (3.5.5) and to substitute their into (3.5.11);

c) to coordinate initial conditions.

The last item means that we should to choose the arbitrary functions $\varphi_1\left(\bar{z}_0\right)$, $\varphi_2\left(\bar{z}_2\right)$, ... by such way the equation is satisfied

$$\bar{z}_0 + \varepsilon u_1\left(\bar{z}_0, 0\right) + \varepsilon^2 u_2\left(\bar{z}_0, 0\right) + \cdots = z_0. \qquad (3.5.14)$$

Thus it is shown that in view of above set forth general idea of the hierarchical approach (see Sections 1–3), the Bogolyubov method indeed can be regarded as a particular variety of hierarchical method. More obviously this thought is illustrated in the diagram represented in Fig. 3.5.1.

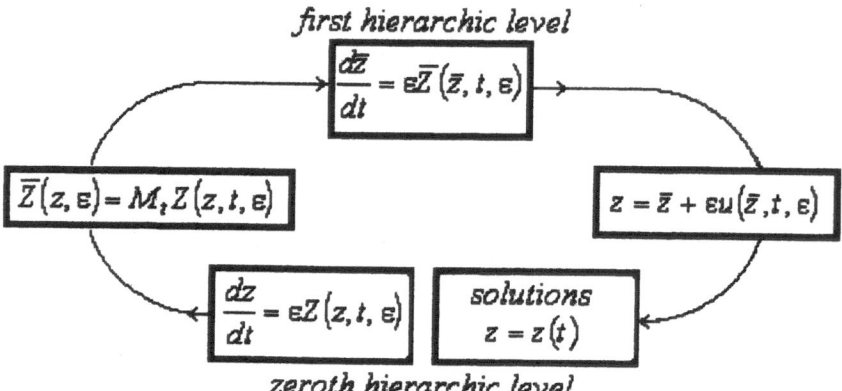

Figure 3.5.1. Illustration of hierarchical nature of the Bogolyubov method.

In principle we might also have some other algorithm for the discussed version of the averaging method [19]. Such algorithm can be realized if the idea of generating equation (see (3.4.2) and comments) is used. (In reference such equations are also called to as *equations of comparison* [19]). In this case the structural operator can be chosen in the following manner

$$\hat{M}_t Z\left(z, t, \varepsilon\right) = \bar{Z}\left(\bar{z}, \varepsilon\right)\big|_{\varepsilon=0} = \bar{Z}\left(\bar{z}, 0\right). \qquad (3.5.15)$$

Respectively, the dynamical equation of the first hierarchical level can be written as

$$\frac{d\bar{z}}{dt} = \varepsilon \bar{Z}\left(\bar{z}, 0\right), \quad \bar{z}\left(0\right) = \bar{z}_0. \qquad (3.5.16)$$

Further calculational scheme is the same. However, it should be mentioned that in a general case the form of asymptotic solution here turns out to be some differ of the solutions (3.5.11)–(3.5.13) [19].

5.2 The Problem of Secular Terms

Unpleased peculiarity of the above discussed versions of the calculational scheme on the average method is presence in asymptotic solutions (3.5.13) *secular terms* of the type $\varphi_s\left(\bar{z}\right) t$. This problem can be solved (completely or particularly) by using of the modernized version of the calculational scheme. Let use to choose the following form of the structural operator

$$\hat{M}_t Z\left(z, t, \varepsilon\right) = Z_0\left(\bar{z}\right) + \varepsilon A_2\left(\bar{z}\right) + \varepsilon^2 A_3\left(\bar{z}\right) + \cdots, \qquad (3.5.17)$$

where A_2, A_3, ... are some unknown (a this stage of the calculational procedure) functions, $Z_0\left(\bar{z}\right)$, as before, is the zeroth term of a Fourier series of type (3.5.8). Then we follow according to the above described calculational scheme. As a result, instead system (3.5.13) we obtain the modernized system

$$\frac{\partial u_1}{\partial t} = Z\left(\bar{z}, t\right) - Z_0\left(\bar{z}\right);$$

$$\frac{\partial u_2}{\partial t} = \left(\frac{\partial Z\left(\bar{z}, t\right)}{\partial \bar{z}}, u_1\right) - \left(\frac{\partial u_1}{\partial \bar{z}}, Z_0\right) - A_2\left(\bar{z}\right);$$

$$\frac{\partial u_3}{\partial t} = \left(\frac{\partial Z\left(\bar{z}, t\right)}{\partial \bar{z}}, u_2\right) + \frac{1}{2!}\left(\left(\frac{\partial^2 Z\left(\bar{z}, t\right)}{\partial \bar{z}^2}, u_1\right), u_1\right)$$

$$- \left(\frac{\partial u_2}{\partial \bar{z}}, Z_0\right) - A_3\left(\bar{z}\right);$$

$$\cdots\cdots\cdots\cdots\cdots\cdots\cdots\cdots\cdots\cdots\cdots\cdots\cdots\cdots\cdots \quad (3.5.18)$$

The next stage of the modernized version of the calculational scheme is the determining of the unknown functions $A_s\left(\bar{z}\right)$. The idea of this procedure follows from comparison of systems (3.5.12) and (3.5.18), respectively. It is understood that the functions $A_s\left(\bar{z}\right)$ can be found, for example, from the condition of elimination of relevant secular terms by corresponding functions $A_s\left(\bar{z}\right)$. As a result of realization of this idea we could obtain the following definitions for these functions

$$A_2\left(\bar{z}\right) = \lim_{T\to\infty}\frac{1}{T}\int\limits_0^T\left[\left(\frac{\partial Z\left(\bar{z}, t\right)}{\partial \bar{z}}, u_1\right) - \left(\frac{\partial u_1}{\partial \bar{z}}, Z_0\right)\right]dt;$$

$$A_3\left(\bar{z}\right) = \lim_{T\to\infty}\frac{1}{T}\int\limits_0^T\left[\left(\frac{\partial Z\left(\bar{z}, t\right)}{\partial \bar{z}}, u_2\right) + \frac{1}{2!}\left(\left(\frac{\partial^2 Z\left(\bar{z}, t\right)}{\partial \bar{z}^2}, u_1\right), u_1\right)\right.$$

$$\left. - \left(\frac{\partial u_2}{\partial \bar{z}}, Z_0\right)\right]dt;$$

$$\cdots\cdots\cdots\cdots\cdots\cdots\cdots\cdots\cdots\cdots\cdots\cdots\cdots\cdots\cdots \quad (3.5.19)$$

The method described above of elimination of *the secular terms* we will use also below in other versions of the averaging method.

6.   METHODS OF AVERAGING. TWO-LEVEL SYSTEMS WITH SLOW AND FAST VARIABLES

6.1   Two-Level Systems with Slow and Fast Variables. General Case

Bogolyubov's standard system of the type (3.5.2) represents the simplest object for application of the averaging methods. Historically such systems have been the first fully fledged realizations of the hierarchical ideas [12, 13]. Later a few other more general versions of such realizations have been developed [6, 14–19]. Let as illustrate one of theses 'more general versions' at an example of so called *system with slow and fast variables*.

The Bogolyubov's standard system (3.5.2) can be regarded as a particular case of the more general system

$$\frac{dz}{dt} = Z\left(z, t, \varepsilon\right).$$ (3.6.1)

Indeed, assuming

$$Z\left(z, t, \varepsilon\right) = \varepsilon Z'\left(z, t, \varepsilon\right)$$ (3.6.2)

we easily come to the standard form (3.5.2). Thus, as is readily seen, Bogolyubov's standard form (3.5.2) is the particular case of general system (3.6.1), when the function $Z\left(z, t, \varepsilon\right)$ is *slowly varying function* on time t (because $\varepsilon << 1$). But such a particular case is not the only one possible. Let us discuss some other particular case. Namely, we have in view the situation, when different velocities of changing of components of different groups of the vector z takes place. In other words, we might say that the obviously expressed dynamical hierarchy of initial variables with respect to the velocities of their changing is a specific characteristic of this case. For simplicity we choose the case of two-level *hierarchy*:

$$z = \{x, y\},$$ (3.6.3)

where x represents the slow part of the vector z, and y represents the fast part of it. The standard system of such kind can be presented in the form

$$\frac{dx}{dt} = \varepsilon X\left(x, y, t, \varepsilon\right);$$

$$\frac{dy}{dt} = \omega\left(x, y, t\right) + \varepsilon Y\left(x, y, t, \varepsilon\right);$$ (3.6.4)

where x, X are the m-dimensional vectors, y, Y and ω are the n-dimensional vectors, $t \in R_1$. The vectors y, Y and ω are determined within some $m + n + 2$-dimensional space

$$G_{m+n+2} = \{(x, y, t, \varepsilon): \ x \in P_m, \ y = G_n, \ t \in R_1, \ \varepsilon \in [0, \hat{\varepsilon}]\}.$$ (3.6.5)

The vector ω comprises, as components, the scalar cyclic frequencies (angular velocities of rotation) $\omega_1, \ldots, \omega_j, \ldots, \omega_n$. Hence we might say that equations (3.6.4) describes a *multi-frequency two-level hierarchical system* with slow and fast variables. Respectively, the vector x is referred to as the *vector of slow variables*, and the vector y is the vector of fast variables.

It is not difficult to show that, as mentioned above, Bogolyubov's standard system (3.5.2), in itself, is the simplest particular case of the system with slow and fast rotating phases. Indeed, we rewrite (3.5,2) in the parametric form (with respect to the parameter $y = t$ and for $z \equiv x$, $Z \equiv X$) to obtain

$$\frac{dx}{dt} = \varepsilon X\left(x, y, \varepsilon\right);$$
$$\frac{dy}{dt} = 1, \tag{3.6.6}$$

i.e., the standard system (3.5.2) can be considered to be equivalent to the standard form with slow and fast phases (3.6.4) in the following particular case

$$X(x, y, t, \varepsilon) \to X(x, y, \varepsilon);$$
$$\omega(x, y, t) = 1;$$
$$Y(x, y, t, \varepsilon) = 0. \tag{3.6.7}$$

Then we should to point out that the general algorithm for obtaining of analytical solutions of the discussed type systems is not known today [19]. Therefore, at a practice some particular cases of the systems similar to (3.6.5) are use, as a rule. The *systems with fast rotating phases* represent one of such particular cases. Owing to their specific features namely these systems are found to be most interesting for practice.

6.2   Two-Level Systems with Fast Rotating Phases

The particular case of the systems with slow variables x and fast phases (3.6.4) is referred as to the system with fast rotating phases if

$$X\left(x, y, t, \varepsilon\right) = X\left(x, y, \varepsilon\right);$$
$$\omega\left(x, y, t\right) = \omega\left(x\right);$$
$$Y\left(x, y, t, \varepsilon\right) = Y\left(x, y, \varepsilon\right). \tag{3.6.8}$$

In this case *the fast variable* y is called as to the *fast rotating phase* $\psi \equiv y$.

Thus the *two-level multi-frequency hierarchical standard system with fast rotating vector phase* can be written as

$$\frac{dx}{dt} = \varepsilon X\left(x, \psi, \varepsilon\right);$$

$$\frac{d\psi}{dt} = \omega\left(x\right) + \varepsilon Y\left(x, \psi, \varepsilon\right). \qquad (3.6.9)$$

Besides that another version for the form of writing of hierarchical standard system with fast rotating phases also will be used in this book below

$$\frac{dx}{dt} = X\left(x, \psi, \xi\right);$$

$$\frac{d\psi}{dt} = \xi\omega\left(x\right) + Y\left(x, \psi, \xi\right), \qquad (3.6.10)$$

where $\xi = 1/\varepsilon$ is the *large scale parameter of the problem*. The various algorithms of asymptotic integration of the systems similar to (3.6.9), (3.6.10) (an some more complex) are set forth below in the following Chapter.

References

[1] M.M. Khapaev. *Asymptotic methods and stability in theory of nonlinear oscillations*. Vysshaja Shkola, Moscow, 1988.

[2] V.V. Kulish. Hierarchical oscillations and averaging methods in nonlinear problems of relativistic electronics. *The International Journal of Infrared and Millimeter Waves*, 18(5):1053–1117, 1997.

[3] V.V. Kulish. Hierarchical approach to nonlinear problems of electrodynamics. *Visnyk Sumskogo Derzshavnogo Universytetu*, 1(7):3–11, 1997.

[4] V.V. Kulish, P.B. Kosel, A.G. Kailyuk. New acceleration principle of charged particles for electronic applications. hierarchical description. *The International Journal of Infrared and Millimeter waves*, 19(1):33–93, 1998.

[5] G.A. Korn, T.W. Korn. *Mathematical handbook for scientists and engineers*. McGraw Hill, NY, 1961.

[6] Y.A. Mitropolsky, B.I. Moseenkov. *Lectures on the asymptotic applications to the solution of partial differential equations*. Mathematics Institute, Kiev, 1968.

[7] V.V. Kulish. *Methods of averaging in non-linear problems of relativistic electrodynamics*. World Scientific Publishers, Atlanta, 1998.

[8] R.K. Dodd, J.C. Eilbeck, J.D. Gibbon, H.C. Morris. *Solutions and nonlinear wave equations*. Academic Press, London, 1982.

[9] M.A. Leontovich. To the problem about propagation of electromagnetic waves in the earth atmosphere. *Izv. Akad. Nauk SSSR*, Phys. Ser.(8):6–24, 1944. ser. Fiz. (Bull. Acad. Sci. USSR, Phys. Ser.).

[10] B. Van der Pohl. *Nonlinear theory of electric oscillations. Russian translation.* Svyazizdat, Moscow, 1935.

[11] V.V. Kulish. *Hierarchical theory of oscillations and waves and its application to nonlinear problems of relativistic electrodynamics.* Causality and locality in modern physics. Kluwer Academic Publishers, Dordrecht/Boston/London, 1998.

[12] N.M. Krylov, N.N. Bogolyubov. *Application of the methods of nonlinear mechanics to the theory of stationary oscillations.* Ukrainian Academic Science Publishers, Kiev, 1934.

[13] N.M. Krylov, N.N. Bogolyubov. *Introduction to nonlinear mechanics.* Ukrainian Academy Science Publishers, Kiev, 1937. English translation: Princeton, New Jersey: Princeton Univ. Press., 1947.

[14] N.N. Bogolubov, Ju.A. Mitropolskii. *Methods of averaging in the theory of nonlinear oscillations.* Publising House Academy of Science of USSR, Moscow, 1963.

[15] N.N. Bogolubov, D.N. Zubarev . Asymptotic approximation method for the system with rotating phases and its application to the motion of charged particles in magnetic fields. *Ukranian Mathem. Zhurn*, 7:201–221, 1955. Ukrainian Mathem. Journal.

[16] V.I. Arnold. Applicability conditions and error estimates for the averaged method applied to the resonant systems. *Dok. Akad. Nauk. SSSR*, 161(1):9, 1965. Sov.Phys.-Doklady.

[17] A.A. Andronov, A.A. Vitt, S.E. Khaikin. *Theory of oscillations.* Fizmatgiz, Moscow, 1959.

[18] N.N. Moiseev. *Asymptotic methods of nonlinear mechanics.* Nauka, Moscow, 1981.

[19] E.A. Grebennikov. *Averaging method in applied problems.* Nauka, Moscow, 1986.

[20] S.S. Kohmanski, V.V. Kulish. To the classic single-particle theory of free electron laser. *Acta Physica Polonica*, A66(6):713–740, 1984.

[21] A.V. Gaponov, L.A. Ostrovskii, M.I. Rabinovich. One-dimensional waves in nonlinear disperse media. *Izv. Vysh. Uchebn.*, 13(2):169–213, 1970. Ser. Radiofizika (Sov. Radiophysics).

[22] M.I. Rabinovich, V.I. Talanov. Four lectures on the theory of nonlinear waves and wave interactions. Leningrad, 1972.

[23] M.I. Rabinovich. On the asymptotic in the theory of distributed system oscillations. *Dok. Akad. Nauk. SSSR*, 191:253–1268, 1971. ser. Fiz. (Sov. Phys.-Doklady).

[24] V. V. Kulish. Nonlinear self-consistent theory of free electron lasers. method of investigation. *Ukrainian Physical Journal*, 36(9):1318–1325, 1991.

[25] V.V. Kulish, A.V. Lysenko. Method of averaged kinetic equation and its use in the nonlinear problems of plasma electrodynamics. *Fizika Plazmy*, 19(2):216–227, 1993. (Sov. Plasma Physics).

[26] V.V. Kulish, S.A. Kuleshov, A.V. Lysenko. Nonlinear self-consistent theory of superheterodyne and free electron lasers. *The International journal of infrared and millimeter waves*, 14(3):451–568, 1993.

[27] V.V. Kulish. Hierarchical method and its application peculiarities in nonlinear problems of relativistic electrodynamics. general theory. *Ukrainian Physical Journal*, 43(4):83–499, 1998.

[28] V.V. Kulish, P.B. Kosel, O.B. Krutko, I.V. Gubanov. Hierarchical method and its application peculiarities in nonlinear problems of relativistic electrodynamics. theory of eh-ubitron accelerator of charged particles. *Ukrainian Physical Journal*, 43(2):33–138, 1998.

[29] V.V. Kulish. Hierarchical method and its application peculiarities in nonlinear problems of relativistic electrodynamics. single-particle model of cyclotron-resonant maser. *Ukrainian Physical Journal*, 43(4):98–402, 1998.

[30] S.S. Kohmanski, V.V. Kulish. To the nonlinear theory of free electron lasers. *Acta Physica Polonica*, A68(5):749–756, 1985.

[31] S.S. Kohmanski, V.V. Kulish. To the nonlinear theory of free electron lasers with multi-frequency pumping. *Acta Physica Polonica*, A68(5):741–748, 1985.

[32] S.S. Kohmanski, V.V. Kulish. Parametric resonance interaction of an electron and the field of electromagnetic waves and the longitudinal magnetic field. *Acta Physica Polonica*, A68(5):725–736, 1985.

[33] V.V. Kulish, I.V. Dzedolik, M.A. Kudinov. Movement of relativistic electrons in periodically reversed electromagnetic field. Deposited in Ukrainian Scientific Research Institute of Thechnical Information, Kiev, 23 jul 1985. Uk-85. Part I. 110 pages.

[34] V.V. Kulish, I.V. Dzedolik. Movement of relativistic electrons in periodically reversed electromagnetic field. Deposited in Ukrainian Scientific Research Institute of Thechnical Information, Kiev, 20 sep 1985. Uk-85. Part II. 54 pages.

[35] V.S. Jakovlev, V.V. Kulish, I.V. Dzedolik, V.G. Motina, S.S. Kohmanski . Generation of energy by relativistic electrons moving in the field of two electromagnetic waves in presence longitudinal magnetic field. PrePrint of University Electrodynamics Academy of Sciences of Ukraine, 1983. 41 pages.

[36] L.I. Mandelshtam and N.D. Papaleksi. On substantiation of an approximate method of solving differential equations. *Zh.Eksp.Teor.Fiz*, 2:220–234, 1934. Sov. Phys.-JETP.

Chapter 4

HIERARCHICAL SYSTEMS
WITH FAST ROTATING PHASES

The hierarchical systems with fast rotating phases are the most 'popular' (amongst other asymptotic hierarchical methods) in electrodynamics applications. Firstly, because the physical models which can be described by such equations, are widespread in a practice. Secondly, because, as mentioned above in Chapter 3, Section 6, analytical algorithms of asymptotic integration of systems of this type can be constructed without excessive difficulties. Taking this into consideration we will pay special attention in this and the following Chapters for studying the systems with fast rotating phases of different types.

1. HIERARCHICAL OSCILLATIONS

1.1 A Few Introductory Words

We begin this Section with discussion of some basic ideas of the theory of hierarchical oscillations in electrodynamic systems. Therefore here we will take an interest, mainly, in the mathematical part of the problem. This problem, as will be understood latter, can quite often be reduced to the asymptotic integration of the hierarchical systems with *fast rotating phases*.

In what follows it should be pointed out that the purely mathematical part in training the theory of *hierarchical systems*, in itself, is not the only source of the principal difficulties for a student or an expert. Let us turn the reader's attention to another side of the discussed object.

As mentioned above in the Preface, this book is intended first of all for students and experts concerned with physical electronics, acceleration devices, plasma electrodynamics, radiophysics, etc.. Initially the idea of the book originated from the lecturing experience gained by the author when delivering specialized lecture courses in vacuum physical electronics. The obtained the experience turned out to be somewhat unexpected. Namely, the author was astonished to discover that students easily cope with the mathematical aspects of the hierarchical method (basic concepts, calculational algorithms, etc.) but find serious difficulties in dealing with the two practical aspects. Moreover, very often the author finds the same situation even in discussions with mature experts. Taking this into consideration, let us briefly comment on the initial nature of these difficulties.

The first difficulty is to formulate the problem in the terms of hierarchical methods. In particular, it is to reduce the initial sets of equations to one of those mentioned above in Chapter 3, Sections 5 and 6 standard forms. As noted earlier in Chapter 3, the asymptotic integration algorithms of this method are valid only for the so called standard systems. The point is that the general procedure of reducing arbitrary systems of equations to standard forms is not known. Quite efficient special approaches, which have been proposed for particular groups of problems, require deep understanding of fine qualitative distinctions in the occurrence of physical processes under consideration. As a result, to formulate the problem adequately, at least in electrodynamics, has turned out to be a skill that is not so easy to acquire. Therefore below we begin exposition of the hierarchical calculational technologies (including those based on the *hierarchical systems* with *fast rotating phases*) from the discussion of the problem of adequate mathematical description (factorization) of electrodynamic systems in the terms of hierarchical methods.

The second difficulty concerns the physical analysis and interpretation of the results obtained. The terminology of the hierarchical method is rather special. The picture of the phenomena to be studied turns out to be described in terms which are not traditional for 'ordinary' electrodynamics. That is why one needs some training in order to understand a physical picture described in 'hierarchical terms'. For example, this might concern an evolution of a slow combination phase of the charged particle. This evolution is really associated with particle nonlinear dynamics in the vicinity of some resonance. Or another example: a certain combination of slowly varying complex wave amplitudes describes self-consistent variations of wave polarizations owed to the nonlinear wave, etc.. A number of special methodical examples are given below interaction (see Chapters 7, 8 and Chapters 9–13, Volume II) for illustration of methods of overcoming these kind of difficulties. The author advises for the reader to study in detail these illustrations for better understanding of methodical and calculational peculiarities of the methods of discussed type.

1.2 Formulation of the Hierarchical Single-Particle Electrodynamic Problem

Taking into consideration the above said in item 4.1.1, we start with discussing of some general features of the hierarchical version of classical single-particle electrodynamic problem. It is the problem of periodical (quasi-periodical) motion of a charged particle in some electromagnetic field in the discussed case.

We accept the superposition that the acting electromagnetic field consists of the wave and non-wave parts, correspondingly. with that the latter (non-wave part) is assumed slowly varying, non-periodic quasi-stationary and quasi-homogeneous fields. Thus the arrangement of the electric and magnetic fields acting on the charged particles can be represented in the form:

$$\vec{E} = \frac{1}{2} \sum_l \left[\vec{E}_l \exp \{im_l p_l\} + \text{c. c.} \right] + \vec{E}_0;$$

$$\vec{B} = \frac{1}{2} \sum_j \left[\vec{B}_j \exp \{im_j p_j\} + \text{c. c.} \right] + \vec{B}_0, \qquad (4.1.1)$$

where $\vec{E}$ and $\vec{B}$ are the electric field intensity and magnetic induction field. $\vec{E}_l = \vec{E}_l(\vec{r}, t)$ and $\vec{B}_j = \vec{B}_j(\vec{r}, t)$ are slowly varying complex amplitudes of electric and magnetic induction fields of partial waves (proper and stimulated); $\vec{E}_0 = \vec{E}_0(\vec{r}, t)$, $\vec{B}_0 = \vec{B}_0(\vec{r}, t)$ are slowly varying quasi-stationary and quasi-homogeneous electric and magnetic induction fields, and m_l, m_j are wave harmonic numbers (m_j, $m_l = \pm 1, \pm 2, \dots$). The quantities

$$p_q = \omega_q t - \vec{k}_q \vec{r}, \quad q = j, l \qquad (4.1.2)$$

are phases, ω_q is the angular frequency, k_q is the wave vector, t is the laboratory time and $\vec{r}$ is the position vector. In the case of proper waves we have the dispersion relations between ω_q and k_q ($\vec{k}_q = \vec{k}_q(\omega_q)$). All these concepts are discussed in Chapter 1 (see Chapter 1, Sections 2 and 3 for more details).

In terms of general hierarchical theory (see Chapters 2 and 3), assigning form (4.1.2) to the fields determines form of the chosen structural operator $M^{(\kappa)}$ (see (2.2.6)).

Further below (in this Volume and in Volume II) we illustrate some practical features of the calculational procedures studied by using examples of various electrodynamic systems. The peculiarity of the latter is that they are always related to some objects like charged particle beams or plasmas fluxes which move within electromagnetic fields. Usually, the single particle and multi-particle (including collective) problems are distinguished.

In this book we are interested in electron, ion or plasma beams of two extreme types: low and high intensity beams, respectively [1–3]. The characteristic feature of beams of the first type is that intensity of Coulomb interaction between the charged particles is relatively low. In fact, such beams are just streams of nearly independent drifting charged particles. Hence we have a possibility to study of a separate particle interaction with the fields. The single-particle theory of motion of charged particles in electromagnetic fields is a result of such study. In turn the single-particle problem is the main point of this theory. The particular case of the multi-particle problem can be realized here by simply summing up of all separate single-particle interactions.

In contrast, the particle–particle Coulomb interactions can be rather intense in high intensity beams. At sufficiently high densities the beams show a collective behavior and they are treated as a quasi-continuous flow of drifting charged (or quasi-neutral) plasmas. Hence, in this case the charged particle beam should be regarded as a whole electrodynamic object.

It is obvious that owing to the differences discussed above of the beams of both types the mathematical descriptions of these objects should olso be different. Below in this Chapter we give hierarchical algorithms for treating of various single-particle systems. Some methods of asymptotic treating of the high density beams will be set forth below in Chapter 5.

Thus the beam's motion of the first type (rarified plasma beams) can be considered as the motion of aggregate of individual particles (Lagrange formalism). In this case, the beam motion can be described by the single-particle equations in the Hamiltonian form (1.4.22):

$$\frac{dH_\alpha}{dt} = \frac{\partial H_\alpha}{\partial t}; \quad \frac{d\vec{P}_\alpha}{dt} = -\frac{\partial H_\alpha}{\partial t}; \quad \frac{d\vec{r}}{dt} = \frac{\partial H_\alpha}{\partial \vec{P}_\alpha}, \qquad (4.1.3)$$

or by the Lorenz equation, which can be regarded as a particular case of the Newton equation (1.4.4) (because $\vec{F} = d\vec{p}/dt = \vec{F}_L$, where $\vec{F}_L$ is the Lorenz force):

$$\frac{d\vec{\beta}_\alpha}{dt} = \frac{q_\alpha}{m_\alpha \gamma_\alpha c} \left\{ \vec{E} + \left[\vec{\beta}_\alpha \vec{B} \right] - \vec{\beta}_\alpha \left(\vec{\beta}_\alpha \vec{E} \right) \right\}. \qquad (4.1.4)$$

Here $H_\alpha = \sqrt{m_\alpha^2 c^4 + c^2 \left(\vec{P}_\alpha - \frac{q_\alpha}{c} \vec{A} \right)} + q_\alpha \varphi$, is the Hamiltonian of a charged particle of type α; q_α is its charge ($q_\alpha = -e$ for electron and

$q_\alpha = +Ze$ for an ion, e is electron charge, Z is charge number of the ion), m α is the rest mass of particle of sort α; $\vec{A}$ is vector-potential and φ is scalar potential of the electromagnetic fields [4]; $\vec{\beta}_\alpha = \vec{v}_\alpha/c$ is the non-dimensional particle velocity; here $\vec{v}_\alpha$ is the velocity, c is light velocity in vacuum, $\vec{P}_\alpha$ is the canonical momentum, $\vec{r}_\alpha$ is the position vector of particle α; $\vec{E}$ and $\vec{B}$ are defined in Chapter 1, Section 4 already; $\gamma_\alpha = \left(1 - \beta_\alpha^2\right)^{-1/2}$ is the relativistic factor (here $\beta_\alpha = \left|\vec{\beta}_\alpha\right|$).

The hierarchical part of the problem consists of two main stages. The first is the reducing the equations (4.1.3) or (4.1.4) to one of standard forms. The second is the asymptotic integration of the standard equations. In the book we study initial sets (4.1.3), (4.1.4) reduced to the standard form hierarchical system with fast rotating phases [5–7]. The simplest two-level versions of such systems we above discussed shortly in Chapter 3, Section 6 (see equations (3.6.9), (3.6.10)). In the general case of m-level hierarchical oscillation systems the standard system with fast rotating phases can be written in the following form [5–7]

$$\frac{dx}{dt} = X\left(x, \psi_1, \ldots, \psi_\kappa, \ldots, \psi_m, \xi_1, \ldots, \xi_m\right);$$

$$\frac{d\psi_1}{dt} = \xi_1 \omega_1\left(x\right) + Y_1\left(x, \psi_1, \ldots, \psi_\kappa, \ldots, \psi_m, \xi_1, \ldots, \xi_m\right);$$

$$\cdots\cdots\cdots\cdots\cdots\cdots\cdots\cdots\cdots\cdots\cdots\cdots\cdots\cdots\cdots\cdots\cdots$$

$$\frac{d\psi_m}{dt} = \xi_m \omega_m\left(x\right) + Y_m\left(x, \psi_1, \ldots, \psi_\kappa, \ldots, \psi_m, \xi_1, \ldots, \xi_m\right), \qquad (4.1.5)$$

where scale parameters $\xi_\kappa = 1/\varepsilon_\kappa$ are the relevant terms of the strong hierarchical series (see for comparison (2.1.3)):

$$\xi_1 >> \xi_2 >> \cdots >> \xi_\kappa >> \cdots >> \xi_m >> 1, \qquad (4.1.6)$$

and x is the vector whose components are slowly varying variables only, ψ_κ are partial vectors of different hierarchy of the vector of fast rotation (revolving) phases ψ, ω_κ are slowly varying part of velocity vector of fast rotating phases, Y_κ are relevant vector-functions, κ is hierarchical level number, and m is the number of the highest hierarchical level.

In view of the setting forth in Section 2, initial equations (4.1.3), (4.1.4) in standard form (4.1.5) mean that some 'hierarchicisation law' (structural operator $M^{(\kappa)}$ (2.2.6)) of the given dynamical system is determined. Real form might be found, from experiments or other considerations.

Within context of the definition (2.2.4), large hierarchical parameters ξ_κ in (4.1.5), (4.1.6) can be determined as

$$\xi_\kappa \sim \left| \frac{d\psi_{\kappa l}}{dt} \right| \Big/ \left| \frac{dx_q}{dt} \right|, \qquad\qquad (4.1.7)$$

where $\psi_{\kappa l}$ is l-th component of vector ψ_κ and x_q is q-th component of vector x. Here we consider the rate of variation of x_q not exceeding rates of variations of other slow variables and the component $\psi_{\kappa l}$ is the 'slowest' of other fast components of the vector ψ_κ. Thus with general definition of hierarchical scale parameter (2.2.4) we affirm that variables $\psi_{\kappa l}$ (in the case $\kappa = 1$) and x_q represent two neighboring different hierarchical levels.

1.3   Classification of Oscillatory Phases and Resonances. Hierarchical Tree

Accordingly with the general principles of the hierarchical approach we should to reduce initial systems (4.1.3) or (4.1.4) to the standard form (4.1.5). Therefore, the first step consists in determining all elements of 'slow vector' x and 'fast phase vector' ψ in (4.1.5). In other words, it consists in classifying total set of variables of systems (4.1.3) and (4.1.4) as slow and fast ones, respectively.

The Hamiltonian $\mathcal{H}$ and variables $\vec{P}$, $\vec{\beta}$ (or $\vec{v} = \vec{\beta}c$) and $\vec{r}$, as a rule, might be considered as slow variables. In contrast, classification of the Lagrange phases of particle oscillations (see about the concepts of Lagrange and Euler variables in Chapter 1, Subsection 4.9) is more complicated. In this case we should separately find the set of phases forming component basis of vectors ψ and x. Besides that, we must divide rotating phases of particle oscillations into those associated with explicit and hidden periods of oscillation (see definitions in Chapter 1, Subsection 2.2). The phases of the first group are related with phases of the particle oscillations in the wave type of fields. For example, they are particle oscillation phases in undulated fields of various types, in the fields of electromagnetic waves or in the fields of proper and stimulated beam waves, and some others (see Chapters 6, 7, 9). The periodicity of these fields is responsible for the periodic character of electron motion with respect to the same wave phases (4.1.2)

$$p_q = \omega_q t - \vec{k}_q \vec{r}. \qquad\qquad (4.1.8)$$

The second case is much more complicated. It can be associated with phases of periodic particle motion in non-periodical quasi-stationary intrinsic beam fields, external focusing and beam-forming fields, etc.. Procedure separating hidden periods of arbitrary functions is known in mathematics. For instance the *Lantzos' method* [8, 9], *Grebennikov's*

method [10] and others. There are also semi-empirical approaches based
on qualitative peculiarities of a studied physical picture [6,7].

We assume the procedure of finding of all particle oscillation phases
(hidden and explicit) has been completed. Analyzing the rates of particle phase varying we find that their total set involves both types of
phases: slow components of vector x and fast components of vector ψ.
Moreover, some nonlinear combinations of two, three or more Lagrange
fast phases produce slowly varying functions. Each slow phase resulting
from this 'combination' action (m-fold, in the general case) corresponds
to some physical mechanism of particle resonance. We distinguish quasilinear and parametric resonance (relevant definitions in Chapter 1, Subsection 2.3). Quasilinear resonances are characterized by two-phase (or
m-phase, in the general case) combinations where one of phases only is
associated with explicit period of the system. The magnitude of stimulated wave force acting on the particle is always linearly dependent on
wave field amplitude in the lowest order in some small parameter. Therefore these resonances are quasilinear . Examples: cyclotron resonance;
various types of synchrotron resonances; etc..

Parametric one-particle resonances correspond to cases when all
phases forming slow phase combination are of wave nature. For instance,
parametric resonances of third-, fourth-, and higher orders [11–18] can
realize in free electron lasers (FELs), in parametric electronic devices
(Adler's lamp, parametric *electron-wave lamps*), etc. [17,18]. In general,
slow $\theta_{\nu g}$ and fast $\psi_{\nu g}$ combined phases of coupled-pair parametric or
quasilinear resonances can be defined as

$$\theta_{\nu g} = \frac{m_\nu n_\nu}{m_g n_g} p_\nu - \sigma_{\nu g} p_g;$$
$$\psi_{\nu g} = \frac{m_\nu n_\nu}{m_g n_g} p_\nu + \sigma_{\nu g} p_g.$$

(4.1.9)

where $\sigma_{\nu g} = \pm 1$ are sign functions, m_ν and m_g are numbers of wave field
harmonics, and n_ν and n_g are numbers of electron oscillation harmonics
in these fields (see definitions (1.2.19), (1.2.20) and relevant discussion).
Hence, the lowest order of relevant amplitudes of stimulated waves in
the case of parametric resonances is quadratic. Therefore, we classify
parametric resonances as nonlinear ones.

According to the hierarchical principles, which are set forth in Chapter 2, we define components of scale parameter tensor according to reciprocal rate of varying of fast to slow combined phases (4.1.9)

$$\xi_{\nu grk} \sim \left| \frac{d\psi_{\nu g}}{dt} \right| \bigg/ \left| \frac{d\theta_{\nu g}}{dt} \right| \gg 1.$$

(4.1.10)

It means that slow $\theta_{\nu g}$ and fast $\psi_{\nu g}$ combined phases are variables of different hierarchical levels. We assume the rates of change of velocities are of the same order (in magnitude) as rates of velocities of other slow variables (see (4.1.7)).

Then we analogously take into account the non-resonant phases of oscillations. The only difference is that instead of slow and fast combined phases we choose ordinary slow and fast phases. On the next stage of calculational, we construct hierarchical series of components of both tensors of scale parameters in the form (4.1.6).

Hierarchical series (4.1.6) are the essential point of the theory called the theory of hierarchical oscillations. In this theory each term of series (2.1.3), (4.1.6) corresponds to a single oscillation or resonance or a group of oscillations and resonances of equivalent hierarchy. In its physical nature this series follows from the assumption accepted above on influences between neighboring hierarchical levels only (see Fig. 2.1.1 and corresponding comments).

For discussed oscillatory hierarchical systems relevant hierarchical trees (similar to those displayed in Fig. 2.2.4, Fig. 2.2.5) can be constructed. Analogously to the case of hierarchical military unit (see Fig. 2.2.4) here we consider for simplicity that each two structural elements of lower hierarchical level form one structural element of the next higher level. In given thought case it means that all resonances are twofold only and all phases of each hierarchical level take part in resonant process. Therefore a phase resonant hierarchical tree can be constructed (Fig. 4.1.1). As well as in the case of military unit (Fig. 2.2.4) we obtain relevant analogy with the Cayley's tree, which are known in the theory of fractal systems, theory of codes, etc. [19, 20]. On the one hand, this circumstance can be treated as confirmation of correctness of developed hierarchical theory. On the other hand, such an analogy is unexpected from physical point of view and even somewhat mystical. Indeed, we have only one charged particle in some electromagnetic field, and, therein obtain the structure coinciding with the characteristic ones for explicit multi-particle objects (Universe, military units, other social and biological systems, etc.)!

Then we discuss the hierarchical tree (4.1.1), from the point of view of the hierarchical principles formulated above. We see that a few 'κ-level phases' (they are two phases only in this particular case) forms each combined phase of the next $(\kappa + 1)$-level. This means that the higher the hierarchical level the fewer are the number of oscillatative phases. The latter can be considered as obvious illustration of the compression hierarchical principle.

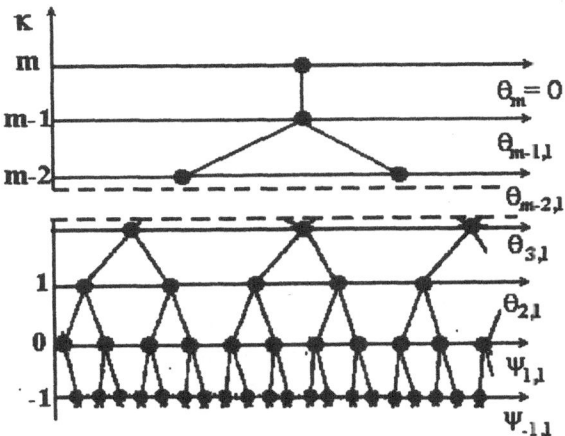

Figure 4.1.1. The simplest case of hierarchical tree consisting of rotating (revolving) charged particle phases. It should be especially pointed out that in this case we have to deal with the dynamical hierarchy, but no structural hierarchy, as it takes place in the cases illustrated in Figs. 2.2.4, 2.2.5. Here κ is hierarchical number of the level, $\psi_{\kappa l}$ and $\theta_{\kappa l}$ are rotating phases with rate of rotation velocities of different hierarchical order (they are shown by dark small circles), m is the highest hierarchical level. We see a full analogy with Cayley's hierarchical tree, which are known in the theory of fractal systems, theory of codes, etc..

The hierarchical resemblance principle holds owing to specific mathematical construction of equations for each hierarchical level. As noted above, all these equations are similar. Characteristic velocities of changing of variables of different levels are higher the lesser number of hierarchical level is (hierarchical analogy with the second thermodynamic principle). It is readily seen that any oscillations are not on the highest hierarchical level: $\theta_m = const, d\theta_m/dt = 0$, $\kappa = m$. It means that on the highest hierarchical level information entropy is equal zero. Hence, hierarchical analogy with the third thermodynamic principle takes place here, too, and so forth.

However, the following observation seems much more interesting. Namely, total energy of the highest level for electronic devices with long time interaction (which possess hierarchical structure like that shown in Fig. 4.1.1) exists in the form of kinetic energy of the initial quasi-stationary beam and the energy of the initial quasi-static electric and magnetic fields only. It is important that in the chosen model relevant sources of electron beam and of the initial quasi-static fields are behind the system. It means that only external quasi-stationary sources can drive the closed hierarchical system through the highest hierarchical level. At that the energy is transmitted in the closed hierarchical system from higher levels to lower one and a reaction (information) of hierarchical levels at this influence transmits in the opposite direction.

It is characteristic that these external sources occupy peculiar place on the top of hierarchical pyramid. Namely, it is the separate point on the top of specific vertical line that connects its with the rest (lower) part of the hierarchical tree. Here we could be draw parallels with the cases of hierarchical tree discussed above in previous Chapter. For example, a similar picture holds in the case of the Universe as a hierarchical structure (see Fig. 2.2.5). Indeed, comparing these two cases we see the stunning and unexpected analogy in the hierarchical structures of Universe and electronic devices: in both cases, we detect the same separate characteristic point at the top of the hierarchical tree. (In hierarchical model of the Universe, this top point corresponds to some unknown energy source of the primary Big Bang). Here also energy flows in the system through the specific vertical line connecting the highest hierarchical level ('Big Bang point') with the hierarchical level 'metagalaxy'. (The fact that in the discussed electrodynamic hierarchical system of Fig. 4.1.1 we consider only pair interactions is not essential.) Moreover, we obtain the same, in principle, situation analyzing military units also, as is seen from comparison of Figs. 2.2.4 and 4.1.1. In the case of Fig. 2.2.4, the point on the top of hierarchical tree is the 'source of power of highest commanders'. Hence we can ask whether we can consider the similar point of Fig. 2.2.5 as the 'source of power of highest commanders'. Unfortunately, today we cannot satisfactorily treat this strange analogy. We only say that we deal with one of obvious manifestations of the discussed in Chapter 2 self-modeling principle. To our regret, such an explanation entails many more new questions than it gives answers. The fact of analogy with the second thermodynamic principle can be illustrated by the same way, too.

With the aid of Fig. 2.2.4 we show that in our model the generation of hierarchical levels with negative numbers $\kappa < 0$ is possible, too. It takes place in the realization of special case of self-consistent systems For instance, it occurs if nonlinear generation of highest resonant harmonics and wave modes is realized. Such generation process really breaks down when imposing real restrictions. The latter can be determined by peculiarities of system design, satisfaction of the conservation principles, etc., or by realization of the effect of hierarchical degeneration.

1.4 Reducing Hierarchical Multi-Level Standard System to the Two-Level Form. The Scheme of Hierarchical Transformations

Thus because of relevant transformations initial system (4.1.3) or (4.1.4) reduces to the hierarchical standard form (4.1.5). Later, we describe the algorithm of asymptotic integration of the system (4.1.5).

Within the definition of κ-th scale parameter (4.1.10), mathematical meaning of hierarchical series (4.1.6) is equivalent to the statement that the system has the hierarchy over velocities of characteristic fast oscillation (rotation) phases of particles, i.e.

$$\left| \frac{d\psi_{1j}}{dt} \right| \gg \left| \frac{d\psi_{2i}}{dt} \right| \gg \left| \frac{d\psi_{\kappa l}}{dt} \right| \gg \cdots \gg \left| \frac{dx_q}{dt} \right|, \qquad (4.1.11)$$

with all evident definitions. Apart from the scale parameter of the first hierarchical level ξ_1 (the leading term in hierarchical series (4.1.6)), all the other fast phases ψ_κ ($\kappa > 1$) are relatively slow quantities. This observation is useful to construct required hierarchical algorithm of asymptotic integration reducing complex total system (4.1.5) to essentially simpler two-level hierarchical form of the type (3.6.10):

$$\frac{dx'}{dt} = X' \left(x', \psi_1, \xi_1 \right);$$

$$\frac{d\psi_1}{dt} = \xi_1 \omega_1 \left(x' \right) + Y_1 \left(x', \psi_1, \xi_1 \right); \qquad (4.1.12)$$

where x' is the expanded vector of slow variables (here all partial fast phases ψ_κ, excepting the first one ψ_1 ($\kappa = 1$), are considered as its components, and so on. We now write the Krylov–Bogolyubov substitution for the system with rotating phases (the dynamical hierarchical operator (2.2.7)). But at that time we do it with respect to corresponding components of the vectors x' and ψ_1:

$$x'_q = \bar{x}'_q + \sum_{n=1}^{\infty} \frac{1}{\xi_1^n} u_q^{(n)} \left(\bar{x}', \bar{\psi}_1 \right);$$

$$\psi_{1l} = \bar{\psi}_{1l} + \sum_{n=1}^{\infty} \frac{1}{\xi_1^n} v_{1l}^{(n)} \left(\bar{x}', \bar{\psi}_1 \right), \qquad (4.1.13)$$

where averaged variables $\bar{x}'_q$ and $\bar{\psi}'_{1l}$ are the components of relevant vectors $\bar{x}' = \{\bar{x}'_q\}$, $\bar{\psi}_1 = \{\bar{\psi}'_{1l}\}$ obtained from truncated (shortened) system of equations for next (first) hierarchical level:

$$\frac{d\bar{x}'}{dt} = \sum_{n=1}^{\infty} \frac{1}{\xi_1^n} A^{(n)} \left(\bar{x}'\right);$$

$$\frac{d\bar{\psi}_1'}{dt} = \xi_1 \omega_1 \left(\bar{x}'\right) + \sum_{n=1}^{\infty} \frac{1}{\xi_1^n} B_1^{(n)} \left(\bar{x}'\right). \tag{4.1.14}$$

The $u_q^{(n)}$, $v_{1l}^{(n)}$, $A^{(n)} = \left\{A_q^{(n)}\right\}$, $B_1^{(n)} = \left\{B_{1l}^{(n)}\right\}$, $\omega_1 \left(\bar{x}'\right)$ are functions calculated below. As we will show later, equations (4.1.12) and (4.1.14) are equivalent with respect to general mathematical structure. It means that hierarchical resemblance principle holds. Analogously, the information compression principle also is satisfied, because equation (4.1.14) describes the simplest physical situation — non-averaged fast vector phases ψ_1 absent in (4.1.14). By virtue of the same cause, hierarchical analogy of the second thermodynamic principle here also holds.

On the next stage of hierarchical analysis, we use discussed hierarchical features of system (4.1.14). Namely, we separate out the following group of fast scalar phases from 'temporary components' of slow variables of the vector $\bar{x}'$. These phases define the dynamics of the vector of partial fast phases ψ_2 (see (4.1.11)). The rest components of vector $\bar{x}'$ and components of averaged phase $\bar{\psi}_{1l}$ serve as a basis forming the vector of slow variables of the next (second) hierarchy:

$$\frac{dx''}{dt} = X'' \left(x'', \bar{\psi}_2, \xi_2\right);$$

$$\frac{d\bar{\psi}_2}{dt} = \xi_2 \omega_2 \left(x''\right) + Y'' \left(x'', \bar{\psi}_2, \xi_2\right);$$

$$x_{q'}'' = \bar{x}_{q'}'' + \sum_{n=1}^{\infty} \frac{1}{\xi_2^n} \bar{u}_{q'}^{(n)} \left(\bar{x}'', \bar{\bar{\psi}}_2\right); \tag{4.1.15}$$

$$\bar{\psi}_{2l'} = \bar{\bar{\psi}}_{2l'} + \sum_{n=1}^{\infty} \frac{1}{\xi_2^n} \bar{v}_{2l'}^{(n)} \left(\bar{x}'', \bar{\psi}_{2l'}\right),$$

where $x'' = \left\{x_{q'}''\right\}$ is the part of the vector $\bar{x}' = \left\{\bar{x}_q'\right\}$ in which the vector of fast phase $\bar{\psi}_2 = \{\psi_{2l}'\}$ is separated. In view of above reasoning the meaning of new functions X'', ω_2, Y_2'', and the others is evident. An algorithm of their calculational is described below.

Then we treat system (4.1.15) in terms of hierarchical algorithm. I.e., we separate out the variables from vector $\bar{x}''$ (this group of variables describes the dynamics of vector phase $\bar{\bar{\psi}}_3$), and so on. The procedure repeats cyclically until no terms remain in hierarchical series (4.1.6) and (4.1.11), i.e., as long as we take into account all the partial vector-phases ψ_κ that form complete vector of the total fast phase ψ. As an eventual result we obtain m-fold averaged (truncated, shortened) equation set ($(m + 1)$ is total number of hierarchical levels) that can be solved easily. This property is the main advantage of discussed version of hierarchical method making it very attractive for applications. Approximate non-averaged solutions of initial equations (4.1.5) are obtained by successive use of relevant inverse transformation formulae (4.1.13).

2. THE CASE OF SIMPLEST TWO-LEVEL SYSTEM WITH ONE ROTATING SCALAR PHASE

2.1 Formulation of the Problem

As shown above in Subsection 1.4, the procedure of asymptotic integration of the general multi-level system (4.1.5) in some special case can be essentially simplified. Namely, it can be reduced to accomplishing a hierarchical chain of integrating of simplified two-level systems (4.1.12), (4.1.14), etc.. Hence the constructing of relevant algorithm of asymptotic integration of a two-level system should represent next step of the discussed calculational technology. We begin discussion of algorithms of such type with the simplest case, when the fast rotating phase $\psi \equiv \psi_1$ in (4.1.12) turns out to be a scalar function. Then we rewrite the equation (4.1.12) in the form through small parameter $\varepsilon = 1/\xi << 1$ (see equation (3.5.9)). As a result we obtain:

$$
\begin{aligned}
\frac{dx}{dt} &= \varepsilon X\left(x, \psi, \varepsilon\right); \\
\frac{d\psi}{dt} &= \omega\left(x\right) + \varepsilon Y\left(x, \psi, \varepsilon\right),
\end{aligned}
\tag{4.2.1}
$$

where $\xi \equiv \xi_1$, $\omega \equiv \omega_1$, $Y \equiv Y_1$, and we have neglected the 'prime' symbol. It is obvious that for scalar phase ψ the function Y is scalar, too.

Without loss of generality we put the period T equal to 2π. Moreover, we assume that functions x, Y, ω, are differentiable with respect to the phase ψ as many times as is the number of terms retained in the asymptotic series expansion.

Later we follow with calculational scheme discussed above in Chapter 3, Section 5. It means that the problem is to find the substitution of the form of the type (3.5.11), which would make it possible to separate slow and fast variables. Taking into consideration the problem of secular terms discussed above (see Chapter 3, Subsection 5.2 for more details) let us write the Krylov–Bogolyubov substitutions as the asymptotic series

$$x = \bar{x} + \sum_{i=1}^{\infty} \varepsilon^i u_i \left(\bar{x}, \bar{\psi} \right) = \bar{x} + \varepsilon u \left(\bar{x}, \bar{\psi}, \varepsilon \right);$$

$$\psi = \bar{\psi} + \sum_{i=1}^{\infty} \varepsilon^i v_i \left(\bar{x}, \bar{\psi} \right) = \bar{\psi} + \varepsilon v \left(\bar{x}, \bar{\psi}, \varepsilon \right),$$

$$(4.2.2)$$

with the unknown functions u_i and v_i to be defined to be in what follows. The equation for the first hierarchical level we also represent in the form of the asymptotic series:

$$\frac{d\bar{x}}{dt} = \sum_{i=1}^{\infty} \varepsilon^i A_i \left(\bar{x} \right) = \varepsilon A \left(\bar{x}, \varepsilon \right);$$

$$\frac{d\bar{\psi}}{dt} = \omega \left(\bar{x} \right) + \sum_{i=1}^{\infty} \varepsilon^i B_i \left(\bar{x} \right) = \omega^* \left(\bar{x}, \varepsilon \right).$$

$$(4.2.3)$$

The coefficients A_i and B_i are unknown, to be found later along with the functions u_i and v_i. In order to specify the problem, we impose addition restrictions on the unknown functions u_i and v_i. Namely, we assume them to be bounded functions of $\bar{\psi}$ for $\bar{\psi} \to \infty$. The similar assumption we accepted above (with respect to time t only — see Chapter 3, Subsection 5.2) for solution of the secular terms problem.

2.2   Algorithm of Asymptotic Integration

In a manner described in Chapter 3, Section 1, we differentiate the representation (4.2.2) with respect to t and make use of the system (4.2.1) and (4.2.3). Moreover, we note that

$$\frac{du_i}{dt} = \frac{\partial u_i}{\partial t} \frac{d\bar{x}}{dt} + \frac{\partial u_i}{\partial t} \frac{d\bar{\psi}}{dt}$$

$$= \frac{\partial u_i}{\partial \bar{x}} \sum_{i=1}^{\infty} \varepsilon^i A_i \left(\bar{x} \right) + \frac{\partial u_i}{dt} \sum_{i=1}^{\infty} \left(\omega \left(\bar{x} \right) + \sum_{i=1}^{\infty} \varepsilon^i B_i \left(\bar{x} \right) \right).$$

$$(4.2.4)$$

After some straightforward algebra the relevant generalized hierarchical equation of the first order of the type (3.1.2) reduces to the expansion

$$\varepsilon A_1\left(\bar{x}\right) + \varepsilon^2 A_2\left(\bar{x}\right) + \cdots + \varepsilon\frac{\partial u_1}{\partial \bar{x}}\left(\varepsilon A_1\left(\bar{x}\right) + \cdots\right)$$

$$+ \varepsilon\frac{\partial u_1}{\partial \bar{\psi}}\left(\omega\left(\bar{x}\right) + \varepsilon B_1\left(\bar{x}\right) + \cdots\right)$$

$$+ \varepsilon^2\frac{\partial u_2}{\partial \bar{\psi}}\left(\varepsilon A_1\left(\bar{x}\right) + \cdots\right) + \varepsilon^2\frac{\partial u_2}{\partial \bar{\psi}}\left(\omega\left(\bar{x}\right) + \varepsilon B_1\left(\bar{x}\right) + \cdots\right)$$

$$= \varepsilon X\left(\bar{x} + \varepsilon u_1 + \cdots, \bar{\psi} + \varepsilon v_1 + \cdots, \varepsilon\right);$$

$$\omega\left(\bar{x}\right) + \varepsilon B_1\left(\bar{x}\right) + \varepsilon^2 B_2\left(\bar{x}\right)$$

$$+ \cdots + \varepsilon\frac{\partial v_1}{\partial \bar{x}}\left(\varepsilon A_1 + \cdots\right) + \varepsilon\frac{\partial v_1}{\partial \bar{\psi}}\left(\omega\left(\bar{x}\right) + \varepsilon B_1\left(\bar{x}\right) + \cdots\right)$$

$$+ \varepsilon^2\frac{\partial v_2}{\partial \bar{x}}\left(\varepsilon A_1 + \cdots\right) + \varepsilon^2\frac{\partial v_2}{\partial \bar{\psi}}\left(\omega\left(\bar{x}\right) + \varepsilon B_1\left(\bar{x}\right) + \cdots\right)$$

$$= \omega\left(\bar{x} + \varepsilon u_1\left(\bar{x}, \bar{\psi}\right) + \cdots\right) + \varepsilon Y\left(\bar{x} + \varepsilon u_1 + \cdots, \bar{\psi} + \varepsilon v_1 + \cdots\right).$$

$$(4.2.5)$$

We equate the coefficients of the equal order terms with respect to ε in the right and left hand parts of (4.2.5) to obtain a system of interconnected equations which determine the required functions, i.e.,

$$\frac{\partial u_1}{\partial \bar{\psi}}\omega\left(\bar{x}\right) = X\left(\bar{x}, \bar{\psi}, 0\right) - A_1\left(\bar{x}\right) \equiv g_1\left(\bar{x}, \bar{\psi}\right) - A_1\left(\bar{x}\right);$$

$$\frac{\partial v_1}{\partial \bar{\psi}}\omega\left(\bar{x}\right) = Y\left(\bar{x}, \bar{\psi}, 0\right) + D_1\left(u_1\right) - B_1\left(\bar{x}\right) \equiv h_1\left(\bar{x}, \bar{\psi}\right) - B_1\left(\bar{x}\right);$$

$$\frac{\partial u_2}{\partial \bar{\psi}}\omega\left(\bar{x}\right) = \frac{\partial X}{\partial \bar{x}}u_1 + \frac{\partial X}{\partial \bar{\psi}}v_1 + \frac{\partial X}{\partial \varepsilon} - \frac{\partial u_1}{\partial \bar{x}}A_1 - \frac{\partial u_1}{\partial \bar{\psi}}B_1 - A_2\left(\bar{x}\right)$$

$$\equiv g_2\left(\bar{x}, \bar{\psi}\right) - A_2\left(\bar{x}\right);$$

$$\frac{\partial v_2}{\partial \bar{\psi}}\omega\left(\bar{x}\right) = \frac{\partial Y}{\partial \bar{x}}u_1 + \frac{\partial Y}{\partial \bar{\psi}}v_1 + \frac{\partial Y}{\partial \varepsilon} - \frac{\partial v_1}{\partial \bar{\psi}}B_1 - \frac{\partial v_1}{\partial \bar{x}}A_1$$

$$+ D_1\left(u_1\right) + D_2\left(u_1\right) - B_2\left(\bar{x}\right)$$

$$\equiv h_2\left(\bar{x}, \bar{\psi}\right) - B_2\left(\bar{x}\right).$$

$$(4.2.6)$$

Here

$$D_1\left(u\right) = \sum_j \frac{\partial}{\partial \bar{x}^{(j)}}u^{(j)}; \quad D_2\left(u\right) = \frac{1}{2}\sum_{j,k} \frac{\partial^2 \omega}{\partial \bar{x}^{(k)}\partial \bar{x}^{(j)}}u^{(j)}u^{(k)},$$

(the functions D_1 and D_2 are obtained by expanding $\omega(\bar{x})$ in Taylor series with respect to ε); the components of the relevant vectors $\bar{x}^{(j)}$, $u_k^{(j)}$, $A_k^{(j)}$ and all derivatives are calculated for $\varepsilon = 0$. It may be easily verified that in the general case the system (4.2.6) may be written as

$$\frac{d\bar{\psi}'_1}{dt} = \xi_1 \omega_1(\bar{x}') + \sum_{n=1}^{\infty} \frac{1}{\xi_1^n} B_1^{(n)}(\bar{x}');$$

$$\omega(\bar{x}) \frac{\partial v_k}{\partial \bar{\psi}} = h_k(\bar{x}, \bar{\psi}) - B_k(\bar{x}).$$

(4.2.7)

We shall find the unknown functions u_k and v_k by means of successive integration of the system (4.2.6). Each subsequent equation is solved taking into account the solution of the previous one, but not vice versa. A complicating point is that equations (4.2.6) and (4.2.7) contain the unknown functions $A_k(\bar{x})$ and B_k. This difficulty may be overcome in view of the condition that both functions u_k and v_k are bounded for $\bar{\psi} \to \infty$.

So let us consider the first equation (4.2.6). We integrate it over $\bar{\psi}$ (bearing in mind that the derivative with respect to $\bar{\psi}$ is partial) to find the formal solution in quadratures to be given by

$$u_1(\bar{x}, \bar{\psi}) = \frac{1}{\omega(\bar{x})} \int_{\bar{\psi}_0}^{\bar{\psi}} \left\{ g_1(\bar{x}, \bar{\psi}) - A_1(\bar{x}) \right\} d\bar{\psi} + \varphi_1^*(\bar{x}),$$

(4.2.8)

where $\varphi_1^*(\bar{x})$ is an unknown function ('integration constant'). When calculating the integrals of the form (4.2.8) here and henceforth, we assume the variable $\bar{x}$ to be constant. We see that the solution (4.2.8) still contains an unknown function $A_1(\bar{x})$. We note that the function g_1 is periodic with respect to $\bar{\psi}$ (by virtue of periodicity $X(\bar{x}, \bar{\psi}, 0)$, see (4.2.6)). Let us calculate the average over the period $T = 2\pi$ of the integrand

$$\left\{ \bar{g}_1(\bar{x}, \bar{\psi}) - \bar{A}_1(\bar{x}) \right\} = \frac{1}{2\pi} \int_{\bar{\psi}}^{\bar{\psi}+2\pi} \left\{ g_1(\bar{x}, \bar{\psi}) - A_1(\bar{x}) \right\} d\bar{\psi} = c(\bar{x}).$$

(4.2.9)

In order to find the value of the function $c(\bar{x})$, we employ the boundedness condition for the function u_k,

$$\lim_{\bar{\psi} \to \infty} u_k(\bar{x}, \bar{\psi}) < \infty.$$

(4.2.10)

We consider the limit

$$\lim_{m \to \infty} u_1 \left(\bar{x}, \bar{\psi}_0 + 2\pi m \right) = \frac{1}{\omega \left(\bar{x} \right)} \lim_{m \to \infty} \left[2\pi c \left(\bar{x} \right) m \right] \qquad (4.2.11)$$

to find that for $m \to \infty$ condition (4.2.10) is satisfied only provided that

$$c \left(\bar{x} \right) = 0. \qquad (4.2.12)$$

Making use of (4.2.12) and (4.2.9) we obtain the required expression for the function A_1, i.e.,

$$A_1 \left(\bar{x} \right) = \bar{g}_1 \left(\bar{x} \right) = \frac{1}{2\pi} \int_0^{2\pi} g \left(\bar{x}, \bar{\psi} \right) d\bar{\psi} = \bar{X} \left(\bar{x} \right). \qquad (4.2.13)$$

This, in turn, makes it possible to find

$$u_1 \left(\bar{x}, \bar{\psi} \right) = \frac{1}{\omega \left(\bar{x} \right)} \left[\int_{\bar{\psi}_0}^{\bar{\psi}} X \left(\bar{x}, \bar{\psi} \right) d\bar{\psi} - \bar{X} \left(\bar{x}, \bar{\psi} \right) \bar{\psi} \right] + \varphi_1^* \left(\bar{x} \right). \qquad (4.2.14)$$

The mathematical structure of the arbitrary function $\varphi_1^* \left(\bar{x} \right)$ suggests that the latter should be chosen in accordance with the initial (boundary) conditions of the problem. For example, if the conditions are

$$x \left(t_0 \right) = \bar{x} \left(t_0 \right); \quad \psi \left(t_0 \right) = \bar{\psi} \left(t_0 \right), \qquad (4.2.15)$$

then $\varphi_1^* \left(\bar{x} \right) = 0$. In principle the solutions can be normalized to the function $\varphi_1^* \left(\bar{x} \right)$ in many other ways. In particular, if the initial system is Hamiltonian, then one can require that the system for the first hierarchical level (4.2.3) should be Hamiltonian too [21].

Now, let us find other unknown functions contained in the transformations (4.2.2) and representations (4.2.3). In a manner similar to solving (4.2.14), we obtain

$$B_1 \left(\bar{x} \right) = \bar{h}_1 = \bar{Y} + D_1 u_1; \qquad (4.2.16)$$

$$v_1 \left(\bar{x}, \bar{\psi} \right) = \frac{1}{\omega \left(\bar{x} \right)} \left\{ \int_{\bar{\psi}_0}^{\bar{\psi}} h_1 \left(\bar{x}, \bar{\psi} \right) d\bar{\psi} - \bar{h}_1 \left(\bar{x} \right) \bar{\psi} \right\} + \psi_1^* \left(\bar{x} \right), \qquad (4.2.17)$$

and so forth. This result may be generalized. Thus we have

$$A_1 = \bar{g}_1; \quad B_1 = \bar{h}_1, \tag{4.2.18}$$

$$u_i\left(\bar{x}, \bar{\psi}\right) = \frac{1}{\omega\left(\bar{x}\right)} \left\{ \int\limits_{\bar{\psi}_0}^{\bar{\psi}} g_i\left(\bar{x}, \bar{\psi}\right) d\bar{\psi} - \bar{g}_i\left(\bar{x}\right)\bar{\psi} \right\} + \varphi_i^*\left(\bar{x}\right); \tag{4.2.19}$$

$$v_i\left(\bar{x}, \bar{\psi}\right) = \frac{1}{\omega\left(\bar{x}\right)} \left\{ \int\limits_{\bar{\psi}_0}^{\bar{\psi}} h_i\left(\bar{x}, \bar{\psi}\right) d\bar{\psi} - \bar{h}_i\left(\bar{x}\right)\bar{\psi} \right\} + \psi_i^*\left(x\right). \tag{4.2.20}$$

Substituting (4.2.19) and (4.2.20) (with regard for (4.2.18)) in (4.2.2) yields the required transformations and, actually, the solution of the problem. Of course, provided the solutions of the system of truncated equations (4.2.3) are known.

Thus as we have already mentioned, the essence of the Bogolyubov method is to integrate the simpler (truncated) equations for the first hierarchical level (4.2.3) rather than complicated equations (4.2.1). In other words, to find the solutions of the truncated equations is sufficient for the asymptotic integration of the initial system (4.2.1) to be completed. The accuracy of the solutions is determined by the number n of terms retained in the series (4.2.2) and (4.2.3). Let us consider this aspect in more detail.

2.3   Accuracy of Approximate Solutions

As well known the accuracy problem is important for application of hierarchical asymptotic methods. Let us illustrate its peculiarities with discussed averaging method.

Suppose number n of terms retained in asymptotic representation similar to (4.1.14) is fixed, i.e.,

$$\frac{d\bar{x}}{dt} = \sum_{i=1}^{n} \varepsilon^i A_i\left(\bar{x}_n\right). \tag{4.2.21}$$

Integrating (4.2.21) over t yields components of vector $\bar{x}_n$. We substitute the latter for the phase $\bar{\psi}_n$ in second equation (4.2.2) to obtain

$$\bar{\psi}_n\left(t\right) = \bar{\psi}_0 + \int\limits_0^t \left\{ \left(\omega\left(\bar{x}_n\right) + \varepsilon B_1\left(\bar{x}_n\right) + \ldots\right) \right\} dt. \tag{4.2.22}$$

Now we treat solutions $\bar{x}_n$, $\bar{\psi}_n$ as approximations of 'true' solutions $\bar{x}$ and $\bar{\psi}$. Then solutions of initial equation (4.2.1) can be written as

$$x\left(t\right) \cong \bar{x}_n\left(t\right) + \sum_{m=1}^{N_1} \varepsilon^m u_m\left(\bar{x}_n, \bar{\psi}_n\right);$$

$$\psi\left(t\right) \cong \bar{\psi}_n\left(t\right) + \sum_{m=1}^{N_1} \varepsilon^m v_m\left(\bar{x}_n, \bar{\psi}_n\right).$$

(4.2.23)

A reasonable question arises of how high is the accuracy of the approximations? We remind the reader that the uncertain degree of accuracy of the approximate expressions (4.2.21), (4.2.22) has been main point of the mathematicians' objections against the Van der Pol method.

Thus let us estimate accuracy of approximations (4.2.23). Since $d\bar{x}/dt$, according to (4.2.21), is of the order of

$$\frac{d\bar{x}}{dt} = \dot{\bar{x}}_n + 0\left(\varepsilon^{n+1}\right),$$

(4.2.24)

then in view of the mathematical structure of equation (4.2.21) the estimate for $d\bar{x}/dt$ can be written as

$$\bar{x} = \bar{x}_n + 0\left(\varepsilon_n\right).$$

(4.2.25)

The asymptotic integration procedure implies that $\omega\left(\bar{x}\right)$ is a slowly varying function of $\bar{x}$. Therefore, its accuracy is given by

$$\omega\left(\bar{x}\right) = \omega\left(\bar{x}_n\right) + 0\left(\varepsilon^n\right).$$

(4.2.26)

We recall that calculating the derivative of a slowly varying function increases the order of magnitude of the result by ε. Therefore integrating a slowly varying function decreases its order of magnitude by ε as well. Hence, we estimate the accuracy of terms contained in the integrand of (4.2.22) as

$$\int_0^t \omega\left(\bar{x}\right) dt = \int_0^t \omega\left(\bar{x}_n\right) dt + 0\left(\varepsilon^{n-1}\right);$$

$$\varepsilon^n \int_0^t B_n\left(\bar{x}\right) dt = \varepsilon^n \int_0^t B_n\left(\bar{x}_n\right) dt + 0\left(\varepsilon^{n-1}\right).$$

Thus for the quadrature (4.2.22) we have

$$\bar{\psi}_n\left(t\right)=\bar{\psi}\left(0\right)+\int_0^t\left\{\omega\left(\bar{x}_n\right)+\varepsilon B_1\left(\bar{x}_n\right)+\ldots+\varepsilon^{n-1}B_{n-1}\left(\bar{x}_n\right)\right\}dt.$$

$$(4.2.27)$$

This means that if in n-th approximation $\bar{x}_n$ is found within the accuracy of ε^n then $\bar{\psi}_n$ is found within ε^{n-1}. This conclusion is important and must be in mind when estimating calculational accuracy. In the lowest approximation ($n=1$), truncated equations are

$$\frac{d\bar{x}}{dt}=\varepsilon\bar{X}\left(\bar{x}\right);\quad x\cong\bar{x};$$

$$\dot{\bar{\psi}}=\omega\left(\bar{x}\right);\quad\psi\cong\bar{\psi},$$

$$(4.2.28)$$

(the so called zeroth approximation — see examples in Chapters 6, 7, 9). In the general approximations of solutions $x\left(t\right)$ and $\psi\left(t\right)$ are

$$x=\bar{x}+\varepsilon u_1\left(\bar{x},\bar{\psi}\right)+\cdots+\varepsilon^{n-1}u_{n-1}\left(\bar{x},\bar{\psi}\right);$$

$$\psi=\bar{\psi}+\varepsilon v_1\left(\bar{x},\bar{\psi}\right)+\cdots+\varepsilon^{n-2}v_{n-2}\left(\bar{x},\bar{\psi}\right).$$

$$(4.2.29)$$

It is important to mention the correspondence between the time interval t_n, for which the solutions thus obtained are valid, and the number of approximation in the averaging method:

$$t_n\leqslant 1/\varepsilon^n=\xi^n.$$

$$(4.2.30)$$

This means that if solutions are obtained in the n-th approximation then physical processes described by them can be adequately interpreted only for interval (4.2.30). Ignoring this fact leads to mistakes (see, for instance, Sections 4 and 5). If some physical effect is predicted theoretically, one has to verify whether criterion (4.2.30) is satisfied. Otherwise the validity of scientific results is doubtful.

Then we turn to the problem of accuracy of Van der Pol solutions (3.4.11), (3.4.12). When equations (4.2.29) (and, respectively, (4.2.28) for $n=1$) are compared to the relevant results, (3.4.11)–(3.4.14), obtained by the Van der Pol method, it may seems that equation (3.4.14) is written with excessive accuracy with respect to (4.1.13). However, this is a premature conclusion. The matter is that, when deriving (3.4.11)–(3.4.14), we assumed $\omega=$ const, i.e., the integral $\int_0^t\omega dt$ can be calculated exactly, the error in the coefficient $B_i\left(\bar{x}_n\right)$ lowers to (ε^n), and so on. As a result we have

$$x = \bar{x} + \varepsilon u_1\left(\bar{x}, \bar{\psi}\right) + \cdots + \varepsilon^{n-1} u_{n-1}\left(\bar{x}, \bar{\psi}\right);$$

$$\psi = \bar{\psi} + \varepsilon v_1\left(\bar{x}, \bar{\psi}\right) + \cdots + \varepsilon^{n-1} u_{n-1}\left(\bar{x}, \bar{\psi}\right);$$

$$\frac{d\bar{x}}{dt} = \varepsilon A_1\left(\bar{x}\right) + \cdots + \varepsilon^n A_n\left(\bar{x}\right);$$

$$\frac{d\bar{\psi}}{dt} = \omega + \varepsilon B_1\left(\bar{x}\right) + \cdots + \varepsilon^n B_n\left(\bar{x}\right).$$

For $n = 1$ (4.2.31) reduces to the Van der Pol results (3.4.11)–(3.4.14), i.e.,

$$x = \bar{x}; \quad \frac{d\bar{x}}{dt} = \varepsilon \bar{X}\left(\bar{x}\right);$$

$$\psi = \bar{\psi}; \quad \frac{d\bar{\psi}}{dt} = \omega + \varepsilon Y\left(\bar{x}\right). \tag{4.2.31}$$

Thus the above mentioned contradiction really does not take a place.

2.4   Asymptotic Integration of Initial Equations by Means of Successive Approximations

In the preceding items we discussed asymptotic integration algorithm for initial set (4.2.1) based on the hypothesis that both functions X and Y are n times differentiable in ψ. Now, following [22, 23] we show this assumption is not fundamental and employed for convenience only.

Again, we consider the set (4.2.1). We write the Krylov–Bogolyubov substitution in the standard form

$$x = \bar{x} + \varepsilon u\left(\bar{x}, \bar{\psi}, \varepsilon\right);$$

$$\psi = \bar{\psi} + \varepsilon v\left(\bar{x}, \bar{\psi}, \varepsilon\right). \tag{4.2.32}$$

The relevant system of equations of the first hierarchical level (4.2.3) is given by

$$\frac{d\bar{x}}{dt} = \varepsilon A\left(\bar{x}\right);$$

$$\frac{d\bar{\psi}}{dt} = \omega\left(\bar{x}\right) + \varepsilon B\left(\bar{x}\right). \tag{4.2.33}$$

Substituting (4.2.32) and (4.2.33) in (4.2.1) yields

$$A\left(\bar{x},\varepsilon\right)+\varepsilon\frac{\partial u}{\partial\bar{x}}A\left(\bar{x},\varepsilon\right)+\frac{\partial u}{\partial\bar{\psi}}\left(\omega\left(\bar{x}\right)+\varepsilon B\left(\bar{x},\varepsilon\right)\right)$$

$$=X\left(\bar{x}+\varepsilon u;\bar{\psi}+\varepsilon v;\varepsilon\right);$$

$$\omega\left(\bar{x}\right)+\varepsilon B\left(\bar{x},\mu\right)+\varepsilon^{2}\frac{\partial v}{\partial\bar{x}}A\left(\bar{x},\varepsilon\right)+\mu\frac{\partial v}{\partial\bar{\psi}}\left(\omega\left(\bar{x}\right)+\varepsilon B\left(\bar{x},\varepsilon\right)\right)$$

$$=\omega\left(\bar{x}+\varepsilon u\right)+\varepsilon Y\left(\bar{x}+\varepsilon u,\bar{\psi}+\varepsilon v,\varepsilon\right).\quad(4.2.34)$$

We analyze (4.2.34) by means of successive approximations. The standard (for this method) iteration procedure implies

$$\frac{\partial u^{(k)}}{\partial\bar{\psi}}\omega\left(\bar{x}\right)=g_{k}-A^{(k)};$$

$$\frac{\partial v^{(k)}}{\partial\bar{\psi}}\omega\left(\bar{x}\right)=h_{k}-B^{(k)},\quad\quad(4.2.35)$$

where k is iteration number,

$$g_{k}=X\left(x+\varepsilon u^{(k-1)},\bar{\psi}+\varepsilon v^{(k-1)},\varepsilon\right)-\varepsilon\frac{\partial u^{(k-1)}}{\partial\bar{x}}A^{(k-1)}$$

$$-\varepsilon\frac{\partial u^{(k-1)}}{\partial\bar{\psi}}B^{(k-1)};$$

$$h_{k}=\frac{\omega\left(\bar{x}+\varepsilon u^{(k-1)}\right)-\omega\left(\bar{x}\right)}{\varepsilon}+Y\left(\bar{x}+\varepsilon\bar{u}^{(k-1)},\bar{\psi}+\varepsilon v^{(k-1)},\mu\right)$$

$$-\varepsilon\frac{\partial u^{(k-1)}}{\partial\bar{x}}A^{(k-1)}-\varepsilon\frac{\partial v^{(k-1)}}{\partial\bar{\psi}}B^{(k-1)}.\quad(4.2.36)$$

We compare systems (4.2.35) and (4.2.7) and find them to be of similar mathematical structure. Repeating above sequence of operations results in solutions similar to (4.2.19), (4.2.20).

2.5   Peculiarities of Asymptotic Hierarchical Calculational Schemes Based on the Fourier Method

We have shown that asymptotic integration of initial system (4.2.1) reduces to solving partial differential equations (4.2.7). The latter can be written in generalized form

$$\frac{\partial F\left(x, \psi\right)}{\partial \psi} = V\left(x, \psi\right) - D\left(x\right). \tag{4.2.37}$$

We recall that $V\left(x, \psi\right)$ is a periodic function of ψ with the period $T = 2\pi$. It is peculiar of the integration procedure applied to the equations of this type that the function $D\left(x\right)$ is unknown and found from the condition that the solution $F\left(x, \psi\right)$ is bounded for $\psi \to \infty$.

The Fourier version of averaging method is the most convenient one. Therefore just it is taken as a basic version further in this book.

We substitute a Fourier series for $V\left(x, \psi\right)$, i.e.,

$$V\left(x, \psi\right) = \sum_{k=-\infty}^{\infty} a_k\left(x\right) e^{ik\psi}, \tag{4.2.38}$$

where a_k is the Fourier amplitude defined in conventional way. Let us consider the zero harmonic of the Fourier expansion (4.2.38). Its amplitude is equal to the mean value of the function $V\left(x, \psi\right)$ for the period T, i.e.,

$$a_0 = \frac{1}{2\pi} \int_{-\pi}^{\pi} V\left(x, \psi\right) d\psi = \bar{V}\left(x\right). \tag{4.2.39}$$

Then equation (4.2.37) can be written as

$$\frac{\partial F}{\partial \psi} = \sum_{k \neq 0} a_k\left(x\right) e^{ik\psi} + \bar{V}\left(x\right) - D\left(x\right). \tag{4.2.40}$$

Let us expand the function F in a Fourier series. Within the context of (4.2.40), we have

$$F\left(x, \psi\right) = \sum_{k=-\infty}^{\infty} b_x\left(x\right) e^{ik\psi} + c\left(x\right)\psi + \varphi\left(x\right), \tag{4.2.41}$$

where $\varphi\left(x\right)$ is the 'integration constant',

$$b_k\left(x\right) = \frac{a_k\left(x\right)}{ik} \ \left(k = 0\right), \quad c\left(x\right) = \bar{V}\left(x\right) - D\left(x\right). \tag{4.2.42}$$

As follows from this result, $c\left(x\right) = 0$. Indeed, function F is bounded for $\psi \to \infty$ only provided the secular term $c\left(x\right)\psi$ contained in (4.2.41) vanishes that is possible for $c\left(x\right) = 0$ only. Thus we obtain the known result from the preceding analysis, i.e.,

$$D\left(x\right) = \bar{V}\left(x\right). \tag{4.2.43}$$

As shown below in this book, the procedure described is efficient when applied to electrodynamics.

3.   CASE OF TWO FAST ROTATING SCALAR PHASES

In Section 1 of this Chapter we have formulated the general calculational hierarchical problem. As was shown, this problem mathematically is the development of an algorithm of asymptotic integration of hierarchical standard systems of the type (4.1.5). It should be mentioned that, in the general case, arbitrary number of components of the total vector of fast rotating phases is characteristic feature of these systems. Above (in Section 2) we discussed the simplest variant of such hierarchical system, where the vector of fast rotating is represented by one scalar phase ($j = 1$) only.

Then we emphasize that systems with number of rotating phases $j > 1$ possess some essentially new peculiar features (in comparison to the discussed case $j = 1$). In the first place this concerns the possibility of realization of resonances in such multi-phases (multi-frequency) objects. Apart from that some specific features also characterize the non-resonant multi-phases systems. Unfortunately the general scheme of asymptotic integration of multi-phases hierarchical systems similar to (4.1.5) (which is given below in Section 4) seems too cumbersome for a beginner. That is why we begin studying of the multi-phases (multi-frequency) hierarchical systems with discussion of simplest system where the phenomenon of resonance can be realized. As analysis has shown, this is the system with two rotating scalar phases and one scalar large parameter.

3.1   Formulation of Problem

Thus we consider a system with two fast scalar phases ψ_j ($j = 1, 2$) and one scalar large parameter $\xi = 1/\varepsilon \gg 1$ given by the following standard form:

$$
\begin{aligned}
\frac{dx}{dt} &= X\left(x, \psi_1, \psi_2, \xi\right); \\
\frac{d\psi_1}{dt} &= \xi\omega_1\left(x\right) + Y_1\left(x, \psi_1, \psi_2, \xi\right); \\
\frac{d\psi_2}{dt} &= \xi\omega_2\left(x\right) + Y_2\left(x, \psi_1, \psi_2, \xi\right),
\end{aligned}
\qquad (4.3.1)
$$

where x and X are vectors, ω_j, Y_j, ψ_j, ξ are scalars. Contrary to the situation studied in the previous Section, now X and Y_j are two-period functions with respect to phases ψ_j with periods

$$T_j = 2\pi/q_j. \tag{4.3.2}$$

In electrodynamic problems coefficients q_j, as a rule, are usually equal to 1. Hence we can put $q_j = 1$ for simplicity. The generalization to the arbitrary case $q_j \neq 1$ is straightforward and does not involve any difficulties (see below the general algorithm described in Section 4).

We write the system of truncated equations (system of the first hierarchical level, according to our terminology) in the form analogous to (4.2.3), i.e.,

$$\frac{d\bar{x}}{dt} = \sum_{n=1}^{\infty} A^{(n)}(\bar{x});$$

$$\frac{d\bar{\psi}_j}{dt} = \xi\omega_j(\bar{x}) + \sum_{n=1}^{\infty} \frac{1}{\xi^n} B_j^{(n)}(\bar{x}). \tag{4.3.3}$$

According to general procedure of the averaging method, we find the change of variables reducing the system (4.3.1) to the form (4.3.3). For this we use the substitution in the standard form given by

$$x = \bar{x} + \sum_{n=1}^{\infty} \frac{1}{\xi} u^{(n)}(\bar{x});$$

$$\psi_j = \bar{\psi}_j + \sum_{n=1}^{\infty} \frac{1}{\xi^n} v_j^{(n)}(\bar{x}). \tag{4.3.4}$$

If solutions of the equation of the first hierarchical level (truncated equations) (4.3.3) are known, then substitutions (4.3.4) play the role of formal solutions of the initial system (4.3.1). Therefore, the main task in this case is to find unknown transformation functions $u^{(n)}(\bar{x})$ and $v_j^{(n)}(\bar{x})$, as well as the auxiliary functions $A^{(n)}(\bar{x})$ and $B_j^{(n)}(\bar{x})$.

3.2   Solutions. Non-Resonant Case

We carry out a procedure similar to that set forth in Section 2 and thus find the first-approximation $(n = 1)$ equations for the transformation functions to be given by

$$\frac{\partial u^{(1)}}{\partial \bar{\psi}}\omega_1(\bar{x}) + \frac{\partial u^{(1)}}{\partial \bar{\psi}_2}\omega_2(\bar{x}) = X(\bar{x}, \bar{\psi}_1, \bar{\psi}_2, \infty) - A^{(1)}(\bar{x});$$

$$\frac{\partial v_1^{(1)}}{\partial \bar{\psi}_1} \omega_1 (\bar{x}) + \frac{\partial v_1^{(1)}}{\partial \bar{\psi}_2} \omega_2 (\bar{x})$$

$$= Y_1 \left(\bar{x}, \bar{\psi}_1, \bar{\psi}_2, \infty \right) + \omega_1 (\bar{x}) u^{(1)} \left(\bar{x}, \bar{\psi}_1, \bar{\psi}_2 \right) - B_1^{(1)} (x) ;$$

$$\frac{\partial v_2^{(1)}}{\partial \bar{\psi}_1} \omega_1 (\bar{x}) + \frac{\partial v_2^{(1)}}{\partial \bar{\psi}_2} \omega_2 (\bar{x})$$

$$= Y_2 \left(\bar{x}, \bar{\psi}_1, \bar{\psi}_2, \infty \right) + \omega_2 (\bar{x}) u^{(1)} \left(\bar{x}, \bar{\psi}_1, \bar{\psi}_2 \right) - B_2^{(1)} (\bar{x}). \quad (4.3.5)$$

It is not difficult to show that in the subsequent approximations, $n > 1$, equations of similar mathematical structure can be obtained. The generalized mathematical construction of the the Krylov–Bogolyubov system of equations can be written similarly to (4.2.37), i.e.,

$$\frac{\partial F \left(\bar{x}, \bar{\psi}_1, \bar{\psi}_2 \right)}{\partial \bar{\psi}_1} \omega_1 (\bar{x}) + \frac{\partial F \left(\bar{x}, \bar{\psi}_1, \bar{\psi}_2 \right)}{\partial \bar{\psi}_2} \omega_2 (\bar{x}) = V \left(\bar{x}, \bar{\psi}_1, \bar{\psi}_2 \right) - D (\bar{x}).$$
$$(4.3.6)$$

As contrast to (4.2.37), however, in the case discussed the fast phase is represented by the two-component vector $(\bar{\psi}_1, \bar{\psi}_2)$, and, correspondingly, we have the two-periodic (with respect to the periods $T_1 = T_2 = 2\pi$) functions F and V. Let us employ the Fourier procedure described above in Subsection 2.4. We expound the functions $u^{(1)}$ and X in double Fourier series, i.e.

$$u^{(1)} \left(\bar{x}, \bar{\psi}_1, \bar{\psi}_2 \right) = \sum_{k=-\infty}^{\infty} \sum_{s=-\infty}^{\infty} b_{ks} (\bar{x}) \exp \left\{ ik\bar{\psi}_1 + is\bar{\psi}_2 \right\}$$

$$+ c (\bar{x}) \bar{\psi}_1 + d (\bar{x}) \bar{\psi}_2, \quad (4.3.7)$$

$$X \left(\bar{x}, \bar{\psi}_1, \bar{\psi}_2 \right) = \sum_{k=-\infty}^{\infty} \sum_{s=-\infty}^{\infty} a_{ks}^{(x)} \exp \left\{ ik\bar{\psi}_1 + is\bar{\psi}_2 \right\}. \quad (4.3.8)$$

Then we substitute (4.3.7) and (4.3.8) in the first equation (4.3.5) and assume that the coefficients of similar exponential functions are equal. Thus we obtain

$$b_{ks} = \frac{a_{ks}^{(x)}}{i \left(k\omega_1 + s\omega_2 \right)}; \quad (4.3.9)$$

$$c\left(\bar{x}\right)\omega_1\left(\bar{x}\right) + d\left(\bar{x}\right)\omega_2\left(\bar{x}\right) = a_{00}^{(x)} - A^{(1)}\left(\bar{x}\right). \tag{4.3.10}$$

The condition that the function $u^{(1)}$ is bounded for $\bar{\psi}_j \to \infty$ is satisfied if the secular terms in (4.3.7) vanish, i.e.,

$$c\left(\bar{x}\right) = 0; \quad d\left(\bar{x}\right) = 0. \tag{4.3.11}$$

In view of (4.3.11), relation (4.3.10) yields

$$A^{(1)} = a_{00}^{(x)}\left(\bar{x}\right) = \bar{X}\left(\bar{x}\right), \tag{4.3.12}$$

where

$$\bar{X}\left(\bar{x}\right) = \frac{1}{T_1 T_2}\int\limits_0^{T_1}\int\limits_0^{T_2} X\left(\bar{x}, \bar{\psi}_1, \bar{\psi}_2, \infty\right) d\bar{\psi}_1 d\bar{\psi}_2. \tag{4.3.13}$$

The other expansion terms of the first approximation can be calculated in the same manner. Thus we obtain

$$
\begin{aligned}
A^{(1)} &= \bar{X}\left(\bar{x}\right); \\
B_1^{(1)} &= \bar{Y}_1 + \omega_1\left(\bar{x}\right)\left\langle u_1\left(\bar{x}, \bar{\psi}_1, \bar{\psi}_2\right)\right\rangle; \\
B_2^{(1)} &= \bar{Y}_2 + \omega_2\left(\bar{x}\right)\left\langle u^{(1)}\left(\bar{x}, \bar{\psi}_1, \bar{\psi}_2\right)\right\rangle; \\
u^{(1)} &= \sum_{s,k\neq 0}\frac{a_{ks}^{(x)}\left(\bar{x}\right)}{i\left(k\omega_1 + s\omega_2\right)}\exp\left\{ik\bar{\psi}_1 + is\bar{\psi}_2\right\} + a_0^{(x)}\left(\bar{x}\right); \\
v_j^{(1)} &= \sum_{k,s\neq 0}\frac{a_{ks}^{(\psi_j)}\left(\bar{x}\right)}{i\left(k\omega_1 + s\omega_2\right)}\exp\left\{ik\bar{\psi}_1 + is\bar{\psi}_2\right\} + a_0^{(\psi_j)}\left(\bar{x}\right),
\end{aligned}
\tag{4.3.14}
$$

where $a_{ks}^{(\psi_j)}$ are the coefficients of the Fourier expansions of the functions $Y_j - \omega_j\left(\bar{x}\right)u^{(1)}$; the arbitrary functions $a_0^{(x)}$, $a_0^{(\psi_j)}$ should be found from the initial (boundary) conditions, $\left\langle u^{(1)}\left(\bar{x}, \bar{\psi}_1, \bar{\psi}_2\right)\right\rangle$ is the twice-averaged (over the phases $\bar{\psi}_1$ and $\bar{\psi}_2$) function $u^{(1)}$:

$$\left\langle u^{(1)}\left(\bar{x}, \bar{\psi}_1, \bar{\psi}_2\right)\right\rangle = \frac{1}{T_1 T_2}\int\limits_0^{T_1}\int\limits_0^{T_2} u^{(1)}\left(\bar{x}, \bar{\psi}_1, \bar{\psi}_2\right) d\bar{\psi}_1 d\bar{\psi}_2.$$

We note that the choice of the functions $a_0^{(x)}$, $a_0^{(\psi_j)}$ does not influence the accuracy of the solutions obtained.

Let us analyze the mathematical structure of the obtained solutions (4.3.14). We see that they contain the specific resonance denominators of the form

$$(k\omega_1(\bar{x}) + s\omega_2(\bar{x})). \qquad (4.3.15)$$

We remind the reader that the problem in the discussed case of two scalar phases has been formulated under the assumption that both phase rotation velocities are commensurable, i.e.,

$$\left|\frac{d\psi_1}{dt}\right| \sim \left|\frac{d\psi_2}{dt}\right| \quad \text{or} \quad |\omega_1(\bar{x})| \sim |\omega_2(\bar{x})|. \qquad (4.3.16)$$

This looks clearer if the following circumstance is taken into consideration. By virtue of the characteristic structure of equation (4.3.1) in the discussed case the difference in the scales of rates of change of the slow, on the one hand, and all fast variables, on the other hand, is determined by the same large parameter of the problem

$$\xi \sim \left|\frac{d\psi_1}{dt}\right| \Big/ \left|\frac{dx_q}{dt}\right| \sim \left|\frac{d\psi_2}{dt}\right| \Big/ \left|\frac{dx_q}{dt}\right| \gg 1, \qquad (4.3.17)$$

where x_q is the q-th component of the vector x. Then we remind the reader that the harmonic number in (4.3.15) is an algebraic quantity, i.e., $k, s = \pm 1, \pm 2, \cdots$. Besides that we remind that the angular velocities of the fast phases rotation (frequencies $\omega_j(\bar{x})$) are algebraic quantities, too. At the same time in principle some denominator in (4.3.15) in some points can be equal to zero, i.e.,

$$k\omega_1(\bar{x}) + s\omega_2(\bar{x}) = 0. \qquad (4.3.18)$$

This is the well known problem of small denominators [10]. Equation (4.3.18) determines the resonance condition (physical essence of this concept is described in Chapter 1, Subsection 2.3 in more detail). It is clear that if this condition is satisfied then the solution (4.3.14) is divergent. This infers that the above calculational procedure is inapplicable in the resonance case (4.3.18). In order to describe the resonance states of the system, the above constructed asymptotic integration algorithm should be some modified. Let us discuss one of methods of such modification.

3.3   Solutions. Resonant Case

Let us find the asymptotic solutions, which would adequately describe the dynamics of the system (4.3.1) in the neighborhood of the resonance point (4.3.18), including this point itself. In order to describe the resonance state, we introduce the it mismatch to be given by

$$h^* (\bar{x}) = k\omega_1 (\bar{x}) + s\omega_2 (\bar{x}) . \qquad (4.3.19)$$

Solving equation (4.3.19) for $h^* (\bar{x}) = 0$ yields the resonant point $\bar{x}^*$. Then we expand the functions $\omega_j (\bar{x})$ in the vicinity of the point $\bar{x}^*$. The analysis of the expansion shows: $h^* \sim 1/\xi$. Thus in the resonance case the large parameter (4.3.17) may be also interpreted as the measure of the approach to the resonance state of the system in the near-resonance region (see also Fig. 1.2.2 and corresponded comments).

The resonances of two types are distinguished - the main resonance and the resonances associated with harmonics. The main resonance occurs for $|k| = |s| = 1$. The harmonic resonances occur for $|k| > 1$, or $|s| > 1$, or when both inequalities are satisfied simultaneously: $|k|, |s| > 1$.

The case of constant frequencies. Let us begin our studying with the case of main resonance for the constant frequencies. It means that the supposition that the frequencies $\omega_j \neq \omega_j (\bar{x})$ are constants is satisfied. If the main resonance occurs in such model, we may write

$$\omega_1 = \omega_2 + (1/\xi) h^*, \qquad (4.3.20)$$

Within the context of (4.3.20), the initial system (4.3.1) may be rewritten as

$$\frac{dx}{dt} = X (x, \psi_1, \psi_2, \xi) ;$$

$$\frac{d\psi_1}{dt} = \xi\omega_1 + Y_1 (x, \psi_1, \psi_2, \xi) ; \qquad (4.3.21)$$

$$\frac{d\psi_2}{dt} = \xi\omega_1 + [Y_2 (x, \psi_1, \psi_2, \xi) + h^*] .$$

We introduce a new variable

$$\theta = \psi_2 - \psi_1, \qquad (4.3.22)$$

which is referred to as the combination phase (see also (1.2.22), (4.1.9) and relevant comments). We differentiate (4.3.22) with respect to time t and make use of (4.3.21). Thus we obtain

$$\frac{d\theta}{dt} = \frac{d\psi_1}{dt} - \frac{d\psi_2}{dt} = Y_2 (x, \psi_1, \psi_2, \xi) + h^* - Y_1 (x, \psi_1, \psi_2, \xi)$$
$$= \Theta (x, \psi_1, \psi_2, \xi) . \quad (4.3.23)$$

Inasmuch as phase rates of change have been assumed to be close (see (4.3.20)), their difference $(d\psi_1/dt) - (d\psi_2/dt)$ can be small. Both the new variable θ and the new function Θ entering (4.3.23) are slowly varying quantities under the resonance conditions. Then we can rewrite the system (4.3.21) taking into account (4.3.22) and (4.3.23). As a result we obtain

$$
\frac{dx}{dt} = X\left(x, \theta, \psi_2, \xi\right);
$$

$$
\frac{d\theta}{dt} = \Theta\left(x, \theta, \psi_2, \xi\right); \tag{4.3.24}
$$

$$
\frac{d\psi_2}{dt} = \xi\omega_1 + \left[Y_2\left(x, \theta, \psi_2, \xi\right) + h^*\right].
$$

It is readily seen from the latter system that, as a result, of the accomplished transformations:

a) the variable θ can be regarded as a component of the new slow variables x';

b) the functions X, Θ, and Y_2 are periodic with respect to one fast phase ψ_2 only (rather than two fast phases ψ_1 and ψ_2 of the system (4.3.1)). In other words, the mathematical properties of the system (4.3.24) are similar to those of the system (4.2.1), in which one should pass to the form initial equations with large parameter ξ (instead of the form with small one ε). Hence we can employ the relevant asymptotic integration algorithm, which is described above in Section 2.

The case of slowly varying frequencies. Further let us turn our attention to the more general case of slowly varying frequencies $\omega_j = \omega_j(x)$. At first we consider the main resonance $|k|, |s| = 1$, for $\omega_j = \omega_j(x)$ bearing in mind that in the general case the functions $\omega_j(x)$ are algebraic quantities. Taking this into account, the main resonance condition similar to (4.3.20) may be written in the case discussed in a form like (4.3.20):

$$
-\sigma\omega_1\left(x\right) = \omega_2\left(x\right) + \left(1/\xi\right)h^*\left(x\right), \tag{4.3.25}
$$

where $\sigma = ks = \pm 1$ is the sign function (because $k, s = \pm 1$, sign $\{\omega_j\} = +1$). Formally, there seems to be no difference between the conditions (4.3.25) (for $\sigma = -1$) and (4.3.20). However, a rather important distinction exists. It is associated with the very possibility that variations of the frequencies $\omega(x)$ can be slow under nonlinear resonance interactions in the system. This means that evolution of the system can be such that the system can slowly get in and out of the resonance state (see also the Fig. 1.2.2 and relevant comments). For each interval of values of x we have to check whether the solutions obtained really contain the resonance denominators (4.3.15). The results of the check up determine whether the non-resonance or resonance asymptotic integration procedure should be employed. In the general case, when solutions of both types occur (associated with different stages of interaction), we have to join the solutions together in terms of some special procedure [7] (see below in Subsection 3.4).

Then we turn to discussion of the algorithm of asymptotic integration in the discussed resonance case.

At the first stage the procedure of separating out the slowly varying combination phase θ could be modified in order generalize it. The general 'multi-phases' procedure of this type is described in monograph [23]. In our case of two scalar phases only it reduces to the following simplified form. We introduce the trivial non-degenerate linear transformation, which relates the fast phase vector $\{\psi_1, \psi_2\}$ to the vector of combination phases $\{\theta, \psi\}$, to be given by

$$\theta = \psi_2 + \sigma\psi_1; \quad \psi = \psi_2 - \sigma\psi_1. \tag{4.3.26}$$

Since the combination phase θ has been shown to be slow and, hence, is associated with a component of the vector of slow variables $x' = \{x, \theta\}$. In turn, the combination phase ψ, by virtue of the definition (4.3.26), must be considered as a fast one. The relations between the old and new variables in such situation are given by the following simplest linear transformations:

$$\psi_1 = \frac{\sigma}{2}(\theta - \psi); \quad \psi_2 = \frac{1}{2}(\theta + \psi). \tag{4.3.27}$$

Then we substitute (4.3.27) into (4.3.1), make use of (4.3.26), and carry out some calculations. As a result we obtain

$$\frac{dx}{dt} = X(x, \theta, \psi, \xi);$$

$$\frac{d\theta}{dt} = [Y_2(x, \theta, \psi, \xi) + \sigma Y_1(x, \theta, \psi, \xi) - h^*(x)];$$

$$\frac{d\psi}{dt} = \xi[2\omega_2(x)] + [Y_2(x, \theta, \psi, \xi) - \sigma Y_1(x, \theta, \psi, \xi) + h^*(x)]. \tag{4.3.28}$$

At the next stage of the calculational procedure we introduce the obvious notation

$$Y_2\left(x,\theta,\psi,\xi\right)+\sigma Y_1\left(x,\theta,\psi,\xi\right)-h^*\left(x\right)=\Theta\left(x,\theta,\psi,\xi\right);$$

$$2\omega_2\left(x\right)=\omega\left(x\right);$$

$$Y_2\left(x,\theta,\psi,\xi\right)-\sigma Y_1\left(x,\theta,\psi,\xi\right)+h^*\left(x\right)=Y\left(x,\theta,\psi,\xi\right);$$

$$x'=\{x,\theta\};\quad X=X'\left(X,\Theta\right). \tag{4.3.29}$$

As a result of the transformations performed we avoid the problems concerned with the small denominations: the system (4.3.28) reduces to the standard form with a single fast rotating phase similar to (4.2.1) (where we should accomplish the change $\varepsilon=1/\xi$ and connected with this relevant transformations), i.e.,

$$\frac{dx'}{dt}=X'\left(x',\psi,\xi\right);$$

$$\frac{d\psi}{dt}=\xi\omega\left(x'\right)+Y\left(x',\psi,\xi\right). \tag{4.3.30}$$

But the algorithm for solving such systems is described above in Section 2. This means that the above formulated resonant problem is solved.

Further we turn our attention to the case of harmonic resonance. Analogously to (4.3.25) the relevant harmonic resonance condition (i.e., in the case $|k|>1$ or $|s|>1$ or $|k|,|s|>1$ simultaneously) may be written as

$$\sigma\frac{k}{s}\omega_1\left(x\right)=\omega_2\left(x\right)+\frac{1}{\xi}h^*\left(x\right). \tag{4.3.31}$$

We define the slow and fast combination phases by the analogous method:

$$\theta=\sigma\frac{k}{s}\psi_2+\psi_1;\quad \psi=\sigma\frac{k}{s}\psi_2-\psi_2, \tag{4.3.32}$$

where, as before, $\sigma=\operatorname{sign}\{\omega_1\omega_2\}$, $k,s>0$. The relation between the old and new variables in now given by

$$\theta=\frac{1}{2}\left(\theta-\psi\right);\quad \psi_2=\sigma\frac{k}{2s}\left(\theta+\psi\right). \tag{4.3.33}$$

We carry out substitutions analogous to those considered in the previous case, and again obtain a system of the form (4.3.30).

It is to be noticed that in practice the definition of harmonic concepts sometimes causes misunderstanding. In electrodynamic problems one has to distinguish the harmonics of wave fields in which the electron moves, from the harmonics of its nonlinear oscillations in these fields (see above the materials of Subsection 1.3). In the asymptotic integration procedure, harmonics of external wave fields are taken into account in the definitions of fast phases, i.e.,

$$\psi_{m1} = m_1\psi_1; \quad \psi_{m2} = m_2\psi_2, \qquad (4.3.34)$$

where $m_j = 1, 2, \ldots$ are the numbers of external wave field harmonics. The mentioned above quantities k and s are harmonics of electron oscillations in these fields. The relevant combination phases (4.3.32) should be then written in the form (4.1.9). The integration procedure for the initial resonance system (4.3.1) remains in principle similar.

4. THE CASE OF MANY ROTATING SCALAR PHASES

4.1 Formulation of the Problem

Generalize the algorithms described above in the case of arbitrary number of the rotating phases (see system (4.1.12) and other relevant materials of Subsection 1.4). Introducing the vector of fast phases in the form $\psi = \{\psi_1, \psi_2, \ldots, \psi_m\}$, we rewrite initial equation (4.1.12) in the following manner

$$\frac{dx}{dt} = X\left(x, \psi, \xi\right);$$

$$\frac{d\psi}{dt} = \xi\omega\left(x\right) + Y\left(x, \psi, \xi\right), \qquad (4.4.1)$$

where all notations are self-evident in the above. We look for a solution to system (4.4.1) in the form of dynamical hierarchical operators (4.1.13):

$$x_q = \bar{x}_q + \sum_{n=1}^{\infty} \frac{1}{\xi^n} u_q^{(n)}(\bar{x}, \bar{\psi}), \quad (q = 1, 2, 3, \ldots, k);$$

$$\psi_j = \bar{\psi}_j + \sum_{n=1}^{\infty} \frac{1}{\xi^n} v_j^{(n)}(\bar{x}, \bar{\psi}), \quad (q = 1, 2, 3, \ldots, m), \qquad (4.4.2)$$

where: x_q, ψ_j are elements of vectors x and ψ, and k, m are numbers of components in the latter. In accepted notations one can rewrite the equations of the first hierarchical level (4.1.14) as

$$\frac{d\bar{x}_q}{dt} = \sum_{n=1}^{\infty} \frac{1}{\xi^n} A_q^{(n)}(\bar{x});$$

$$\frac{d\bar{\psi}_j}{dt} = \xi\omega_j(\bar{x}) + \sum_{n=1}^{\infty} \frac{1}{\xi^n} B_j^{(n)}(\bar{x}). \qquad (4.4.3)$$

4.2   Algorithm of Asymptotic Integration

Strict determination of unknown functions in (4.4.2), (4.4.3) is ambiguous. This is owed to arbitrariness in attributing different terms of the series. Following the above calculational schemes discussed we can eliminate this arbitrariness assuming all $u_q^{(n)}$ and $v_l^{(n)}$ are free of null (in $\bar{\psi}_l$) Fourier harmonics. Thus we postulate the whole averaged motion is described by $\bar{x}_q$ and $\bar{\psi}_l$.

Let us resume the differentiation of (4.4.2) and, taking into account (4.4.3), substitute the obtained result into (4.4.1), equating the coefficients of equal powers ξ^{-1}. Therefore we obtain the infinite sequence of relations

$$\sum_{j=1}^{m} \frac{\partial u_q^{(1)}}{\partial \psi_j} \omega_j = X_q - A_q^{(0)}; \qquad (4.4.4)$$

$$\sum_{j=1}^{m} \frac{\partial u_q^{(2)}}{\partial \psi_j} \omega_j = \sum_{i=1}^{k} \left(\frac{\partial X_q}{\partial x_i} u_i^{(1)} - \frac{\partial u_q^{(1)}}{\partial x_i} A_i^{(0)} \right)$$
$$+ \sum_{j=1}^{m} \left(\frac{\partial X_q}{\partial \psi_j} v_j^{(1)} - \frac{\partial u_q^{(1)}}{\partial \psi_j} B_j^{(0)} \right) - A_q^{(1)}, \quad (4.4.5)$$

$$\cdots\cdots\cdots\cdots\cdots\cdots\cdots\cdots\cdots\cdots\cdots\cdots\cdots\cdots$$

$$\sum_{l=1}^{m} \frac{\partial v_j^{(1)}}{\partial \psi_l} \omega_j = \sum_{i=1}^{k} \frac{\partial \omega_j}{\partial x_i} u_i^{(1)} + Y_j - B_j^{(0)} \equiv \psi_j(x, \psi) - B_j^{(0)}; \qquad (4.4.6)$$

$$\sum_{l=1}^{m} \frac{\partial v_l^{(2)}}{\partial \psi_l} \omega_j = \sum_{i=1}^{k} \frac{\partial \omega_j}{\partial x_i} u_i^{(2)} + \frac{1}{2} \sum_{i,l=1}^{k} \frac{\partial^2 \omega_j}{\partial x_i \partial x_l} u_i^{(1)} u_l^{(1)} + \sum_{i=1}^{k} \frac{\partial Y_j}{\partial x_i} u_i^{(1)}$$
$$+ \sum_{l=1}^{m} \frac{\partial Y_l}{\partial \psi_l} v_l^{(1)} - \sum_{i=1}^{k} \frac{\partial v_j^{(1)}}{\partial x_i} A_q^{(0)} - \sum_{l=1}^{m} \frac{\partial v_j^{(1)}}{\partial \psi_l} B_l^{(0)} - B_j^{(1)}; \quad (4.4.7)$$

. . . .

Hereafter, the averaging sign is omitted for simplicity. Taking into account (4.4.4), it is easy to see that functions x_q are periodic in ψ_l. Using this, we expand them into Fourier series of multiplicity m:

$$X_q = \sum_{sk...p} a_q^{sk...p}(x) \exp\left[2\pi i \left(\frac{s\,\psi_1}{T_1} + \cdots + \frac{p\,\psi_m}{T_m}\right)\right]. \qquad (4.4.8)$$

Here T_j is the period associated with phase ψ_j $(j = 1, 2, \ldots, m)$, $s, k, \ldots, p$ are the relevant Fourier harmonics. Proceeding from similar considerations as well as (3.1.18), we represent unknown function $u_q^{(n)}$ as

$$u_q^{(1)} = \sum_{sk...p} b_q^{sk...p}(x) \exp\left[2\pi i \left(\frac{s\,\psi_1}{T_1} + \cdots + \frac{p\,\psi_m}{T_m}\right)\right] + \sum_{j=1}^{m} C_{qj}(x)\psi_j. \qquad (4.4.9)$$

Upon substituting (4.4.8), (4.4.8) into (4.4.4) and equating the coefficients with equal exponents, we obtain expressions for amplitudes $b_q^{sk...p}(x)$ and functions $A_q^{(0)}$:

$$b_q^{sk...p}(x) = a_q^{sk...p}(x) \left[2\pi i \left(\frac{s\,\psi_1}{T_1} + \cdots + \frac{p\,\psi_m}{T_m}\right)\right]^{-1}, \qquad (4.4.10)$$

$$A_q^{(0)} = a_q^{00...0} - \sum_{j=1}^{m} C_{qj}(x)\,\omega_j, \qquad (4.4.11)$$

where functions $C_{qj}(x)$ in limits on $u_q^{(1)}$ satisfy normalization condition

$$C_{qj} = 0. \qquad (4.4.12)$$

Accordingly, definitions for coefficients $A_q^{(0)}$ (4.4.11) can be rewritten as

$$A_q^{(0)} = a_q^{00...0}(x) = \frac{1}{T_1 T_2 \cdots T_m} \int_0^{T_1} \cdots \int_0^{T_m} X_q \, d\psi_1 \ldots d\psi_m. \qquad (4.4.13)$$

Taking into account the condition of absence of null harmonics on ψ_j and $u_q^{(1)}$, for the latter we finally formulate the definition different from (4.4.9) only by exclusion of terms in s, k, ..., $p \neq 0$. Similarly to (4.4.11), the second approximation equation is solved, etc.. From the structure (4.4.10), in particular, it follows that

$$\frac{s\,\omega_1}{T_1} + \cdots + \frac{p\,\omega_m}{T_m} \neq 0, \qquad (4.4.14)$$

i.e., there are no resonances amongst components of the vector of fast phases. (Failure of (4.4.14) means that at the stage of classification of phases (and their linear combinations) not all slow phases have been singled out). Following the procedure, we obtain expressions for other unknown functions, in particular:

$$u_q^{(1)} = \sum_{sk\ldots p} \frac{1}{2\,\pi\,i} \frac{a_q^{sk\ldots p}(x)}{\frac{s\,\omega_1}{T_1} + \cdots + \frac{p\,\omega_m}{T_m}} \exp\left(2\,\pi\,i\left(\frac{s\,\omega_1}{T_1} + \cdots + \frac{p\,\omega_m}{T_m}\right)\right),$$

$$(4.4.15)$$

$$v_j^{(1)} = \sum_{sk\ldots p} \frac{1}{2\,\pi\,i} \frac{d_j^{sk\ldots p}(x)}{\frac{s\,\omega_1}{T_1} + \cdots + \frac{p\,\omega_m}{T_m}} \exp\left(2\,\pi\,i\left(\frac{s\,\omega_1}{T_1} + \cdots + \frac{p\,\omega_m}{T_m}\right)\right);$$

$$(4.4.16)$$

$$A_q^{(1)} = \left\langle \sum_{i=1}^{k} \frac{\partial X_q}{\partial x_i} u_i^{(1)} + \sum_{j=1}^{m} \frac{\partial X_q}{\partial \psi_j} v_j^{(1)} \right\rangle; \qquad (4.4.17)$$

$$B_1^{(0)} = \langle Y_1 \rangle, \qquad (4.4.18)$$

$$B_j^{(1)} = \left\langle \sum_{i,l=1}^{k} \left[\frac{1}{2}\frac{\partial^2 \omega_j}{\partial x_i \partial x_l} u_i^{(1)} u_l^{(1)} \frac{\partial Y_j}{\partial x_i} v_l^{(1)}\right] + \sum_{l=1}^{m} \frac{\partial Y_j}{\partial \psi_l} v_l^{(1)} \right\rangle, \qquad (4.4.19)$$

where $\langle \ldots \rangle$ means the averaging in all fast phases (similar to (4.4.13)); $d_j^{sk\ldots p}(x)$ are factors of expansions in Fourier series of the functions of this form.

5. ALGORITHM FOR SEWING TOGETHER RESONANT AND NON-RESONANT SOLUTIONS

5.1 Essence of the Problem

Thus depending on satisfaction of a condition like to (4.4.14) we could have two of principle different calculational situations in the considered oscillatative systems. The first is the non-resonant case. It takes place if condition (4.4.14) is satisfied. The calculational algorithms described above in Sections 2 and 4 are dedicated for the studying of systems of this type namely. The other situation corresponds to the resonant case. In such a case the relevant resonant condition can be constructed also (see, for example, Subsection 3.3).

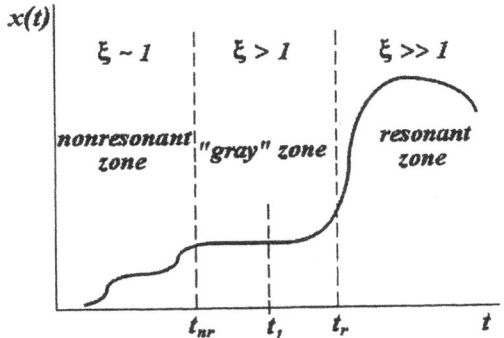

Figure 4.5.1. Illustration of calculational situation in the system with non-resonant and resonant states. Here the interval of interaction that is characterized by the parameter $\xi \sim \left|\frac{d\psi}{dt}\right| / \left|\frac{d\theta}{dt}\right| \sim 1$ corresponds to 'non-resonant zone' (here ψ and θ are some fast and slow combinative phases, respectively — see also definition (4.1.10) and relevant comments), the resonant character of the solution $x(t)$ we have on the interval $\xi \gg 1$, and the intermediate interval between these types of solutions ($\xi > 1$) is treated as a peculiar 'gray zone'.

Up to now we have discussed the models, which are characterized by resonant or non-resonant states, but no both they in the same model. However, the situations, when both (i.e., the non-resonant as well as the resonant) states could be realized in the same system are also rather typical for practice. Most often this takes place in the nonlinear dynamical systems, which pass consequently through non-resonant and resonant states during their evolution on time or space (see Fig. 4.5.1 and illustration example shown in Fig. 1.2.2). It is readily seen that we can separate three characteristic zones for the complete solution $x(t)$ in the discussed case (see Fig. 4.5.1).

The first is the zone which corresponds to non-resonant part of the solution. Characteristic feature of the latter is that the considered physical process here carries an explicitly non-resonant nature. According to the general theory, which was set forth above in Subsection 2.3, Chapter 1, such non-resonant states can be determined by some conditions such as the following

$$\xi \sim \left| \frac{d\psi}{dt} \right| \Big/ \left| \frac{d\theta}{dt} \right| \sim 1, \tag{4.5.1}$$

where ψ and θ are corresponded fast and slow combinative phases (see definition (4.1.10) and relevant comments). As shown in preceding Section 4, condition (4.5.1) can be assumed to be equivalent to condition (4.1.14):

$$\frac{s\,\omega_1}{T_1} + \cdots + \frac{p\,\omega_m}{T_m} \gg \frac{1}{\xi} h^*. \tag{4.5.2}$$

where h^* is the so called resonant mismatch (see also (4.3.20), (4.3.25), (4.3.31), and relevant explanations).

The resonant part of solution (the *resonant zone*) is characterized by satisfying of some resonant condition. In the framework of developed hierarchical theory such condition could be formulated in one of the forms like to (1.2.13), (1.2.18), (1.2.21) or (4.1.7). For instance:

$$\xi \sim \left| \frac{d\psi}{dt} \right| \Big/ \left| \frac{d\theta}{dt} \right| \gg 1 \tag{4.5.3}$$

or

$$\frac{s\,\omega_1}{T_1} + \cdots + \frac{p\,\omega_m}{T_m} \cong \frac{1}{\xi} h^*, \tag{4.5.4}$$

correspondingly.

In view of the above, it is obvious that in the general case the non-resonant and resonant solutions have essentially different mathematical structures, because they are obtained in different mathematical ways. Thus the problem of sewing them should be solved in order to obtain the complete solution $x(t)$.

5.2 Sewing Together of Resonant and Non-Resonant Solutions

A specific feature of the sewing procedure in the case of hierarchical asymptotic methods is the presence a peculiar 'gray' zone between the non-resonant and resonant zones (see Fig. 4.5.1). Strictly speaking, neither non-resonant ($\xi \sim 1$) nor resonant $\xi \gg 1$ solutions are invalid within this zone. But, on the other hand, it is obvious that the non-resonant solutions should transform into the resonant ones within the 'gray' zone. In principle, more than one such 'gray' zones can exist in a general multi-resonance case. We may assume that some point t_j, where the non-resonant solutions transform into resonant one immediately, should exist for each such j-th 'gray' zone . It is obvious, that we should have one such point t_1 only in the simplest one-resonant particular case illustrated in Fig. 4.5.1.

In what follows, let us turn to the problem of formulating the sewing condition. It is assumed that a k-dimensional vector of slow variables $x = x\{x_1, \ldots, x_k\}$, and some m-dimensional vector of fast rotating phases $\psi = \psi\{\psi_1, \ldots, \psi_m\}$ characterize the considered system. Accordingly with the above said, we should satisfy the following $2(k+m)$ sewing conditions for realization of the sewing procedure in any 'gray' zone:

$$
\begin{aligned}
x_{nr}(t)|_{t=t_j} &= x_r(t)|_{t=t_j} \,, \\
\psi_{nr}(t)|_{t=t_j} &= \psi_r(t)|_{t=t_j} \,, \\
dx_{nr}(t)/dt|_{t=t_j} &= dx_r(t)/dt|_{t=t_j} \,, \\
d\psi_{nr}(t)/dt|_{t=t_j} &= d\psi_r(t)/dt|_{t=t_j} \,.
\end{aligned}
\tag{4.5.5}
$$

Here the subscript 'nr' corresponds to the non-resonant parts of the complete solution, and the subscript 'r' describes relevant variables of the resonant solutions. As before t_j is the time when transformation of non-resonant solutions into resonant ones within a j-th zone occurs (see Fig. 4.5.1). Taking into consideration the properties discussed above of the resonant and non-resonant condition, the following 'gray' condition

$$
|s\,\omega_1 + \cdots + p\omega_m| \cong \alpha \tag{4.5.6}
$$

can be formulated (here $\alpha = |s\,\omega_1 + \ldots + p\omega_m| \sim 1/\xi\,(t=t_j) = \alpha\,(t_j)$ is some constant). Therewith, the condition

$$
|s\,\omega(x(t))_1 + \cdots + p\omega(x(t))_m| > \alpha, \tag{4.5.7}
$$

is satisfied for the time interval $t < t_j$, and the condition

$$
|s\,\omega(x(t))_1 + \cdots + p\omega(x(t))_m| < \alpha. \tag{4.5.8}
$$

is valid for the time interval $t > t_j$.

The system's evolution in the inverse direction (i.e., from the resonant state to the non-resonant one) can be described in an analogous manner.

Then let as remember that some unknown functions $a_0^{(x)}$ and $a_0^{(\psi_j)}$ (see solutions (4.3.14), for instance) were obtained for the above described procedures for constructing resonant and non-resonant solutions. It is important to draw the reader's attention that in the discussed 'sewing' case the relevant scheme of determining explicit view of these functions is some different. (Because the combinative phase θ is a component of the vector of slow variables in the resonant case, and, at the same time, analogous fast phase plays a role of component of the vector of fast rotating phases in the non-resonant case — see Sections 2–4 in detail). Hence this circumstance should be taken into account in 'sewing' conditions (4.5.5). As a result of utilization of the calculational schemes described in Sections 2–4 and the above noted peculiarities, conditions (4.5.5) can be transformed into corresponding system of equations with respect to the functions $a_0^{(x)}$ and $a_0^{(\psi_j)}$.

Thus we obtain two sets of the vector arbitrary functions $a_0^{(x)}$ and $a_0^{(\psi_j)}$ as a result of the accomplished calculations (one set we have for the resonant solutions, and the other set related with the non-resonant ones). One of them could be obtained by the method described in Sections 2–4. The second set is determined by 'sewing' condition of (4.5.5). This gives the method of solving the considered 'sewing' problem.

It might be seem that the above described calculational algorithm look too abstract for immediate practical realization. Let us illustrate it further in more detail by the simple example of the '*stimulated Duffing equation*' (see below equation (5.4.3) and corresponding commentaries, and [10]).

5.3   Example for the Solution 'Sewing': The 'Stimulated' Duffing Equation

The *Duffing equation* for stimulated oscillations, as is well known (see, for instance, [10]) can be written in the form:

$$\frac{d^2y}{dt^2} + \omega^2 y = -\frac{1}{\xi}y^3 + R\cos\lambda t, \qquad (4.5.9)$$

where ω, λ, and R are some constants, $\xi \gg 1$, as earlier, is the large problem parameter. Let as transform equation (4.5.9), using the Van der Pol variables $x(t)$ and $\psi(t)$ (see Chapter 3, Subsection 4.1):

$$\begin{aligned}
y(t) &= x(t)\cos\psi(t), \\
\dot{y}(t) &= -x(t)\omega\sin\psi(t).
\end{aligned} \qquad (4.5.10)$$

As a result the following equations for $x(t)$ and $\psi(t)$ (equations of the zeroth hierarchical level) could be obtained:

$$\frac{dx}{dt} = \frac{x^3}{4\xi\omega}\sin 2\psi + \frac{x^3}{8\xi\omega}\sin 4\psi - \frac{R}{\omega}\cos \lambda t \sin \psi,$$

$$\frac{d\psi}{dt} = \omega + \frac{3x^2}{8\xi\omega} + \frac{x^2}{2\xi\omega}\cos 2\psi + \frac{x^2}{8\xi\omega}\cos 4\psi - \frac{R}{x\omega}\cos \lambda t \cos \psi. \quad (4.5.11)$$

Accomplishing the obvious exchanging of variables $\psi_1 = \psi$, $\psi_2 = \lambda t$ it is not difficult to rewrite equations (4.5.11) in the following more symmetric form:

$$\frac{dx}{dt} = \frac{x^3}{4\xi\omega}\sin 2\psi_1 + \frac{x^3}{8\xi\omega}\sin 4\psi_1 - \frac{R}{\omega}\cos \psi_2 \sin \psi_1,$$

$$\frac{d\psi_1}{dt} = \omega + \frac{3x^2}{8\xi\omega} + \frac{x^2}{2\xi\omega}\cos 2\psi_1 + \frac{x^2}{8\xi\omega}\cos 4\psi_1 - \frac{R}{x\omega}\cos \psi_2 \cos \psi_1,$$

$$\frac{d\psi_2}{dt} = \lambda. \quad (4.5.12)$$

It is readily be convinced that equations (4.5.12) could be regard as a standard system with scalar slow variable x and two fast rotating phases ψ_1 and ψ_2. The analogous (but more general) system had been studied in Section 3 (see equations (4.3.1)). Hence all calculational schemes described there for considering system (4.5.12) can be used here. Later we will make use this circumstance in our analysis. At that according to that set forth in Section 3, we assume that non-resonant as well as resonant states could be realized in the same (considered) physical system that is described by equations (4.5.12). This means that behavior of the system at some time-intervals might be described by the resonant solutions, and, therein, the non-resonant solutions characterize the system evolution at other time-intervals. So let us begin our studying with discussion of the non-resonant states (or state) of the system.

Non-resonant part of the complete solution. It is considered that the non-resonant condition like to (4.5.2) is realized ($m = 2$, $T_1 = T_2 = 2\pi$ — see (4.3.2) and corresponding comments)

$$(s\omega + k\lambda) \geqslant \frac{1}{\xi}h^*, \text{ i.e. } (s\omega + k\lambda) \neq 0, \quad (4.5.13)$$

where s, k are any integers for which $1 \leqslant |s| + |k| \leqslant N$, $N \geqslant 2$.

Accordingly with the algorithm described in Section 3 the Bogolyubov substitutions (4.3.4)

$$x = \bar{x} + \sum_{n=1}^{\infty} \frac{1}{\xi^n} u_x^{(n)} \left(\bar{x}, \bar{\psi}_1, \bar{\psi}_2 \right);$$

$$\psi_1 = \bar{\psi}_1 + \sum_{n=1}^{\infty} \frac{1}{\xi^n} v_1^{(n)} \left(\bar{x}, \bar{\psi}_1, \bar{\psi}_2 \right); \qquad (4.5.14)$$

$$\psi_2 = \bar{\psi}_2 + \sum_{n=1}^{\infty} \frac{1}{\xi^n} v_2^{(n)} \left(\bar{x}, \bar{\psi}_1, \bar{\psi}_2 \right)$$

should be accomplished at the next stage of our calculational procedure. Here the relevant averaging values are determined by the equations like to (4.3.3):

$$\frac{d\bar{x}}{dt} = \sum_{n=1}^{\infty} \frac{1}{\xi^n} A_x^{(n)} (\bar{x});$$

$$\frac{d\bar{\psi}_1}{dt} = \omega(\bar{x}) + \sum_{n=1}^{\infty} \frac{1}{\xi^n} B_1^{(n)} (\bar{x}); \qquad (4.5.15)$$

$$\frac{d\bar{\psi}_2}{dt} = \lambda + \sum_{n=1}^{\infty} \frac{1}{\xi^n} B_2^{(n)} (\bar{x}).$$

Let us limit ourselves by calculating the functions $A_x^{(1)}$, $B_1^{(1)}$, $B_2^{(1)}$, $u_x^{(1)}$, $v_1^{(1)}$, $v_2^{(1)}$ in the first approximation only. As a result we can find for the function $(1/\xi) A_x^{(1)}$:

$$\frac{1}{\xi} A_x^{(1)} = \frac{1}{(2\pi)^2} \int_0^{2\pi} \int_0^{2\pi} \left\{ \frac{\bar{x}^3}{4\xi\omega} \sin 2\bar{\psi}_1 + \frac{\bar{x}^3}{8\xi\omega} \sin 4\bar{\psi}_1 \right.$$

$$\left. - \frac{R}{\omega} \cos \bar{\psi}_2 \sin \bar{\psi}_1 \right\} d\bar{\psi}_1 d\bar{\psi}_2 = 0. \quad (4.5.16)$$

The function $u_x^{(1)}$ we obtain from the equation

$$\frac{1}{\xi} \left(\frac{\partial u_x^{(1)}}{\partial \bar{\psi}_1} \omega(\bar{x}) + \frac{\partial u_x^{(1)}}{\partial \bar{\psi}_2} \lambda \right)$$

$$= \frac{\bar{x}^3}{4\xi\omega} \sin 2\bar{\psi}_1 + \frac{\bar{x}^3}{8\xi\omega} \sin 4\bar{\psi}_1 - \frac{R}{\omega} \cos \bar{\psi}_2 \sin \bar{\psi}_1. \quad (4.5.17)$$

After calculations it is not difficult to obtain also (see (4.3.14)):

$$
\frac{1}{\xi} u_{\bar{x}}^{(1)} = -\frac{\bar{x}^3}{8\xi\omega^2} \cos 2\bar{\psi}_1 - \frac{\bar{x}^3}{32\xi\omega^2} \cos 4\bar{\psi}_1
$$
$$
+ \frac{R}{2\omega} \left[\frac{\cos(\bar{\psi}_1 + \bar{\psi}_2)}{\omega + \lambda} + \frac{\cos(\bar{\psi}_1 - \bar{\psi}_2)}{\omega - \lambda} \right] + a_0^{(x)}(\bar{x}). \quad (4.5.18)
$$

In what follows, we can accept for simplicity: $a_0^{(x)}(\bar{x}) = 0$. Analogously, relevant calculations for the functions $B_1^{(1)}$ and $B_2^{(1)}$ yield:

$$
\frac{1}{\xi} B_1^{(1)} = \frac{1}{(2\pi)^2} \int_0^{2\pi} \int_0^{2\pi} \left\{ \frac{3\bar{x}^2}{8\xi\omega} + \frac{\bar{x}^2}{2\xi\omega} \cos 2\bar{\psi}_1 + \frac{\bar{x}^2}{8\xi\omega} \cos 4\bar{\psi}_1 \right.
$$
$$
\left. - \frac{R}{\bar{x}\omega} \cos \bar{\psi}_2 \cos \bar{\psi}_1 \right\} d\bar{\psi}_1 d\bar{\psi}_2 = \frac{3\bar{x}^2}{8\xi\omega}, \quad (4.5.19)
$$

$$
\frac{1}{\xi} B_2^{(1)} = \frac{1}{(2\pi)^2} \int_0^{2\pi} \int_0^{2\pi} \{0\} d\bar{\psi}_1 d\bar{\psi}_2 = 0, \quad (4.5.20)
$$

The equation

$$
\frac{1}{\xi} \left(\frac{\partial v_1^{(1)}}{\partial \bar{\psi}_1} \omega(\bar{x}) + \frac{\partial v_1^{(1)}}{\partial \bar{\psi}_2} \lambda \right)
$$
$$
= \frac{\bar{x}^2}{2\xi\omega} \cos 2\bar{\psi}_1 + \frac{\bar{x}^2}{8\xi\omega} \cos 4\bar{\psi}_1 - \frac{R}{\bar{x}\omega} \cos \bar{\psi}_2 \cos \bar{\psi}_1 \quad (4.5.21)
$$

could be found for the determining of the function $v_1^{(1)}$. Solving it we obtain:

$$
\frac{1}{\xi} v_1^{(1)} = \frac{\bar{x}^2}{4\xi\omega^2} \sin 2\bar{\psi}_1 + \frac{\bar{x}^2}{32\xi\omega^2} \sin 4\bar{\psi}_1
$$
$$
- \frac{R}{2\bar{x}\omega} \left[\frac{\sin(\bar{\psi}_1 + \bar{\psi}_2)}{\omega + \lambda} + \frac{\sin(\bar{\psi}_1 - \bar{\psi}_2)}{\omega - \lambda} \right] + a_0^{(\psi_j)}(\bar{x}). \quad (4.5.22)
$$

We also accept, as well as in the case (4.5.18): $a_0^{(\psi_j)}(\bar{x}) = 0$.
Accomplishing the analogous calculations it is readily be obtained:

$$
\frac{1}{\xi} v_2^{(1)} = 0. \quad (4.5.23)
$$

Resonant part of the solution. Let us consider the case of main resonance $s = 1$, $k = -1$ (see (4.3.25)). Then, the resonant condition (4.5.4) can be rewritten in the form ($T_1 = T_2 = 2\pi$):

$$(\omega - \lambda) \approx 0. \tag{4.5.24}$$

Further, using the procedure of reducing the resonant problem to the non-resonant one, which is described above in Subsection 3.3, we should pass to the slow combinative phase (4.3.22)

$$\theta = s\psi_1 + k\psi_2 = \psi_1 - \psi_2, \tag{4.5.25}$$

and to accomplish the exchanging

$$(\psi_1, \psi_2) \rightarrow (\theta, \psi), \quad \psi_2 = \psi, \quad \psi_1 = \psi + \theta. \tag{4.5.26}$$

As a result, equations (4.5.12) could be transformed into the following form:

$$\frac{dx}{dt} = \frac{x^3}{4\xi\omega} \sin 2(\psi + \theta) + \frac{x^3}{8\xi\omega} \sin 4(\psi + \theta) - \frac{R}{\omega} \cos\psi \sin(\psi + \theta),$$

$$\frac{d\theta}{dt} = \omega - \lambda + \frac{3x^2}{8\xi\omega} + \frac{x^2}{2\xi\omega} \cos 2(\psi + \theta) + \frac{x^2}{8\xi\omega} \cos 4(\psi + \theta)$$

$$- \frac{R}{x\omega} \cos\psi \cos(\psi + \theta),$$

$$\frac{d\psi}{dt} = \lambda, \tag{4.5.27}$$

It is obviously that the new variable θ is a slowly varying function and the system (4.5.27) carries the non-resonant nature. Therefore, for the finding of asymptotic solutions of system (4.5.27) we can apply the 'non-resonant' algorithm described in Subsection 3.2. So Krylov–Bogolyubov substitutions (4.3.4)

$$x = \bar{x} + \sum_{n=1}^{\infty} \frac{1}{\xi^n} u_x^{(n)} \left(\bar{x}, \bar{\theta}, \bar{\psi}\right);$$

$$\theta = \bar{\theta} + \sum_{n=1}^{\infty} \frac{1}{\xi^n} u_\theta^{(n)} \left(\bar{x}, \bar{\theta}, \bar{\psi}\right); \tag{4.5.28}$$

$$\psi = \bar{\psi} + \sum_{n=1}^{\infty} \frac{1}{\xi^n} v^{(n)} \left(\bar{x}, \bar{\theta}, \bar{\psi}\right)$$

can be consider as a first step of this stage of our calculational procedure. The second stage, analogously to the non-resonant case, is the
constructing the equations of the firs hierarchical level (i.e., equations
for averaging values)

$$
\frac{d\bar{x}}{dt} = \sum_{n=1}^{\infty} \frac{1}{\xi^n} A_x^{(n)}(\bar{x}, \bar{\theta});
$$

$$
\frac{d\bar{\theta}}{dt} = \sum_{n=1}^{\infty} \frac{1}{\xi^n} A_\theta^{(n)}(\bar{x}, \bar{\theta}); \tag{4.5.29}
$$

$$
\frac{d\bar{\psi}}{dt} = \lambda + \sum_{n=1}^{\infty} \frac{1}{\xi^n} B^{(n)}(\bar{x}).
$$

After corresponding calculations we obtain the following expressions
for the function $A_x^{(1)}$, $A_\theta^{(1)}$, $B^{(1)}$ (the first Bogolyubov approximation
$(n = 1)$):

$$
\frac{1}{\xi} A_x^{(1)} = \frac{1}{2\pi} \int_0^{2\pi} \left\{ \frac{\bar{x}^3}{4\xi\omega} \sin 2(\bar{\theta} + \bar{\psi}) + \frac{\bar{x}^3}{8\xi\omega} \sin 4(\bar{\theta} + \bar{\psi}) \right.
$$
$$
\left. - \frac{R}{\omega} \cos\bar{\psi} \sin(\bar{\theta} + \bar{\psi}) \right\} d\bar{\psi} = -\frac{R}{2\omega} \sin\bar{\theta}. \tag{4.5.30}
$$

$$
\frac{1}{\xi} A_\theta^{(1)} = \frac{1}{2\pi} \int_0^{2\pi} \left(\omega - \lambda + \frac{3\bar{x}^2}{8\xi\omega} + \frac{\bar{x}^2}{2\xi\omega} \cos 2(\bar{\psi} + \bar{\theta}) + \frac{\bar{x}^2}{8\xi\omega} \cos 4(\bar{\psi} + \bar{\theta}) \right.
$$
$$
\left. - \frac{R}{\bar{x}\omega} \cos\bar{\psi} \cos(\bar{\psi} + \bar{\theta}) \right) d\bar{\psi}
$$
$$
= \omega - \lambda + \frac{3\bar{x}^2}{8\xi\omega} - \frac{R}{2\bar{x}\omega} \cos\bar{\theta}. \tag{4.5.31}
$$

The Krylov–Bogolyubov equation

$$
\frac{1}{\xi} \left(\frac{\partial u_x^{(1)}}{\partial \bar{\psi}} \lambda \right) = \frac{\bar{x}^3}{4\xi\omega} \sin 2(\bar{\theta} + \bar{\psi}) + \frac{\bar{x}^3}{8\xi\omega} \sin 4(\bar{\theta} + \bar{\psi}) - \frac{R}{2\omega} \sin(\bar{\theta} + 2\bar{\psi})
$$
$$
\tag{4.5.32}
$$

can be found for the function $u_x^{(1)}$. Solving (4.5.32) we have:

$$\frac{1}{\xi}u_x^{(1)} = -\frac{\bar{x}^3}{8\xi\omega\lambda}\cos 2(\bar{\theta}+\bar{\psi}) - \frac{\bar{x}^3}{32\xi\omega\lambda}\cos 4(\bar{\theta}+\bar{\psi})$$

$$+ \frac{R}{4\omega\lambda}\cos(\bar{\theta}+2\bar{\psi}) + \frac{1}{\xi}a_0^{(x)}\left(\bar{x},\bar{\theta}\right). \quad (4.5.33)$$

Analogously, we obtain for the functions $u_\theta^{(1)}, B^{(1)}$, and $v^{(1)}$:

$$\frac{1}{\xi}\left(\frac{\partial u_\theta^{(1)}}{\partial\bar{\psi}}\lambda\right) = \frac{\bar{x}^2}{2\xi\omega}\cos 2(\bar{\psi}+\bar{\theta}) + \frac{\bar{x}^2}{8\xi\omega}\cos 4(\bar{\psi}+\bar{\theta})$$

$$- \frac{R}{2\bar{x}\omega}\cos(2\bar{\psi}+\bar{\theta}); \quad (4.5.34)$$

$$\frac{1}{\xi}u_\theta^{(1)} = \frac{\bar{x}^2}{4\xi\omega\lambda}\sin 2(\bar{\psi}+\bar{\theta}) + \frac{\bar{x}^2}{32\xi\omega\lambda}\sin 4(\bar{\psi}+\bar{\theta})$$

$$- \frac{R}{4\bar{x}\omega\lambda}\sin(2\bar{\psi}+\bar{\theta}) + \frac{1}{\xi}a_0^{(\theta)}(\bar{x},\bar{\theta}). \quad (4.5.35)$$

$$\frac{1}{\xi}B^{(1)} = \frac{1}{2\pi}\int_0^{2\pi}\{0\}\,d\bar{\psi} = 0. \quad (4.5.36)$$

$$\frac{1}{\xi}v^{(1)} = 0 + \frac{1}{\xi}a_0^{(\psi)}(\bar{x},\bar{\theta}) \quad (4.5.37)$$

Thus we have obtained the solutions for resonant and non-resonant parts of the complete solution. The next stage of the considered calculational procedure is 'sewing' of these solutions within the 'gray' zone (see Fig. 4.5.1).

The sewing the resonant and non-resonant parts of the solution. Let as suppose that the frequency λ is a slowly varying on time value (i.e., $\lambda = \lambda(t)$). This means that the considered physical system passes through non-resonant and resonant states during its temporary evolution. Therefore, we should to 'sew' the resonant and non-resonant parts of solution for obtaining some discontinuous complete only solution. The key point of corresponding 'sewing' algorithm is the 'sewing' condition (4.5.5). In the case considered condition (4.5.5) can be rewritten in the form:

$$x_{nr}(t)|_{t=t_1} = x_r(t)|_{t=t_1} \; ;$$
$$\psi_{1nr}(t)|_{t=t_1} = \psi_{1r}(t)|_{t=t_1} \; ;$$
$$\psi_{2nr}(t)|_{t=t_1} = \psi_{2r}(t)|_{t=t_1} \; ;$$
$$dx_{nr}(t)/dt|_{t=t_1} = dx_r(t)/dt|_{t=t_1} \; ; \qquad (4.5.38)$$
$$d\psi_{1nr}(t)/dt|_{t=t_1} = d\psi_{1r}(t)/dt|_{t=t_1} \; ;$$
$$d\psi_{2nr}(t,\xi)/dt|_{t=t_1} = d\psi_{2r}(t,\xi)/dt|_{t=t_1} \; ,$$

where, in view of definitions used in (4.5.5) all designations are self-evident.

However, let us turn the reader attention to one important circumstance. The resonant solutions (4.5.30)–(4.5.37) are written by the using the concept of slow $\theta = \psi_1 - \psi_2$ and fast $\psi = \psi_1 + \psi_2$ combinative phases. On the one hand, we did not need to utilize analogous combinative phases for description of the non-resonant part of complete solution. Instead this we had used the fast phases $\psi_{1,2}$ immediately. We have some contradiction of designations for the resonant and non-resonant parts of solution. For the sake of levelling the designations for description of both parts of solutions further we introduce the concept of fast combinative phase $\theta = \psi_1 - \psi_2$ (apart from the fast phase ψ) for the non-resonant part (!) of solution. This gives a possibility to accomplish the transformations like to $(\psi_1, \psi_2) \rightarrow (\theta, \psi)$ in non-resonant solutions (4.5.16)–(4.5.23). The 'sewing' conditions (4.5.38) reduce to the form:

$$x_{nr}(t)|_{t=t_1} = x_r(t)|_{t=t_1} \; ,$$
$$\theta_{nr}(t)|_{t=t_1} = \theta_r(t)|_{t=t_1} \; ,$$
$$\psi_{nr}(t)|_{t=t_1} = \psi_r(t)|_{t=t_1} \; ,$$
$$dx_{nr}(t)/dt|_{t=t_1} = dx_r(t)/dt|_{t=t_1} \; , \qquad (4.5.39)$$
$$d\theta_{nr}(t)/dt|_{t=t_1} = d\theta_r(t)/dt|_{t=t_1} \; .$$
$$d\psi_{nr}(t,\xi)/dt|_{t=t_1} = d\psi_r(t,\xi)/dt|_{t=t_1} \; .$$

Substituting the above obtained solutions for the non-resonant and resonant parts into (4.5.39) we eventually obtain the system of equations with respect to the unknown functions $a_0^{(x)}(\bar{x},\bar{\theta})$, $a_0^{(\theta)}(\bar{x},\bar{\theta})$, and $a_0^{(\psi)}(\bar{x},\bar{\theta})$. Let us determine these functions in the first Bogolyubov approximation. Utilizing (4.5.39), Krylov–Bogolyubov substitutions (4.5.14), and solutions obtained for non-resonant and resonant parts, it is not difficult to construct the system for determining the unknown functions $a_0^{(x)}(\bar{x},\bar{\theta})$, $a_0^{(\theta)}(\bar{x},\bar{\theta})$, and $a_0^{(\psi)}(\bar{x},\bar{\theta})$:

$$\bar{x} + \frac{1}{\xi} u_x^{(1)}\Big|_{nr,\,t=t_1} = \bar{x} - \frac{\bar{x}^3}{8\xi\omega^2}\cos 2(\theta + \psi) - \frac{\bar{x}^3}{32\xi\omega^2}\cos 4(\theta + \psi)$$

$$+ \frac{R}{2\omega}\left[\frac{\cos(\theta + 2\psi)}{\omega + \lambda} + \frac{\cos(\theta)}{\omega - \lambda}\right]$$

$$= \bar{x} + \frac{1}{\xi} u_x^{(1)}\Big|_{r,\,t=t_1} = \bar{x} - \frac{\bar{x}^3}{8\xi\omega\lambda}\cos 2(\bar{\theta} + \bar{\psi})$$

$$- \frac{\bar{x}^3}{32\xi\omega\lambda}\cos 4(\bar{\theta} + \bar{\psi}) + \frac{R}{4\omega\lambda}\cos(\bar{\theta} + 2\bar{\psi}) + \frac{1}{\xi}a_0^{(x)}(\bar{x}, \bar{\theta}). \quad (4.5.40)$$

$$\theta_{nr}(t)|_{t=t_1} = \bar{\psi}_1 + \frac{1}{\xi}v_1^{(1)} - \bar{\psi}_2 - \frac{1}{\xi}v_2^{(1)}\Big|_{nr,\,t=t_1}$$

$$= \bar{\theta} + \frac{\bar{x}^2}{4\xi\omega^2}\sin 2(\bar{\psi} + \bar{\theta}) + \frac{\bar{x}^2}{32\xi\omega^2}\sin 4(\bar{\psi} + \bar{\theta})$$

$$- \frac{R}{2\bar{x}\omega}\left[\frac{\sin(2\bar{\psi} + \bar{\theta})}{\omega + \lambda} + \frac{\sin(\bar{\theta})}{\omega - \lambda}\right] = \bar{\theta} + \frac{1}{\xi}u_\theta^{(1)}\Big|_{r,\,t=t_1}$$

$$= \bar{\theta} + \frac{\bar{x}^2}{4\xi\omega\lambda}\sin 2(\bar{\psi} + \bar{\theta}) + \frac{\bar{x}^2}{32\xi\omega\lambda}\sin 4(\bar{\psi} + \bar{\theta})$$

$$- \frac{R}{4\bar{x}\omega\lambda}\sin(2\bar{\psi} + \bar{\theta}) + \frac{1}{\xi}a_0^{(\theta)}(\bar{x}, \bar{\theta}); \quad (4.5.41)$$

$$\psi_{nr}(t)|_{t=t_1} = \bar{\psi}_2 + \frac{1}{\xi}v_2^{(1)}\Big|_{nr,\,t=t_1} = \bar{\psi} = \psi_r(t)|_{t=t_1} = \bar{\psi} + \frac{1}{\xi}v^{(1)}\Big|_{r,\,t=t_1}$$

$$= \bar{\psi} + \frac{1}{\xi}a_0^{(\psi)}(\bar{x}, \bar{\theta}). \quad (4.5.42)$$

It should be pointed out that here we confine ourselves by accounting only three first conditions (4.5.39). This is explained by that accounting of other three conditions (4.5.39) (which contain derivatives) do not exert remarkable influence at the complete solution, obtained in the first Bogolyubov approximation.

Further we use the observation (see Fig. 4.5.1) that we can accept for the 'sewing time' $t = t_1$:

$$\bar{x}|_{nr} = \bar{x}|_r = \bar{x};$$

$$\bar{\theta}\big|_{nr} = \bar{\theta}\big|_r = \bar{\theta};$$
$$\bar{\psi}_{2nr} = \bar{\psi}_r = \bar{\psi};$$
$$\omega - \lambda = 1/\xi, \qquad\qquad (4.5.43)$$

(here we assume that the frequencies ω and λ, as well as other values, are non-dimension ones). As a result of accomplished transformations we can write relevant expressions for the looked for functions $a_0^{(x)}(\bar{x}, \bar{\theta})$, $a_0^{(\theta)}(\bar{x}, \bar{\theta})$, and $a_0^{(\psi)}(\bar{x}, \bar{\theta})$:

$$\frac{1}{\xi} a_0^{(x)}(\bar{x}, \bar{\theta}) = \frac{R\cos(\bar{\theta})}{2\omega(\omega - \lambda)}\bigg|_{t=t_1};$$
$$\frac{1}{\xi} a_0^{(\theta)}(\bar{x}, \bar{\theta}) = -\frac{R\sin(\bar{\theta})}{2\bar{x}\omega(\omega - \lambda)}\bigg|_{t=t_1};$$
$$\frac{1}{\xi} a_0^{(\psi)}(\bar{x}, \bar{\theta}) = 0. \qquad\qquad (4.5.44)$$

In what follows, we substitute obtained expressions (4.5.44) into resonant solutions (4.5.33), (4.5.35), and (4.5.37). In turn let us substitute the got by such manner expressions into Krylov–Bogolyubov substitutions (4.5.14). This yields the resonant part of solutions which are 'sewed' with non-resonant ones in the point $t = t_1$. The conditions (5.4.3) could be used for determining of this point.

Thus the above stated problem of 'sewing' resonant and non-resonant solutions is solved.

References

[1] A.A. Ruhadze, L.S. Bogdankevich, S.E. Rosinkii, V.G. Ruhlin. *Physics of high-current relativistic beams*. Atomizdat, Moscow, 1980.

[2] R.C. Davidson. *Theory of nonlinear plasmas*. Benjamin, Reading, Mass, 1974.

[3] A.G. Sitenko, V.M. Malnev. *Principles of the plasma theory*. Naukova Dumka, Kiev, 1994.

[4] L.P. Landau, E.M. Liftshitz. *Theory of field*. Nauka, Moscow, 1974.

[5] V.V. Kulish. Nonlinear self-consistent theory of free electron lasers. method of investigation. *Ukrainian Physical Journal*, 36(9):1318–1325, 1991.

[6] V.V. Kulish, A.V. Lysenko. Method of averaged kinetic equation and its use in the nonlinear problems of plasma electrodynamics. *Fizika Plazmy*, 19(2):216–227, 1993. Sov. Plasma Physics.

[7] V.V. Kulish, S.A. Kuleshov, A.V. Lysenko. Nonlinear self-consistent theory of superheterodyne and free electron lasers. *The International journal of infrared and millimeter waves*, 14(3):451–568, 1993.

[8] M.G. Serebriannikov, A.A. Pervozvansky. *Discovery of hidden periodicities.* Nauka, Moscow, 1965.

[9] L. Lancos. *Practical methods of applied analysis.* Fizmatgiz, Moscow, 1961.

[10] E.A. Grebennikov. *Introduction to the theory of the reasonable systems.* Izdatelstvo MGU, Moscow, 1987.

[11] A.P. Sukhorukov. *Nonlinear wave interactions in optics and radiophysics.* Nauka, Moscow, 1988.

[12] N. Bloembergen. *Nonlinear optics.* Benjamin, New York, 1965.

[13] J. Weiland, H. Wilhelmsson. *Coherent nonlinear interactions of waves in plasmas.* Pergamon Press, Oxford, 1977.

[14] L.A. Vainstein, V.A. Solnzev. *Lectures on Microwave electronics.* Sov. Radio, Moscow, 1973.

[15] V.I. Gaiduk, K.I. Palatov, D.M. Petrov. *Principles of microwave physical electronics.* Sov. Radio, Moscow, 1971.

[16] A.F. Alexandrov, L.S. Bogdankevich, A.A. Ruhadze. *Principles of plasma electrodynamics.* Vyschja Shkola, Moscow, 1978.

[17] A.N. Kondratenko, V.M. Kuklin. *Principles of plasma electronics.* Energoatomizdat, Moscow, 1988.

[18] B.E. Zshelezovskii. *Electron-beam parametrical microwave amplifiers.* Nauka, Moscow, 1971.

[19] A.I. Olemskoi, A.Ya. Flat. Application of the factual concept in the condensed-matter physics. *Physics-Uspekhy*, 163(12):1–104, 1993.

[20] R. Rammal, G. Toulouse, M.A. Virasoro. Ultrametricity for physicists. *Reviews of Modern Physics*, 58(3):765–788, 1986.

[21] V.V. Kulish, P.B. Kosel, A.G. Kailyuk. New acceleration principle of charged particles for electronic applications. *The International Journal of Infrared and Millimeter Waves*, 19(1):33–93, 1998.

[22] N.N. Moiseev. *Asymptotic methods of nonlinear mechanics.* Nauka, Moscow, 1981.

[23] V.V. Kulish. *Methods of averaging in non-linear problems of relativistic electrodynamics.* World Federation Publishers, Atlanta, 1998.

Chapter 5

HIERARCHICAL SYSTEMS WITH FAST ROTATING PHASES. EXAMPLES OF PRACTICAL APPLICATIONS

Previously in this book we have set forth various versions of the asymptotic hierarchical algorithms and relevant calculational schemes. But experience shows that most of these calculational schemes are not simple for an average reader to understanding. Moreover, as the author is convinced, most professional physicists have heard about averaging methods, but only some of them know how to use them at practice. The main cause of such a situation is some psychological unusualness of the structure, the basic idea as well as the special 'language' of the methods. This could explain the main cause of the fact that hierarchical (including averaging) methods now are not too widespread in physics, in general, and in electrodynamics in particular. On the other hand, the methods discussed demonstrate an obvious highly practical effectiveness. Personal author pedagogical experience shows that the best teaching effect can be achieved in such situation only when all above discussed methodical ideas were illustrated by means of many calculational details and evident applied examples.

Observation formulated above predetermined the main idea and the structure of the book, as a whole. Indeed, the most part of its contents comprises titles of examples, which demonstrate various practical calculational peculiarities of the methods. In particular, Volume II, as a whole, serves this purpose. Besides that, some illustration examples are given in this (first) Volume. The difference is only that the examples described in Volume II are mainly intended for professional physicists, postgraduate, and master level students, who have essential preliminary training in physics and mathematics. In contrast to this, the illustration examples given in Volume I are intended for readers whose training is not so advanced. Therefore characteristic feature of the examples given below is the relative simplicity of the considered models and presence of a lot of illustrative calculational details. The model *'charged particles in the field of standing electromagnetic waves'*, which is studied below in this Chapter, is chosen as a basis for such simplified illustration example.

1. GENERAL PROPERTIES OF THE MODEL: 'A CHARGED PARTICLE IN THE FIELD OF A STANDING ELECTROMAGNETIC WAVE' AND EXAMPLES OF ITS PRACTICAL REALIZATION

1.1 Systems for Transformation of Optical Signals into Microwave Signals as a Convenient Illustrative Examples

The systems which can be described in framework of the model 'a charged particle in the field of a *standing electromagnetic wave'* are widespread in electronics, radiophysics, and plasma electrodynamics. For instance, one might find them amongst various systems of laser plasma diagnostic and control, systems for transformation, generation, and amplification of electromagnetic waves [1–8], etc.. For instance, as it will be shown further in Chapter 10, Volume II, the models of free electron lasers (FELs) in some cases can be reduced to such model [4–8].

So let us discuss the system for transformation of optical signals into microwave signals [5, 9–11], as a convenient illustration example. As the analysis shows, the chosen model can represent this system in satisfactory way. The devices of this type can be used as receivers of electromagnetic signals of sub-millimeter-infrared (IR) ranges, as systems for dynamical parameter control of especially intensive laser beams in vacuum, etc.. It is interesting to note that the solid state analogs of this device are also known [11–13]. They are IR receivers of electromagnetic signals constructed on the basis of some types of piezo-semiconductors. Moreover, they could be studied by using the discussed hierarchical methods. So, the calculational technology set forth below can, in principle, be applied for analysis of some solid state systems, too.

Let us restrict ourselves by discussion of only two versions of this system design. These versions are presented in Figs. 5.1.1 and 5.1.2. We begin our discussion with the design-version represented in Fig. 5.1.1.

The system shown in Fig. 5.1.1, which intended for transformation of submillimeter-IR range signals into microwave signals, operates in the following way. Transforming electromagnetic signal 1 with frequency ω_1 enters the signal input 2. Heterodyne signal 10 with frequency ω_2 enters the heterodyne input 9. Both inputs are placed in such manner that the transforming and heterodyne signals have the possibility to propagate within the same system working bulk in mutually opposite directions along the system optical axis. Electron gun 3 generates non-relativistic electron beam 5, which also moves along this axis. So electron beam 5 moves in the working bulk of the device in the field of two opposite directed transverse running electromagnetic waves Namely, they are transforming 1 and heterodyne 10 waves with frequencies ω_1 and ω_2, respectively. Owing to nonlinear properties of plasmas of beam 5 longitudinal Langmuir electron wave with the differential frequency $\omega_3 = \omega_1 - \omega_2$ (difference frequency electron wave) is excited. This wave also propagates in the direction of electron beam motion. The frequencies ω_1 and ω_2 (that belong to submillimeter-IR range) are chosen close to each other, so that the differential frequency $\omega_3 = \omega_1 - \omega_2$ lies within the microwave range (i.e., $\omega_3 << \omega_{1,2}$).

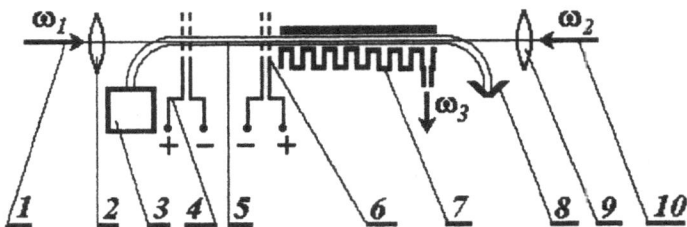

Figure 5.1.1. Design scheme of the system for optical signals transformation into microwave signals (version 1). Here 1 is the transforming optical signal, 2 is the input system for the transforming of the optical signal, 3 is the electron gun, 4 is the retarding system of electrodes for the electron beam, 5 is the electron beam, 6 is the system of electrodes for acceleration of the electron beam, 7 is the retarding system for the transformed microwave signal, 8 is the collector for the utilized electron beam, 9 is the input system for the heterodyne optical signal, 10 is the heterodyne signal, ω_1 is the frequency of the transforming optical signal, ω_2 is the frequency of the heterodyne signal, $\omega_3 = \omega_1 - \omega_2$ is the frequency of the transformed microwave signal.

The process of excitation of an electron beam wave, according to principles of the physical electronics [1–4], could be interpreted as the *effect of electron beam modulation*. The beam modulation in the discussed case occurs under action of the field of two oppositely propagated transverse electromagnetic waves. *The Langmuir electron waves*, to the physical point of view, are analogous to the usual sound waves in a gas, because the beam plasmas could be regarded as a peculiar flux of a charged gas.

Thus some (first) part of the device really works as a *modulation section*. Besides that, the device contains some other design elemets, such as retarding system 7. The difference frequency electron wave ω_3, which is excited within the modulation, transforms into the electromagnetic microwave signal with the same frequency ω_3 in retarding system 7. This signal we regard as an *output transformed electromagnetic signal* for the discussed device.

As the analysis shows (see, for example, [1–4]), effectiveness of any modulation process always depends strongly on certain correlation between different characteristic dynamical parameters of a system. In our case they are magnitudes of the frequencies $\omega_{1,2}$, on the one hand, and the velocity v_0 of the beam as a whole, on the other hand. Qualitatively this peculiarity can be illustrated in the following way. Propagating in the opposite directions running electromagnetic waves ω_1 and ω_2 form a *drifting standing electromagnetic wave*. It is well known that amplitude maximums and minimums of an 'ordinary' (i.e., no drifting) standing waves always *stand* on some fixed spatial points. In contrast, the analogous points of the drifting standing wave *drift* along the system optical axis with the *phase velocity*

$$v_{ph} = \frac{\omega_1 - \omega_2}{\omega_1 + \omega_2}c, \qquad (5.1.1)$$

(see for more details about the concept of drifting standing wave in Chapter 10, Volume II). Here c is, as before, the velocity of light in vacuum (the general concept of the phase velocity is discussed also in Chapter 1, Section 3). It can easily be seen that in our case the phase velocity $v_{ph} < c$ always, because, as mentioned above, $\omega_3 \neq \omega_{1,2}$. Moreover, for $\omega_3 << \omega_{1,2}$ (i.e., for $\omega_1 \sim \omega_2$) the case of non-relativistic drifting velocities ($v_{ph} << c$) can occur. In the particle case $\omega_1 = \omega_2$, the velocity $v_{ph} = 0$, and hence the drifting standing wave transforms into the 'usual' standing wave.

Then, we remind that in the considered device the electron beam 5 also moves, as a whole, along the optical axis. At that the direction of the drifting wave in the case considered is the same as the direction of electron beam 5. Such direction of the drifting wave is attained by the mentioned special selection of the frequencies $\omega_{1,2}$ ($\omega_1 > \omega_2$). In particular, the case is possible when the so called *strong synchronism* appears. This takes place when the drifting wave velocity v_{ph}, on the one hand, and the electron beam velocity v_0, on the other hand, are found to be equal

$$v_{ph} = v_0 = \frac{\omega_1 - \omega_2}{\omega_1 + \omega_2}c. \qquad (5.1.2)$$

As the analysis show [5, 11], the highest effectiveness of the electron beam modulation (i.e., excitation of the difference frequency electron wave ω_3) could be achieved in this case namely. After analyzing the condition (5.1.2) one can be convinced that this is the condition for realization of the so called *parametrical resonance*. (In detail the concept of parametrical resonance is discussed earlier in Chapter 1, Subsection 2.3 and Chapter 4, Subsection 1.3). Such work regime is chosen as a basic for the modulation section of the device shown in Fig. 5.1.1.

It should be mentioned that the theoretical model discussed of the device modulation section formally coincides, in principle, with the simplest model of *free electron laser* (FEL) with Dopplertron (including, laser) pumping [14, 15] (see also Chapters 10–13 Volume II). The heterodyne electromagnetic wave can be interpreted as the peculiar pumping wave, which is purposed in FELs for achieving additional amplification for the signal (transforming) wave ω_1. However, this amplification in the case considered is really inessential because the typical set of parameter for the considered transformer (the electron beam here is non-relativistic and weak-current). We shall neglect the influence of this amplification effect.

In what follows we turn the reader's attention to the fact that another design element, whose work effectiveness also depend's essentially on the beam velocity v_0, presents in the device. This is the retarding system 7 which, as mentioned above, transforms the electron wave ω_3 into the transformed output electromagnetic signal with the same frequency ω_3. In fact, this design part can be treated as a specific section of the Travelling Wave Tube (TWT) (about this type of electron devices see, for instance, in [1, 2]). Hence a corresponding synchronism condition can also be formulated for this transformer section that includes retarding system 7. It can be written as an approximate equality of the phase velocity of the transformed electromagnetic wave v_{ph3} and the beam velocity v_0:

$$v_{ph3} \approx v_0. \tag{5.1.3}$$

Then, we note that very often requirements for the beam modulation process (5.1.2) and for the maximum of transformation effectiveness of the retarding system 7 (5.1.3) cannot be satisfied simultaneously. Two special pairs of electrodes 4 and 6, correspondingly, are introduced in the device shown in Fig. 5.1.1 for correction of this drawback. Earlier we talked about the modulation section but never concretizes its design. These electrodes, and part of the electron beam between them, form the modulation section discussed above. Owing to the electrodes we have a possibility to change beam velocity v_0 in the working bulk of the modulation section and, consequently, to provide satisfaction of optimal conditions (5.1.2). Or, in other words, we can provide in this way optimal conditions for excitation of the difference frequency electron wave ω_3 in modulation section. At, satisfying the optimal condition for transformation of this electron wave into the electromagnetic microwave signal within the retarded system 7 (5.1.3) is provided by virtue of the last pair of electrodes 6. Usually the case of essentially differing synchronous conditions (5.1.2) and (5.1.3) is realized in the design versions intended for work in the submillimeter-IR range. It should be mentioned especially that the presence of running electromagnetic waves 1 and 10 (see Fig. 5.1.1) within the working bulk of retarding system 7 in such situations does not have any significant influence at effectiveness of the transformation process.

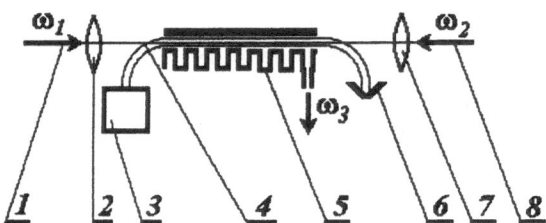

Figure 5.1.2. Design scheme of the system for transformation of optical signals into microwave signals (version 2). Here 1 is the transforming optical signal, 2 is the input system for the transforming optical signal, 3 is the electron gun, 4 is the electron beam, 5 is the retarding system for the transformed microwave signal, 6 is the collector for the utilized electron beam, 7 is the input system for the heterodyne optical signal, 8 is the heterodyne signal, ω_1 is the frequency of the transforming optical signal, ω_2 is the frequency of the heterodyne signal, $\omega_3 = \omega_1 - \omega_2$ is the transformed microwave electromagnetic signal.

So the characteristic feature of the discussed design-version (see Fig. 5.1.1) is the possibility of *spatial separating* of the processes of beam modulation and electron wave transformation, respectively. Hence both these processes can be studied separately owing to this feature. Later we shall widely use the latter peculiarity.

The distinctive, in principle, situation takes place for the second type of the discussed device, whose design scheme is shown in Fig. 5.1.2. Its characteristic feature is the possibility of simultaneous satisfaction of the conditions (5.1.2) and (5.1.3) in the same device working bulk. This means that, from an engineering point of view we can omit here the electrode systems like 4 and 6 in Fig. 5.1.1. As a result, the design as a whole simplifies.

Thus the difference between the devices shown in Fig. 5.1.2 and Fig. 5.1.1 is only the following. The systems of electrodes for controlling of the beam velocity v_0 are not foreseen in the device in Fig. 5.1.2, because of primordial coinciding both synchronous conditions (5.1.2) and (5.1.3), respectively. This means automatically that the modulation and transformation processes in this case occur simultaneously and this occurs in the same working bulk of the device. Or, more exactly, both processes happen within the working bulk of the retarded system 5. However, the found design peculiarity changes essentially the general physical situation in the device. Namely, in contrast to the previous design version (see Fig. 5.1.1) we do not have in this case any possibility for spatial separation of both discussed processes (modulation and transformation, respectively). Both these processes in the second case represented in Fig. 5.1.2 cannot be studied separately.

Generalizing the above we can state that in spite of the seeming design resemblance of both represented (in Fig. 5.1.1 and Fig. 5.1.2) devices they are characterized by essentially different operation physics. This means automatically that their theoretical models are also different. In the first case we have to do with the connection in series of two separate models-sections. The modulation and transformation processes, respectively, are characteristic for them. Therein the theoretical model: '*an electron beam in the field of two oppositely propagating electromagnetic waves*' corresponds to the modulation (i.e., first) section. Whereas the model '*electron beam in the field of three electromagnetic waves*' are characteristic for the second case. All this will be taken into consideration during analysis of physical processes in the discussed systems.

As follows from what was said above, all working physical mechanisms in the discussed devices really possess obviously multi-particle nature. At the same time, the analysis shown that used usually electron beams could be classified as a *weakly intensive*. This implies that reciprocal particle interactions (Coulomb's and magnetic, in the general case) can be neglected in the framework of such system theory. Hence, complete physical picture here is formed as a result of simple supposition of many individual no interacted between themselves particles, which, in turn, interact with the external heterodyne and signal electromagnetic wave fields. The concept of single-particle interaction electrons with the electromagnetic fields is accepted further as a basic one.

Thus we begin our illustration process with the simplest *single particle problems*. Or, in more detail, we will illustrate the hierarchical asymptotic calculational technology to the single particle theory of the devices discussed above. Therefore as already mentioned above, we have a possibility of construct two different, in principle, theoretical models in the discussed situation. The first one describes physical processes, which take place in the modulation section of the system, like that is shown in Fig. 5.1.1. (Let us recall once more that we classify it as the model with electron oscillations in the field of two oppositely directed running electromagnetic waves with frequencies $\omega_{1,2}$). The second one is the model with electron oscillations of in the field of three running electromagnetic waves (with frequencies $\omega_{1,2}$ and ω_3, respectively).

1.2   Formulation of the Problem of Electron Motion in the Field of Two Oppositely Propagating Electromagnetic Waves

Let us consider in more detail the dynamics of electron motion in the single-particle model of the first type. According to reasons discussed above we assume that the synchronous condition (5.1.2) is satisfied. Then we perform the transition to the co-ordinate system K', which moves along z-axis with the phase velocity of the drifting standing wave v_{ph}. This means that the drifting standing wave in the case of system K' transforms into an 'ordinary' standing wave, whose maximums and minimums of oscillations amplitude are spatially fixed. In turn, the fact of satisfying condition (5.1.2) means that the velocity of the electron beam, as a whole, is equal to zero in the co-ordinate system K' ($v_0' = 0$). Hence concerning the initial averaged velocity of a single chosen electron in the system K' we can assume that it is also equal to zero.

We describe the electron motion in the field of two oppositely running electromagnetic waves in terms of the Hamilton equation (see above Subsection 1.4.7)

$$\frac{d\mathcal{H}}{dt} = \frac{\partial\mathcal{H}}{\partial t}; \quad \frac{d\vec{\mathcal{P}}}{dt} = -\frac{\partial\mathcal{H}}{\partial \vec{r}}; \quad \frac{d\vec{r}}{dt} = \frac{\partial\mathcal{H}}{\partial \vec{\mathcal{P}}}, \qquad (5.1.4)$$

where $\mathcal{H} = \sqrt{m^2 c^4 + c^2 \left(\vec{\mathcal{P}} + \frac{e}{c}\vec{A}\right)^2}$ and $\vec{\mathcal{P}}$ are the Hamiltonian and canonical momentum of an electron in the electromagnetic field of standing electromagnetic wave [16], t is the laboratory time, $\vec{r} = \{x, y, z\}$ is the position vector (radius-vector) of the electron. Here and below everywhere we omit the primes that indicate that the quantities under consideration are written in the system K'.

As has been already mentioned above, any standing wave can be represented in the form of two running electromagnetic waves propagating in mutually opposite directions. In our case we suppose for simplicity that these running waves are plane and linearly polarized:

$$\vec{A} = \sum_{j=1}^{2} \frac{c\mathcal{E}_j}{\omega_j} \sin(p_j + \varphi_j) \cdot \vec{e}_x, \qquad (5.1.5)$$

where $\vec{A}$ is the vector-potential of the standing wave; $\mathcal{E}_j$ are the amplitudes of electric components of the running waves; $p_j = \omega_j t - s_j k_j z$ are their wave phases; $s_j = \text{sign}\left\{\vec{k}_j \vec{n}\right\} = \pm 1$ are the sign functions which determine the directions of the running wave propagation along z-axis (in our case of two opposite running waves: $s_1 = +1$, $s_2 = -1$), $\vec{k}_j$ are the wave vectors, $\vec{n}$ is the unit vector along the z-axis; ω_j are the cyclic frequencies of the running wave, $k_j = \left|\vec{k}\right|$ are the wave numbers (analogously with the situation with frequencies, we consider $k_1 = k_2 = k$); φ_j are the initial phases oscillations of the running waves; $\vec{e}_x$ is the unit vector along the x-axis; ω_j are the cyclic frequencies of the running wave. Here we should mention that in the framework of the chosen coordinate system K' the cyclic frequencies of both running electromagnetic waves are found to be equal: $\omega_1 = \omega_2 = \omega$, whereas they are not equal ($\omega_1 > \omega_2$) in the laboratory coordinate system K. The physical sense of this mathematical result will be discussed below in Volume II. Here we only note that Doppler effect can explain this phenomenon. The transforming signal wave (which propagates along the positive direction of the z-axis — see below in Fig. 5.1.3) reduces its frequency at the transition into the coordinate system K'. At the same time, oppositely directed heterodyne wave raises its frequency. As a result in the system K' both frequencies turn out to be equal.

The graphical illustration of the considered model-scheme is given in Fig. 5.1.3. Relevant definitions for all used concepts were given before in Sections 1.2 and 1.3.

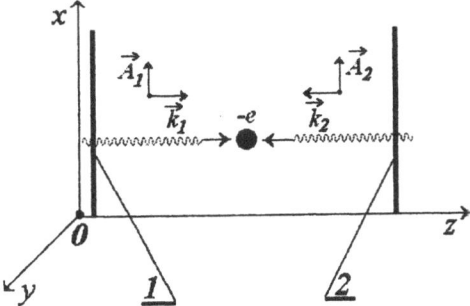

Figure 5.1.3. The illustration of the single-particle model of the modulation section of the device shown in the Fig. 5.1.1. Here 1 is the left boundary of the model, 2 is the right boundary of the model, x, y, and z are the coordinate axes, $\vec{A}_j$ ($j = 1, 2$) are the vector-potentials, $\vec{k}_j$ are the wave vectors of the transforming ($j = 1$) and heterodyne ($j = 2$) plane linearly polarized electromagnetic waves, $-e$ is the charged particle (in the considered specific case it is an electron).

Analyzing the initial system (5.1.4) from the point of view of definition (5.1.5), one can surely be convinced that equation (5.1.4) possesses an *integral of motion* (conservation law) with respect to transverse component of the canonical momentum:

$$\vec{\mathcal{P}}_\perp = \text{const}. \tag{5.1.6}$$

Indeed, inasmuch as any explicit dependencies of the Hamiltonian $\mathcal{H}(x,y,z)$ a determined by the explicit dependency

$$\vec{A} = \vec{A}(p_j) = A(z,t), \tag{5.1.7}$$

really the Hamiltonian $\mathcal{H}$ explicitly depends on the spatial coordinate z and laboratory time t only

$$\mathcal{H} = \mathcal{H}(z,t). \tag{5.1.8}$$

Due to this the partial derivatives $\partial\mathcal{H}/\partial x$ and $\partial\mathcal{H}/\partial y$ turn out to be equal to zero. Then, taking into consideration this and the second of equations (5.1.4) we can write the equation for the transverse component of the canonical momentum:

$$\frac{d\vec{\mathcal{P}}_\perp}{dt} = -\frac{\partial\mathcal{H}(z,t)}{\partial\vec{r}_\perp} = 0, \quad \text{i.e. } \vec{\mathcal{P}}_\perp = \text{const}, \tag{5.1.9}$$

where $\vec{r}_\perp = x\vec{e}_x + y\vec{e}_y$ is the transverse radius-vector, $\vec{e}_y$ is the unit vector along the axis y. The other values were determined above.

It is obvious that for known initial transverse momentum $\vec{\mathcal{P}}_{\perp 0} = \vec{\mathcal{P}}_\perp$ relevant exact solution for the transverse coordinates $\vec{r}_\perp$ can be found easily:

$$\frac{d\vec{r}_\perp}{dt} = \frac{\partial\mathcal{H}\left(\vec{\mathcal{P}}_{\perp 0}\right)}{\partial\vec{\mathcal{P}}_{\perp 0}}; \tag{5.1.10}$$

$$\vec{r}_\perp = \int_0^t \frac{\partial}{\partial\vec{\mathcal{P}}_{\perp 0}}\left\{\sqrt{m^2c^4 + c^2\left(\vec{\mathcal{P}}_{\perp 0} + \mathcal{P}_z\vec{n} + \frac{e}{c}\vec{A}(z,t)\right)^2}\right\}dt. \tag{5.1.11}$$

However, concerning the longitudinal component of motion, the calculational situation is found to be essentially more complex. The point is that there is no any possibility to obtain any exact solutions $z(t)$ and $\mathcal{P}_z(t)$ in this case. One of ways for obtaining of such asymptotic (i.e., approximate) solutions will be shown below.

2. MODEL 'AN ELECTRON IN THE FIELD OF TWO OPPOSITELY DIRECTED ELECTROMAGNETIC WAVES' AS A TWO-LEVEL HIERARCHICAL OSCILLATATIVE SYSTEM

2.1 Reducing Initial Motion Equations to the Standard Forms with Two Rotation Phases

Now we shall try to apply the hierarchical method for solution of the longitudinal part of the discussed electron motion problem. We start with classification of the variables and reducing the initial set of equations to some standard hierarchical form.

Taking into account results (5.1.9)–(5.1.11), the initial system (5.1.4), can be rewritten in the following manner:

$$\frac{d\mathcal{H}}{dt} = \frac{\partial \mathcal{H}}{\partial t}; \quad \frac{d\mathcal{P}_z}{dt} = -\frac{\partial \mathcal{H}}{\partial z}; \quad \frac{dz}{dt} = \frac{\partial \mathcal{H}}{\partial \mathcal{P}_z}, \qquad (5.2.1)$$

Analyzing system (5.2.1) together with definition (5.1.5), one can be convinced that here two fast rotating phases of the electron oscillations could be separated. Formally (in view of periodicity of the right hand parts of equations (5.2.1)) these phases coincide with the oscillation phases of the running electromagnetic plane waves:

$$\psi_1 = p_1 = \omega t - kz, \quad \psi_2 = p_2 = \omega t + kz. \qquad (5.2.2)$$

Next, let us turn to the problem of identification of slow variables. The variables $\mathcal{H}$, $\mathcal{P}_z$, and z should be considered as the slow ones from physical reasons. For instance, we take into consideration that according to the well known definition the velocity of changing of the electron energy in time is the power of acting forces (see Section 1.4). It is obvious that this power in a real physical situation must be always limited. Analogously, the velocity of changing of the canonical momentum is the force that acts on the electron and it also should be limited, and so on.

Bearing in the mind the above mentioned reasoning, let us perform some transformations in system (5.2.1). First, we transform definition for field of the standing wave (5.1.5) in the complex form, taking into account the definitions (5.2.2):

$$\vec{A} = A\vec{e}_x = \frac{1}{2i} \left[A_1 \exp\{i\psi_1\} - A_1^* \exp\{-i\psi_1\} \right.$$
$$\left. + A_2 \exp\{i\psi_2\} - A_2^* \exp\{-i\psi_2\} \right] \vec{e}_x, \quad (5.2.3)$$

where $i = \sqrt{-1}$ is the imaginary unit, the values

$$A_j = \frac{c\mathcal{E}_j}{\omega} \exp\{i\varphi_j\} \qquad (5.2.4)$$

are the complex amplitudes (see corresponding definitions above in Section 1.2). Let us assume that the phases $\psi_{1,2}$ in (5.2.3) are the electron oscillation phases. Then, substituting (5.2.3), (5.2.4) in the definition for Hamiltonian (see comments for equations (5.1.4)), we reduce the first two equations of system (5.2.1) to the form:

$$\frac{d\mathcal{H}}{dt} = \frac{ec\mathcal{P}_{x0}}{\mathcal{H}} \frac{\partial A(\psi_1, \psi_2)}{\partial t} + \frac{e}{2\mathcal{H}} \frac{\partial A^2(\psi_1, \psi_2)}{\partial t};$$
$$\frac{d\mathcal{P}_z}{dt} = -\frac{ec\mathcal{P}_{x0}}{\mathcal{H}} \frac{\partial A(\psi_1, \psi_2)}{\partial z} - \frac{e}{2\mathcal{H}} \frac{\partial A^2(\psi_1, \psi_2)}{\partial z}. \qquad (5.2.5)$$

In what follows, differentiating equations (5.2.2) by t and after some relatively simple transformation we obtain

$$\frac{d\psi_1}{dt} = \omega - k\frac{dz}{dt} = \omega - k\frac{c^2\mathcal{P}_z}{\mathcal{H}}; \qquad (5.2.6)$$

$$\frac{d\psi_2}{dt} = \omega + k\frac{dz}{dt} = \omega + k\frac{c^2\mathcal{P}_z}{\mathcal{H}} \qquad (5.2.7)$$

Here the well known relationships of relativistic dynamics [17]:

$$\mathcal{H} = E = \frac{mc^2}{\sqrt{1 - (v_x^2 + v_y^2 + v_z^2)/c^2}};$$

$$\mathcal{P}_z = \frac{mv_z}{\sqrt{1 - (v_x^2 + v_y^2 + v_z^2)/c^2}}. \qquad (5.2.8)$$

are used. Here $v_x = dx/dt$, $v_y = dy/dt$, $v_z = dz/dt$ are the components of the vector of electron velocity $\vec{v}$. We have also used the property that the Hamiltonian is the particle energy expressed through canonical variables (see relevant definitions in Section 1.4 and standard textbooks on the General Physics (see, fore instance, [16]).

Thus analyzing the mathematic structure of system (5.2.5)–(5.2.7) we conclude that the initial system (5.2.1) can be reduced to the standard form with two fast rotating phases like (4.3.1)

$$\frac{dx}{dt} = X\left(x, \psi_1, \psi_2, \xi\right);$$

$$\frac{d\psi_1}{dt} = \xi\omega_1\left(x\right) + Y_1\left(x, \psi_1, \psi_2, \xi\right); \qquad (5.2.9)$$

$$\frac{d\psi_2}{dt} = \xi\omega_2\left(x\right) + Y_2\left(x, \psi_1, \psi_2, \xi\right),$$

where the vector x is constructed from the slow variables: $x = \{\mathcal{P}_z, \mathcal{H}\}$; the vector-function X is constructed from left hand sides of equations (5.2.5). As it is easily seen (compare the last two of equations (5.2.9) and (5.2.6), (5.2.7), correspondingly) that we have two possible versions of determining of functions $\xi\omega_j$ and $Y_j(...)$ in the considered specific case. We can accept for the first version the concept of slowly varying nonlinear frequency

$$\xi\omega_j\left(x\right) = \omega \pm k\frac{c^2\mathcal{P}_z}{\mathcal{H}}.$$

In the case of the second version we can write

$$\xi\omega_j\left(x\right) \equiv \omega; \; Y_j\left(x, \psi_1, \psi_2, \xi\right) = \pm k\frac{c^2\mathcal{P}_z}{\mathcal{H}}. \qquad (5.2.10)$$

Let us chose version (5.2.10) as the more convenient for our purposes. But it is interesting to point out that, as the analysis shows, the choice of that or other above discussed versions does not exert any influence at the terminal calculational result.

It is necessary turn the reader attention also at some another specific calculational feature of discussed type problems. Namely, here, according to the definition given above of problem large parameter ξ, we should determine it as a ratio of fast to slow variables. We have two different variants for such definition:

$$\xi \sim \left|\frac{d\psi_{1,2}}{dt}\right| \Big/ \left|\frac{d\mathcal{P}_z}{\mathcal{P}_{z0}dt}\right| >> 1 \text{ and } \xi \sim \left|\frac{d\psi_{1,2}}{dt}\right| \Big/ \left|\frac{d\mathcal{H}}{\mathcal{H}_0 dt}\right| >> 1, \quad (5.2.11)$$

respectively. Here the functions $d\psi_{1,2}/dt$, $d\mathcal{P}_z/dt$, and $d\mathcal{H}/dt$ are determined by the right hand parts of equations (5.2.5)–(5.2.7). The smaller of parameters (5.2.11) is chosen as the problem large parameter ξ.

Thus it is clearly seen that the problem large parameter ξ in our particular case *turns out to be a slowly varying function on time*. That is why in calculational practice the large parameters in an explicit form are, as a rule, not separated out. Usually their peculiar 'hidden' form is used, when they are present in relevant equations in an implicit form. The relationships like ones in (5.2.11) are used for estimation of order of relevant terms in equation right hand sides. These relationships are also used for numerical current control of accuracy of the used calculational asymptotic procedures at their different stages.

It should be remind that the algorithm of asymptotic integration of systems similar to (5.2.9) has been described before in Chapter 4, Section 3. So, let us find relevant solutions using this algorithm, and perform their short physical analysis.

2.2   Zeroth Hierarchical Level. Parametrical Resonance

Analyzing the discussed system we can be convinced that the resonant states could be realized here (see condition (5.1.2) and corresponding comments). The latter conclusion is one of the sequences of synchronous condition (5.1.2), which had been imposed above. As has already been mentioned, this condition coincides with relevant result, which can be got from the parametrical resonance condition (see corresponding discussion in Subsections 1.2.3 and 4.1.3, respectively). Following the general algorithm of asymptotic integration (see above Section 4.3)), we should introduce in such situation the slow and fast combinative phases (4.3.26)

$$\theta = \psi_1 - \psi_2 = p_1 - p_2 = -2kz = -\frac{2\omega}{c}z, \qquad (5.2.12)$$

$$\psi = \psi_1 + \psi_2 = p_1 + p_2 = 2\omega t. \qquad (5.2.13)$$

It is obvious that the phase θ is a *slow variable* (because the electron coordinate z is a slowly varying on time function), whereas the phase ψ is fast one. After performing differentiation of equation (5.2.12) by time, and taking into account the relationship

$$\frac{dz}{dt} = v_z = \frac{c^2 \mathcal{P}_z}{\mathcal{H}}, \qquad (5.2.14)$$

(which, in turn, follows from expressions (5.2.8)) it is not difficult to obtain the corresponding differential equation for the slow combinative phase θ:

$$\frac{d\theta}{dt} = -2\omega c \frac{\mathcal{P}_z}{\mathcal{H}}. \qquad (5.2.15)$$

Analogously the differential equation for the fast combinative phase could be got in the similar way:

$$\frac{d\psi}{dt} = 2\omega. \tag{5.2.16}$$

As a result of the accomplished transformation we reduce the initial standard system (5.2.5)–(5.2.7) to the standard system with one fast rotating phase:

$$
\begin{aligned}
\frac{d\mathcal{H}}{dt} &= \frac{ec\mathcal{P}_{x0}}{\mathcal{H}} \frac{\partial A(\theta, \psi)}{\partial t} + \frac{e}{2\mathcal{H}} \frac{\partial A^2(\theta, \psi)}{\partial t}; \\
\frac{d\mathcal{P}_z}{dt} &= -\frac{ec\mathcal{P}_{x0}}{\mathcal{H}} \frac{\partial A(\theta, \psi)}{\partial z} - \frac{e}{2\mathcal{H}} \frac{\partial A^2(\theta, \psi)}{\partial z}; \\
\frac{d\theta}{dt} &= -2\omega c \frac{\mathcal{P}_z}{\mathcal{H}}; \\
\frac{d\psi}{dt} &= 2\omega.
\end{aligned}
\tag{5.2.17}
$$

Indeed, having constructed new extended vector of slow variables $x' = \{\mathcal{H}, \mathcal{P}_z, \theta\}$ and new vector-function X' we can formulate equation system (5.2.17) in the 'classical' standard form with one fast rotating phase (4.3.30) ψ

$$
\begin{aligned}
\frac{dx'}{dt} &= X'\left(x', \psi, \xi\right); \\
\frac{d\psi}{dt} &= \xi\omega' + Y\left(x', \psi, \xi\right),
\end{aligned}
\tag{5.2.18}
$$

where

$$
\begin{aligned}
\xi\omega' &\equiv 2\omega; \\
Y\left(x', \psi, \xi\right) &= 0, \\
\xi &\sim \left|\frac{d\psi}{dt}\right| \Big/ \left|\frac{d\theta}{dt}\right| \gg 1.
\end{aligned}
\tag{5.2.19}
$$

It is interesting to point out that the system obtained (5.2.17) can be classified as a *system with constant rotation frequency* (the difference between the systems with changing and constant frequencies is discussed in detail above in Chapter 4, Section 3).

2.3 Passage to First Hierarchical Level. Nonlinear Pendulum

According to the basic idea of the hierarchical method we should pass from zero to the first hierarchical levels at the next stage of our calculational procedure. For this the following the algorithm described above in Chapter 4, Section 3, we use a transformations similar to (3.3.4)

$$x' = \bar{x}' + \sum_{n=1}^{\infty} \frac{1}{\xi^n} u^{(n)}\left(\bar{x}'\right);$$

$$\psi = \bar{\psi} + \sum_{n=1}^{\infty} \frac{1}{\xi^n} v^{(n)}\left(\bar{x}'\right),$$

(5.2.20)

where we should construct relevant truncated equation of the type (see (3.3.3)) for determining the averaged values

$$\frac{d\bar{x}'}{dt} = \sum_{n=1}^{\infty} \frac{1}{\xi^n} A^{(n)}\left(\bar{x}'\right);$$

$$\frac{d\bar{\psi}}{dt} = \xi\omega + \sum_{n=1}^{\infty} \frac{1}{\xi^n} B^{(n)}\left(\bar{x}'\right).$$

(5.2.21)

Thus the main problem on this step is determining the unknown transformation functions $u^{(n)}\left(\bar{x}\right)$, $v^{(n)}\left(\bar{x}\right)$ and the functions $A^{(n)}\left(\bar{x}\right)$, $B^{(n)}\left(\bar{x}\right)$ in right hand parts of truncated equations (5.2.21). Let us solve this problem following with the algorithm discussed above. Therein we confine ourselves by calculations in the first approximation only of the averaging method ($n = 1$). Therefore, we will keep only the terms in right hand parts of equations not higher than cubic order, with respect to amplitudes of the running electromagnetic waves $\mathcal{E}_j$. The latter supposition can be substantiated by the following observation. The values

$$\varepsilon_j = e\mathcal{E}_j/mc\omega\gamma = v_{\perp j}/c$$

(5.2.22)

describe the amplitude of oscillations of the normalized (on the light velocity c) transverse electron velocity $v_{\perp j}$ in the field of the j-th running electromagnetic wave (here $\gamma = \mathcal{H}/mc^2$ is the normalized electron energy — *relativistic factor*)). In the case when the non-relativistic case of electron motion takes place ($v_j \ll c$) the following evident criterion is satisfied

$$\varepsilon_j \ll 1.$$

(5.2.23)

In relativistic electrodynamics the values ε_j are referred to as *acceleration parameters*. In principle, the definition (5.2.22) we can also regard as a convenient formula for normalization of the amplitudes $\mathcal{E}_j$.

Hence, speculating about the 'keeping in right hand parts of equations the terms only not higher than cubic order, with respect to amplitudes of the running electromagnetic waves $\mathcal{E}_j$', we actually have in mind the accelerator orders of accelerative parameters ε_j. Therefore, these parameters are used as relevant expansion parameters in many electrodynamic problems.

Thus following with the above mentioned algorithm, we can formally write the solutions of the problem (5.2.17) in the form of Krylov–Bogolyubov substitutions

$$\begin{aligned}
\mathcal{H} &= \bar{\mathcal{H}} + u_{\mathcal{H}}(\bar{x}, \bar{\psi}), \\
\mathcal{P}_z &= \bar{P}_z + u_{P_z}(\bar{x}, \bar{\psi}), \\
\theta &= \bar{\theta} + u_\theta(\bar{x}, \bar{\psi}), \\
\psi &= \bar{\psi} + u_\psi(\bar{x}, \bar{\psi}),
\end{aligned} \tag{5.2.24}$$

where the averaged values can be found from the truncated equations

$$\frac{d\bar{\mathcal{H}}}{dt} = \frac{e^2}{2\bar{\mathcal{H}}} \frac{\partial \bar{A}^2}{\partial t}; \tag{5.2.25}$$

$$\frac{d\bar{P}_z}{dt} = -\frac{e^2}{2\bar{\mathcal{H}}} \frac{\partial \bar{A}^2}{\partial z}; \tag{5.2.26}$$

$$\vec{\mathcal{P}}_\perp = \vec{P}_\perp = \text{const}; \tag{5.2.27}$$

$$\frac{d\vec{r}_\perp}{dt} = \frac{c^2 \vec{P}_\perp}{\bar{\mathcal{H}}}; \tag{5.2.28}$$

$$\frac{d\bar{\theta}}{dt} = -2\omega c \frac{\bar{P}_z}{\bar{\mathcal{H}}} = -2\frac{\omega}{c} \frac{d\bar{z}}{dt}; \tag{5.2.29}$$

$$\frac{d\bar{\psi}}{dt} = 2\omega. \tag{5.2.30}$$

The characteristic expression for the function $e^2 \bar{A}^2$ can be in the following form:

$$e^2 \bar{A}^2 = e^2 \bar{A}^2 (\bar{\theta}, \bar{\mathcal{H}}) = \bar{\mathcal{H}}^2 \left[\frac{1}{2}(\varepsilon_1^2 + \varepsilon_2^2) + \varepsilon_1 \varepsilon_2 \cos(\bar{\theta} + \bar{\alpha}) \right], \tag{5.2.31}$$

where $\varepsilon_j = \bar{\varepsilon}_j = e\mathcal{E}_j/mc\omega \bar{\gamma} = e|A_j|/\bar{\mathcal{H}}$, $\bar{\gamma} = \bar{\mathcal{H}}/mc^2$, $\bar{\alpha} = \varphi_1 - \varphi_2$.

Analyzing equation (5.2.25) and taking into account definition (5.2.31) one could easily be convinced that the averaged Hamiltonian $\bar{\mathcal{H}}$ (i.e., the Hamiltonian of the first hierarchical level) is *conserved in time*

$$\bar{\mathcal{H}} \cong \sqrt{m^2 c^4 + c^2 \mathcal{P}_{\perp 0}^2 + c^2 \mathcal{P}_z^2 + e^2 \bar{A}^2} + 0\left(\varepsilon_j^4\right)$$

$$\cong \bar{\mathcal{H}}\left(t = t_0\right) = \bar{\mathcal{H}}_0 = \text{const}, \quad (5.2.32)$$

i.e. the expression (5.2.32) we can regard as the *motion integral* of the first hierarchical level. This can be explained by the fact that the function $e^2 \bar{A}^2$, according to (5.2.31), does not depend on time explicitly. Therein averaged combination phase $\bar{\theta}$ depends linearly on the averaged longitudinal coordinate $\bar{z}$

$$\bar{\theta} = -2 \frac{\omega}{c} \bar{z} \qquad (5.2.33)$$

which easily can be obtained by simple averaging of definition (5.2.12).

Than, we differentiate equation (4.2.29) by time t, taking into consideration motion integral (5.2.32), expression (5.2.33), and equation (5.2.26). As a result we have:

$$\frac{d^2 \bar{\theta}}{dt^2} = -2 \frac{\omega c}{\bar{\mathcal{H}}} \frac{d\bar{\mathcal{P}}_z}{dt} = -2 \frac{\omega c}{\bar{\mathcal{H}}} \left(\frac{e^2}{2\bar{\mathcal{H}}} \frac{2\omega}{c} \frac{\partial \bar{A}^2}{\partial \bar{\theta}} \right) = -2\omega^2 \varepsilon_1 \varepsilon_2 \sin\left(\bar{\theta} + \bar{\alpha}\right).$$

$$(5.2.34)$$

Introducing the definition

$$\omega_0 = \omega \sqrt{2\varepsilon_1 \varepsilon_2} \qquad (5.2.35)$$

we can write down result (5.2.34) in the more elegant form:

$$\frac{d^2 \bar{\theta}}{dt^2} + \omega_0^2 \sin(\bar{\theta} + \bar{\alpha}) = 0. \qquad (5.2.36)$$

Let us discuss brief the results we obtained. Comparing (5.2.36) and (1.2.5), it is not difficult to see that equation (4.2.35) formally reproduces the analogous equation for the nonlinear pendulum (see Chapter 1, Subsection 2.1 for more details). But let us recall that the equation (1.2.5) has been obtained for the case of a material point motion in some *potential well*. The homogeneous *gravitational field* determines the forming of this potential well in that case (see Fig. 1.2.1 and relevant commentaries). We also recall that, as is well known, the gravitational fields related to so called *potential fields*. Besides that, we can regard this field as a *proper field of the zeroth hierarchical level* of the system presented in Fig. 1.2.1.

Following with this analogy we can introduce formally for the model considered the concept of *quasi-potential field*. In contrast to the case with 'true nonlinear pendulum' (see Fig. 1.2.1) this field belong to the first hierarchical level. Action of this field provides nonlinear electron oscillations on the first hierarchical level, which is similar with oscillations of the nonlinear pendulum on the zeroth level in potential gravitational field (see Fig. 1.2.1).

It should be mentioned that the concept of proper field of a κ-th hierarchical level looks, at list, unusual and somewhat whimsical. Therefore, let us discuss it in more detail. We revert for this to the discussion about hierarchical nature of our Universe (see Chapter 2, Sections 1 and 2; and also Tables 2.1.1, 2.1.2, Fig. 2.2.5, and corresponding discussion). Besides that, we will compare some hierarchical properties of the Universe and the considered model 'an electron in the field of electromagnetic standing wave', as hierarchical dynamical systems. It is not difficult to see that general situation in both systems is similar, which in itself, looks incredibly in principle. Analogously to the situation in the considered model, *some proper set of fields also exists on each Universe hierarchical level*. But, in contrast to the considered model, all proper fields on all Universe hierarchical levels carry explicitly expressed fundamental nature. For example, weak and strong interactions are proper for the hierarchical level 'elementary particles', electromagnetic fields are proper for the atoms and molecules, gravitational field is proper for the planet and star systems, and (it is possible) for the galactic systems, and so on.

But let us once more turn to the topic of fictitious (effective) and fundamental fields in the physics. The quasi-potential field is a fictitious (effective) one. It is obvious that it cannot be considered as the fundamental one. We regard the quasi-potential field as one of many displaying of a 'hidden' acting the electromagnetic wave fields (that belong, as noted, to the zeroth hierarchical level) on the first hierarchical level.

However, we should recognize that, talking about the fundamental nature of Universe fields, we have in the mind the fact that they are really some postulates only. Indeed, all these fields appear as a result of many experimental observations and they can not be obtained strictly in framework of some general theoretical approach. (We avoid here discussion of basic problems of the unified field theory). Or, in other words, by this we would like say the following: it is not known surely today how the proper (fundamental) fields of higher hierarchical levels are connected with proper (also fundamental) fields of the lower hierarchical levels. The specific peculiarity of the hierarchical structure given by the considered model 'electron in the field of standing electromagnetic field' is the following. We achieved here the *revealing of some specific physical mechanism of forming the proper field of first hierarchical level (quasi-potential field)* by the fields of the zeroth level (electromagnetic wave fields). I.e., we reveal, in fact, the specific mechanism of correlation between the fields of the zeroth (fundamental fields) and the first (non-fundamental ones) hierarchical levels.

Basing on the above formulated observation, let us propose the following hypothesis. We suppose that all fundamental fields in the Universe are formed analogously with the above-described mechanism of forming quasi-potential field in our particular model with electron in the standing wave. This means that all proper (fundamental) fields of higher hierarchical levels in the Universe appear as a result of displaying of acting of proper (also fundamental) fields of the lower hierarchical levels. It could be assumed that the mechanism of this displaying could be described by some 'generalized averaging procedure', similar to that used above for obtaining the quasi-potential field. We expect also that the two main problems will arise in this case. The first is connected with the finding of proper sets of dynamical variables for each hierarchical level. The second is the constructing of relevant dynamical operators, which regulate the rules of transacting between the hierarchical levels. The hierarchical principles discussed earlier in Chapter 2 could serve as a basis for solving of these problems.

Beyond doubt, the topic about the developing of a 'new hierarchical unified field theory' is very interesting. But let us again revert to the 'sinful earth', i.e., to our model 'an electron in the field of standing electromagnetic wave'.

2.4   Nonlinear Pendulum.
The Miller–Gaponov Potential

Let us temporarily interrupt the discussion of considered calculational problem. Instead this we will turn to some interesting physical peculiarities of our model. Namely, we will discuss the concept of *effective potentials (Miller–Gaponov potentials)*.

A.M. Miller and his student A.V. Gaponov (future Soviet academician) [18, 19] were the first who turned attention to the discussed peculiarity of motion of charged particles in the field of two opposite directed electromagnetic waves. They have proposed the concept of *effective potential* field for description of the quasi-potential field. Corresponding definition for the potential of such field can easily be obtained if we recall the well known correlation between a force $\vec{F}$ acting on the charged particle and *the potential function* $U(\vec{r}, t)$ (see expression (1.4.17) and relevant commentaries)

$$\vec{F} = -\operatorname{grad} U(r,t) = -\left(\vec{\nabla}U\right). \tag{5.2.37}$$

In what follows we turn to equation (5.2.36) and the second law of dynamics (see (1.4.4))

$$\vec{F} = \frac{d\vec{P}}{dt}. \tag{5.2.38}$$

Taking into account (5.2.26), (5.2.37), and (5.2.38) we can introduce the concept of the potential function (potential energy) of electron in the quasi-potential field U_{MG}. Accordingly with (5.2.37), the effective force can be expressed via the potential function in the following way:

$$F_z = \frac{d\vec{P}_z}{dt} = -\frac{e^2}{2\bar{\mathcal{H}}}\frac{\partial \bar{A}^2}{\partial z} = -\frac{\partial U_{MG}}{\partial z} \tag{5.2.39}$$

Integration by z in (5.2.39) yields

$$U_{MG} = \frac{e^2}{2\bar{\mathcal{H}}_0}\bar{A}^2 + \Theta\left(\vec{P},t\right), \tag{5.2.40}$$

where $\Theta\left(\vec{P},t\right)$ is some arbitrary function. As relevant analysis show. it could be accepted for standard boundary and initial conditions for the potential function U_{MG}

$$\Theta\left(\vec{P},t\right) = U_0 = \text{const}.$$

Taking into consideration expression (5.2.31), the potential function U_{MG} in the considered case can be written eventually in the following form:

$$U_{MG} = \bar{\mathcal{H}}_0\varepsilon_1\varepsilon_2 \cos\left(\bar{\theta} + \bar{\alpha}\right) + \frac{1}{2}\bar{\mathcal{H}}_0\left(\varepsilon_1^2 + \varepsilon_2^2\right) \tag{5.2.41}$$

i.e., here we obtained for the constant U_0 the expression

$$U_0 = \frac{\bar{\mathcal{H}}_0}{2}\left(\varepsilon_1^2 + \varepsilon_2^2\right). \tag{5.2.42}$$

It should be mentioned that sometimes in references the potential function U_{MG} is referred as *Miller–Gaponov potential*. To be exact, this is not quite correct, because according to classical definition, the potential of an electromagnetic field should be defined in the following manner: $\varphi = U/q$ (in our case $q = -e$). But, really it is not important because the noted problem have purely philological nature. Therefore, here and on we will understand definition (5.2.41) as the Miller–Gaponov potential. We note that in tradition references the Miller–Gaponov potential wells are called the *buckets*.

2.5   Nonlinear Pendulum. The First Motion Integral

Let us continue the discussion of the calculational peculiarities of the considered model.

The problem of nonlinear pendulum is classical and it is represented rather widely in references. We confine ourselves further by only the discussion of some characteristic methodical steps of the solution procedure and short analysis of the obtained solutions.

We begin with the so called *first integral of motion* that can be obtained from the equation of nonlinear pendulum (5.2.36). Let us multiply for this equation (5.2.36) by the value $d\bar\theta/dt$. It is not difficult get after obvious simple transformations:

$$\frac{d}{dt}\left[\frac{1}{2}\left(\frac{d\bar\theta}{dt}\right)^2 + \omega_0^2 \cos\left(\bar\theta + \bar\alpha\right)\right] = 0 \qquad (5.2.43)$$

or, after integration, the sought first motion integral can be found:

$$\frac{1}{2}\left(\frac{d\bar\theta}{dt}\right)^2 + \omega_0^2 \cos\left(\bar\theta + \bar\alpha\right) = H = \text{const}. \qquad (5.2.44)$$

Let us clarify the physical sense of this integral and the constancy of H in our case. Substituting definition (5.2.13) into (5.2.44) and multiplying this equation by the relativistic electron mass $m_r = \bar{\mathcal{H}}_0/c^2$ and the coefficient $c^2/4\omega^2$ we transform expression (5.2.44) to the form of the energy conservation law:

$$\frac{m_r \bar{v}_z^2}{2} + U_{MG} = \bar{\mathcal{H}}_0 \frac{H}{4\omega^2} + \bar{\mathcal{H}}_0 \frac{(\varepsilon_1^2 + \varepsilon_2^2)}{2} = \bar{\mathcal{H}}_0 - mc^2 = const. \quad (5.2.45)$$

Thus as can be obviously seen, the first motion integral (5.2.44) by itself, from physical point of view, has the sense of the energy conservation law for the first hierarchical level (5.2.32). To obtain equation (5.2.45) we also used the relationships (5.2.13), (5.2.41), and the definition of the averaged longitudinal electron velocity $\bar{v}_z = d\bar{z}/dt$.

2.6   Nonlinear Pendulum. Exact Solutions and Analysis

Thus as shown above, equation (5.2.44) can be regarded as a normalized form of conservation law (5.2.45). Hence it is convenient to introduce formally the concept of *normalized 'energy'*

$$H = 4\omega_0^2 \left(1 - \frac{\varepsilon_1^2 + \varepsilon_2^2}{2} - \frac{mc^2}{\bar{\mathcal{H}}_0} \right) \tag{5.2.46}$$

and the *normalized 'potential function'* [19]

$$V = \omega_0^2 \cos(\bar{\theta} + \bar{\alpha}). \tag{5.2.47}$$

As it can be easily seen, function $V\left(\bar{\theta} + \bar{\alpha}\right)$ represents the sequence of maxima and minima of normalized potential energy of the electron (see Fig. 5.2.1a). We also might say that this is a periodic sequence of the *potential wells*. Such type wells are referred to as the *buckets* in the applied electrodynamics. Following this tradition, we will use such terminology further in this book.

Then let us find and analyze the solutions of equations (5.2.36). The initial conditions are

$$t = t_0; \quad \bar{\theta} + \bar{\alpha} = \pi + 2\pi n; \quad n = 0, \pm 1, \pm 2, \cdots). \tag{5.2.48}$$

We introduce the *energy parameter* by means of the expression [19]

$$\kappa = \left[\frac{1}{2} \left(1 + \frac{H}{\omega_0^2} \right) \right]^{1/2}. \tag{5.2.49}$$

The using of the energy parameter κ is very convenient for analysis of the physical processes in the model 'an oscillating electron in the bucket'. Dynamics of these processes is illustrated in obvious way in Fig. 5.2.1. The dependence of the normalized potential function V on the oscillation phase $\left(\bar{\theta} + \bar{\alpha}\right)$ is shown in Fig. 5.2.1a), and the phase portrait of the electron motion is pictured in Fig. 5.2.1b).

The three typical cases of electron motion are illustrated there. E.g., the case $\kappa < 1$ corresponds to the oscillations of electrons localized in the bucket (i.e., trapped), $\kappa > 1$ is associated with the electron *drift (transit) motion*, and $\kappa = 1$ gives rise to the motion along the *separatrix* (see Fig. 2.2.1b). The solutions of the equation of nonlinear pendulum similar to (5.2.36) are well known (see, for example, [19]). In our designations they may be written as

$$\frac{d\bar{\theta}}{dt} = \dot{\bar{\theta}} = \pm 2\kappa \cdot \omega_0 \begin{cases} cn\left[\omega_0(t - t_0), \kappa\right], & (\kappa < 1); \\ dn\left[\omega_0(t - t_0), 1/\kappa\right], & (\kappa \geq 1). \end{cases} \tag{5.2.50}$$

Making use of the Fourier expansions for the elliptic functions $cn[\ldots]$ and $dn[\ldots]$, we rewrite solutions (5.2.50) in the form

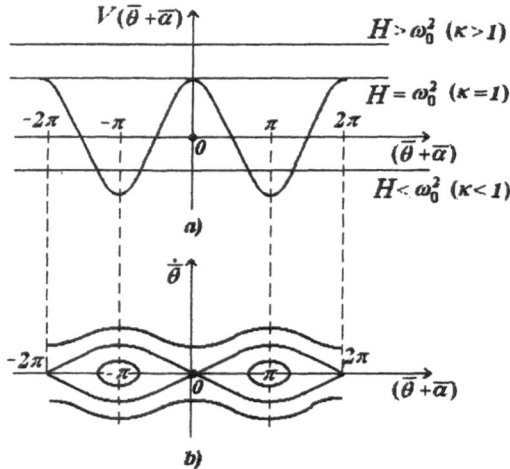

Figure 5.2.1. Dependence of the normalized potential function $V(\bar{\theta} + \bar{\alpha})$ (a) and the phase portrait of the electron motion in such quasi-potential field (b).

$$\frac{d\bar{\theta}}{dt} = \pm 8\omega_b \begin{cases} \sum_{n=1}^{\infty} \frac{(-1)^{n+1} a^{n-1/2}}{1 + a^{2n-1}} \cos\left[(2n-1)\omega_b (t - t_0)\right], & (\kappa \leq 1); \\ \frac{1}{4} + \sum_{n=1}^{\infty} \frac{a^n}{1 + a^{2n}} \cos\left[2n\omega_b(t - t_0)\right], & (\kappa \geq 1). \end{cases}$$

$$(5.2.51)$$

which determines the dependence of $\bar{\theta}$ after being integrated over t. Here, the following designation has been introduced:

$$a = \exp(-\pi F'/F), \qquad (5.2.52)$$

where

$$F' = F(\frac{\pi}{2}, \sqrt{1 - \bar{\kappa}^2}), \quad F = F(\frac{\pi}{2}, \bar{\kappa})$$

are full elliptic integrals. The nonlinear electron oscillation frequency is then given by [19]

$$\omega_b = \omega_b (H) = \frac{\pi}{2} \omega_0 \begin{cases} \dfrac{1}{F(\frac{\pi}{2}; \kappa)}, & (\kappa \leq 1); \\ \dfrac{k}{F(\frac{\pi}{2}; 1/\kappa)}, & (\kappa \geq 1). \end{cases}$$

$$(5.2.53)$$

We can see that the oscillation frequency ω_b of an electron in the bucket depends on its normalized energy H, i.e., the electron in this case is a *non-isochronous* oscillator. We shall show in what follows that this peculiarity can lead to some nontrivial consequences.

Thus the problem of electron motion in the field of two oppositely directed electromagnetic waves for the first hierarchical level is reduced to the well known problem of nonlinear pendulum. Therefore, all conclusions and solutions thereof can be employed to analyze the processes of interest in the model under consideration.

Let us continue the discussion of the characteristic features of electron motion on the first hierarchical level. For the sake of convenience we introduce the number

$$N = \frac{\omega_0}{\omega_b(H)} = \frac{2}{\pi} F(\frac{\pi}{2}; \kappa); \quad (\kappa \leq 1). \tag{5.2.54}$$

Apart from that we employ the asymptotic of the elliptic integral $F\left(\frac{\pi}{2}, \kappa\right)$ that is described by the expressions

$$F(\frac{\pi}{2}; \kappa) = \begin{cases} \dfrac{\pi}{2}, & (\kappa << 1) \\[2mm] \dfrac{1}{2} \ln \dfrac{32 H_s}{H_s - H}, & (1 - \kappa^2 << 1), \end{cases} \tag{5.2.55}$$

where $H_s = \omega_0^2$ is the energy of the electron that moves along the separatrix. We make use of (5.2.55) for $\kappa << 1$ to find that

$$\frac{d\bar{\theta}}{dt} \cong 2\kappa\omega_0 \cos(\omega_0 t + \alpha'), \tag{5.2.56}$$

Hence electron oscillations are harmonic near the bucket bottom which corresponds to the case of small oscillations. As normalized energy H increases, the nonlinear nature of the energy dependence of the oscillation frequency $\omega_b(H)$ is revealed to greater extent (see (5.2.54)). As a result the motion is appreciably modified. Near the separatrix, we have $\kappa^2 \to 1$, $H \to H_s$, and the oscillation frequency tends to zero: $\omega_b(H) \to 0$. The plot of velocity versus time, $\dot{\bar{\theta}}(t) = d\bar{\theta}(t)/dt$, resembles a sequence of solitons (see Fig. 5.2.2).

The distance between the neighboring solitons (bumps) is approximately equal to $2\pi/\omega$, and each soliton width is close to $2\pi/\omega_0$. This means that the number N, defined by equation (5.2.54), is a measure of relative pulse duration of the sequence of pulses–solitons. The analysis of the Fourier amplitudes of electron oscillation harmonics (making use of the solutions of (5.2.51)) reveals rich oscillations spectrum in the vicinity

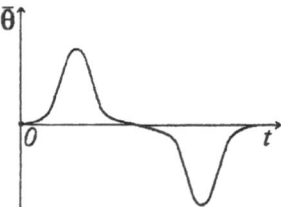

Figure 5.2.2.   The dependence of velocity of changing of the phase $\dot{\bar{\theta}} = d\bar{\theta}/dt$ in the vicinity of the separatrix).

of separatrix. The amplitudes of many first harmonics here being almost equal (see Fig. 5.2.3) [19]. It should be noted that in this case the number N describes the number of characteristic oscillation harmonics. The two above mentioned marginal cases $\kappa^2 \ll 1$ and $\kappa^2 \to 1$ make it possible to analyze the electron oscillation dynamics in the bucket for the whole energy range from $H = 0$ to $H = H_s$. In the case of drift motion (for $H > H_s$), the picture looks symmetric, too. Electron oscillations near the separatrix have many harmonics as well. When moving from the separatrix the oscillation type changes towards the harmonic law.

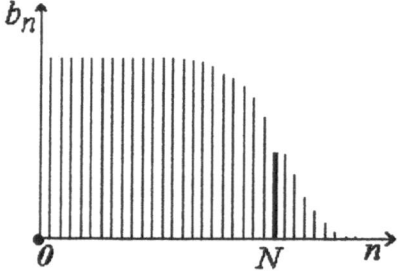

Figure 5.2.3.   The Fourier spectrum of the velocity $\dot{\bar{\theta}} = d\bar{\theta}/dt$ in vicinity of the separatrix. Here b_n is the Fourier amplitude with the harmonic number n.

Thus summarizing the discussed above results, we can conclude that a proper, rather peculiar, and somewhat whimsical, physical picture characterizes the first hierarchical level of the considered system. Generalizing this observation we can claim that each hierarchical level of any dynamical hierarchical system always possesses a proper specific physics. Other obvious illustrations of this conclusion we can obtain analyzing such hierarchical object as, for example, our Universe (see materials Chapter 2, Section 1 for more details). We should stress that general physical picture of zero (initial) hierarchical level is formed as a superposition of the physical pictures of all hierarchical levels. Let us illustrate this phenomenon further using the illustration example of the considered model.

2.7 Full Solutions of the Initial System

Taking into consideration the Bogolyubov substitutions (4.3.34) and solutions (5.2.50), and after corresponding calculations, we can write the full solutions of the considered problem for the zeroth (initial) hierarchical level:

$$
\mathcal{P}_z = \bar{\mathcal{P}}_z + \bar{\mathcal{P}}_x[\varepsilon_1 \sin((\bar{\psi} + \bar{\theta})/2 + \bar{\varphi}_1) - \varepsilon_2 \sin((\bar{\psi} - \bar{\theta})/2 + \bar{\varphi}_2)]
$$
$$
+ mc\bar{\gamma}[\varepsilon_2^2 \cos(\bar{\psi} - \bar{\theta} + 2\bar{\varphi}_2) - \varepsilon_1^2 \cos(\bar{\psi} + \bar{\theta} + 2\bar{\varphi}_1)]/4; \quad (5.2.57)
$$

$$
\mathcal{H} = \bar{\mathcal{H}} + \bar{\mathcal{P}}_x c[\varepsilon_1 \sin((\bar{\psi} + \bar{\theta})/2 + \bar{\varphi}_1) + \varepsilon_2 \sin((\bar{\psi} - \bar{\theta})/2 + \bar{\varphi}_2)]
$$
$$
- mc^2\bar{\gamma}[\varepsilon_2^2 \cos(\bar{\psi} - \bar{\theta} + 2\bar{\varphi}_2) + \varepsilon_1^2 \cos(\bar{\psi} + \bar{\theta} + 2\bar{\varphi}_1)
$$
$$
+ 2\varepsilon_1\varepsilon_2 \cos\left(\bar{\psi} + \varphi_1 + \varphi_2\right)]/4; \quad (5.2.58)
$$

$$
\mathcal{P}_\perp = \bar{\mathcal{P}}_\perp; \quad (5.2.59)
$$

$$
\theta = \bar{\theta}; \quad (5.2.60)
$$

$$
\psi = \bar{\psi}, \quad (5.2.61)
$$

where solutions for the averaged values are obtained in the form

$$
\bar{\mathcal{H}} = \bar{\mathcal{H}}_0 = \text{const}; \quad (5.2.62)
$$

$$
\bar{\mathcal{P}}_\perp = \bar{\mathcal{P}}_{\perp 0} = \text{const}; \quad (5.2.63)
$$

$$
\bar{\mathcal{P}}_z = -\frac{\bar{\mathcal{H}}}{2kc^2} \cdot \frac{d\theta}{dt} = \mp \frac{\bar{\mathcal{H}}}{kc^2} \cdot \kappa \cdot \omega_0 \left\{ \begin{array}{ll} cn\left[\omega_0(t - t_0), \kappa\right], & (\kappa < 1); \\ dn\left[\omega_0(t - t_0), 1/\kappa\right], & (\kappa \geq 1); \end{array} \right.
$$
$$
(5.2.64)
$$

$$
\bar{\theta} = \pm 2\kappa \cdot \omega_0 \int_{t_0}^{t} \left\{ \begin{array}{ll} cn\left[\omega_0\left(t - t_0\right), \kappa\right] \\ dn\left[\omega_0\left(t - t_0\right), 1/\kappa\right] \end{array} \right\} dt; \quad \begin{array}{l} (\kappa < 1); \\ (\kappa \geq 1). \end{array} \quad (5.2.65)
$$

$$\bar{\psi} = 2\omega t; \tag{5.2.66}$$

It is easily seen that the mathematical arrangement of solutions (5.2.57)–(5.2.66) obviously describes the slow electron oscillations (that correspond to the first hierarchical level) on background of the fast oscillations (the zeroth hierarchical level). This means that the total oscillation process itself is found to be rather complex. In spite of the use of the hierarchical approach giving a real possibility for separating the oscillations of different hierarchical level. This allows to study these oscillations separately, that essentially simplifies the general physical analysis.

Following the general idea of the hierarchical method (see Fig. 1.1.2), the above used calculational can be shown as it was done in Fig. 5.2.4.

At the first sight, according to the scheme in Fig. 5.2.4 the model studied should be classified as a two-level hierarchical oscillatative system (zeroth and first levels, respectively) only. However, that is not really correct because the system comprises one more (the highest) hierarchical level, too. This level was not used in the described above calculational scheme, because the equations of the first hierarchical level allow exact solutions. In the hypothetical case we did not find such solutions. Relevant asymptotic solutions should be constructed in this case with accounting the second hierarchical level (see below Subsection 5.3). This means that we would separate additionally *new fast oscillatative phase* in such calculational scheme and construct corresponding equations of the following (second) hierarchical level and so on.

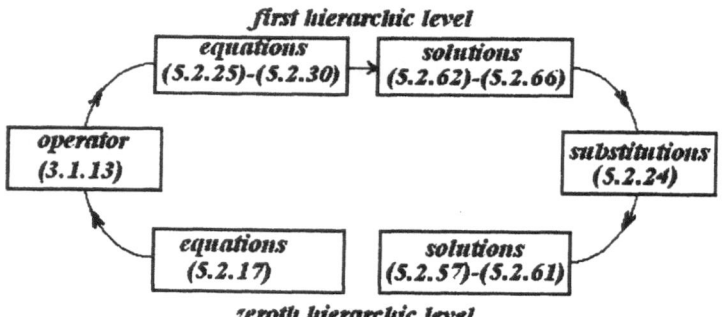

Figure 5.2.4. Graphical illustration of the two-level hierarchical calculational scheme which has been used above for solution of the motion problem in the considered model.

As has already been mentioned above, this second hierarchical level is used nowhere in the discussed calculational procedure before. According to the given explanations this had happened because in the special case considered we have a possibility of obtaining exact solutions of the first

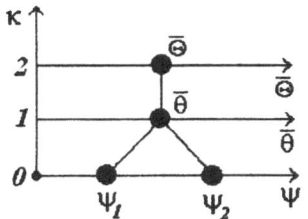

Figure 5.2.5. Graphical illustration of the considered model 'an electron in the field of two oppositely directed electromagnetic waves' as a three-level hierarchical oscillatative system (hierarchical tree).Here κ is the number of a hierarchical level, $\psi_j = p_{1,2}$ are the fast phases of the electromagnetic waves, $\bar{\theta} = \bar{p}_1 - \bar{p}_2$ is the averaged slow combinative phase of the first hierarchical level, $\bar{\Theta}$ is the averaged 'super-slow' phase of the second hierarchical level (in the considered case the velocity of changing of this phase equals to zero: $d\bar{\Theta}/dt = 0$ — see also Fig. 4.1.1 and corresponding commentaries).

hierarchical level problem. This means that we need to use in this particular case the second hierarchical level for obtaining the full solutions of the initial problem. But in fact the situations of such type (when the problem of some lower hierarchical level has an exact solution) are not typical in practice. Moreover, they are rather rare. That is why further in the following Subsection we will give another version of calculational scheme for the considered problem. At that time we will conditionally assume that exact solutions for the first hierarchical level are not known. We will look in this case for approximate solutions which can be obtained by using of the hierarchical method.

3. MODEL 'AN ELECTRON IN THE FIELD OF TWO OPPOSITELY DIRECTED ELECTROMAGNETIC WAVES' AS A THREE-LEVEL HIERARCHICAL OSCILLATATIVE SYSTEM

3.1 Transition to the Second Hierarchical Level

Thus later we shall study the model with two oppositely directed electromagnetic waves (i.e., standing electromagnetic wave) which we regard as a three-level hierarchical oscillatative system. Let us solve once more the considered problem taking into consideration its three-level hierarchical nature.

The general calculational schemes of both hierarchical models (two- and three-level, respectively), in principle, coincide. However, exact coincidence takes place until we reach the stage of obtaining averaged equations (5.2.25)-(5.2.30) only (see Fig. 5.2.5). Then the difference concerns the method of solution of these equations. We have used in the first case the *method of exact solutions*. In the second case (considered below) we shall look for approximate solutions of these equations (5.2.25)–(5.2.30). We will apply for this the hierarchical calculational scheme again for their asymptotic integration. Therefore, we can omit the first step of the discussed calculational procedure in this case and begin the further study with equation of nonlinear pendulum (5.2.36) immediately.

Taking into account expansions in series (1.2.7), we introduce the auxiliary function $\Phi\left(\bar{\theta}\right)$:

$$\varepsilon\Phi\left(\bar{\theta}\right) = \omega_0^2\left(\bar{\theta} - \sin\bar{\theta}\right), \qquad (5.3.1)$$

where $\varepsilon = 1/\xi \ll 1$ is the small parameter of the problem (really, as corresponding analysis shows, the small parameter $\varepsilon \sim \varepsilon_j^2$, where ε_j is the bigger of the accelerative parameters (5.2.23)). Separating the function $\sin\bar{\theta}$ in definition (5.3.1), and substituting it into (5.2.36), we can reduce this equation to the form

$$\frac{d^2\bar{\theta}}{dt^2} + \omega_0^2\bar{\theta} = \varepsilon\Phi\left(\bar{\theta}\right). \qquad (5.3.2)$$

Here in (5.3.1) and (5.3.2) we performed the transition to the so called *extended combinative phase* $\bar{\theta} + \alpha \rightarrow \bar{\theta}'$, neglecting further the sign of prime.

Then we must carry out the four following steps for accomplishing the hierarchical procedure of asymptotic integration:

1) to separate oscillatative phases of the system;

2) to reduce the equation (5.3.2) to the standard form with rotating phases similar to (5.2.18);

3) to construct the approximate solutions for the next (second, in this case) hierarchical level;

4) to transform the obtained solutions into the lower (first) hierarchical level.

We take into consideration that corresponding formulae for transformation of the first level solutions to the zeroth (initial) hierarchical level are given by relationships (5.2.57)–(5.2.61).

Concerning the procedure of separation of rotating phases on the first hierarchical level, we should note the following. First, here we have only one rotating phase. Second, this phase, in contrast with the situation on the zeroth hierarchical level, turns out to be hidden (we had explicit phases (5.2.2) in the preceding hierarchical level).

The Van der Pol variables (see in Chapter 3, Subsection 4.1) are applied for solution of the first problem (i.e., for the separating oscillatative

phases). As it was shown in Chapter 3, Subsection 4.1, they originate
from the solutions of a generating (linear) equation similar to (3.4.2):

$$\bar\theta = \Theta \cos\hat\psi; \hat\psi = \omega_0 (t - t_0),\qquad(5.3.3)$$

i.e., we consider that due to weak nonlinear nature of oscillations we can
consider that period of nonlinear oscillations is close to a period of linear
system like (3.4.2).

We consider the amplitude Θ to be slowly (super-slowly) changing
phase. Then, following the scheme described in Chapter 3, Subsection 4.1, we can transform equation (5.3.2) into the form that is similar
to (3.4.8), (3.4.9):

$$\frac{d\Theta}{dt} = -\varepsilon\omega_0 \left[\Theta\cos\hat\psi - \sin\left(\Theta\cos\hat\psi\right)\right]\sin\hat\psi;$$

$$\frac{d\hat\psi}{dt} = \omega_0 - \frac{\varepsilon\omega_0}{\Theta}\left[\Theta\cos\hat\psi - \sin\left(\Theta\cos\hat\psi\right)\right]\cos\hat\psi.$$

$$(5.3.4)$$

Equations (5.3.2) and (5.3.4), from mathematical point of view, are
equivalent. However, it can be easily seen that, in contrast to (5.3.2),
system (5.3.4) is written in the hierarchical standard form like in (4.2.1).
Consequently, further we can use the standard procedure for its asymptotic integration that is described earlier in Chapter 4, Section 2. According to this procedure we should be looking for formal solutions of
the problem (Bogolyubov substitutions) in the form like in (4.2.2):

$$\Theta = \bar\Theta + \varepsilon u_{1\Theta}\left(\bar\Theta, \bar{\hat\psi}\right) + \cdots$$

$$\hat\psi = \bar{\hat\psi} + \varepsilon v_{1\psi}\left(\bar\Theta, \bar{\hat\psi}\right) + \cdots,\qquad(5.3.5)$$

where all notations are standard.

It is easy to see in (5.3.5) that the variable $\bar\Theta$ in the considered case
plays the role of some 'super-slow' oscillation phase for the second hierarchical level. This means that the studied system indeed consists of
(at least) three hierarchical levels (the zeroth, first and second levels,
correspondingly — see Fig. 5.2.5). But the question arises: is this third
level higher for our system? Positive answer for this question could be
obtained we will prove that $d\bar\Theta/dt = 0$ (hierarchical analogue of the
third thermodynamic principle — see Chapter 2, Subsections 2.1 and
4.1.3). Below (in the following Subsection) we will show that this really
takes place for our model.

The truncated equations (equations for the second hierarchical level)

$$\frac{d\bar{\Theta}}{dt} = \frac{\varepsilon\omega_0}{2\pi} \int\limits_0^{2\pi} \sin\left(\bar{\Theta}\cos\bar{\bar{\psi}}\right)\sin\bar{\bar{\psi}}d\bar{\bar{\psi}};$$

$$\frac{d\bar{\bar{\psi}}}{dt} = \omega_0 + \frac{\varepsilon\omega_0}{2\pi\bar{\Theta}} \int\limits_0^{2\pi} \sin\left(\bar{\Theta}\cos\bar{\bar{\psi}}\right)\cos\bar{\bar{\psi}}d\bar{\bar{\psi}}.$$

(5.3.6)

can be constructed for determining of the averaged variables in substitutions (5.3.5). The integration on the right hand side of equations (5.3.6) can be easily accomplished if the integrands are represented as expansions in a Fourier–Bessel series. Unfortunately, exact analytical solutions for the obtained systems, as a whole, were not found yet. We are forced in this situation to confine ourselves by the analysis of some simplified particular cases only. Let us consider such a simplified example, which is known in reference as the *case of small oscillations*

$$\bar{\theta} << 1$$

(5.3.7)

(for more detail concerning the classification of oscillatative models see in Chapter 1, Subsection 2.1).

3.2   Duffing Oscillator

We can expand the function $\sin\bar{\theta}$ in (5.3.1) in the case (5.3.3) in a power series

$$\sin\bar{\theta} \approx \bar{\theta} - \frac{1}{3!}\bar{\theta}^3 + \frac{\bar{\theta}^5}{5!} - \cdots.$$

(5.3.8)

Let us confine ourselves by considering the cube term in (5.3.8) only. In this case we can transform equation of nonlinear pendulum (5.3.2) in the following manner

$$\frac{d^2\bar{\theta}}{dt^2} + \omega_0^2\bar{\theta} = \frac{\omega_0^2}{6}\bar{\theta}^3.$$

(5.3.9)

Then we perform some auxiliary transformations. They are introduced as the *normalized combinative phase* x and the *normalized time* τ

$$x = \frac{c\omega_0}{l}\bar{\theta}; \tau = \frac{c}{l}t,$$

(5.3.10)

where l is the characteristic size of the interaction region (for instance, distance between the electrodes 4 and 6 in Fig. 5.1.1), c is the light velocity in vacuum. Equation (5.3.9) after simple transformations can be rewritten in the form

$$\frac{dx}{d\tau} + \hat{\omega}^2 x = -\mu x^3,$$ (5.3.11)

where

$$\hat{\omega} = c\omega_0/l$$ (5.3.12)

is the *normalized frequency*, and

$$\mu = -l/6c\omega_0$$ (5.3.13)

is the *Duffing parameter*. As noted in Chapter 1, Subsection 2.1, equation (5.3.11) in references is known as the *Duffing equation*.

Using the Van der Pol variables

$$x = a \cos \psi'$$ (5.3.14)

(here all notations are self-evident considering what was said above) we transform equation (5.3.11) into the standard form similar to (3.4.8), (3.4.9):

$$\frac{da}{d\tau} = \mu \frac{a^3}{4\hat{\omega}} \sin 2\psi' + \mu \frac{a^3}{8\hat{\omega}} \sin 4\psi';$$

$$\frac{d\psi'}{d\tau} = \omega + \mu \frac{3a^2}{8\hat{\omega}} + \mu \frac{a^2}{2\hat{\omega}} \cos 2\psi' + \mu \frac{a^2}{8\hat{\omega}} \cos 4\psi'.$$ (5.3.15)

It is easy to be convinced that the hierarchical small parameter of the problem ε

$$\varepsilon \sim \left| \frac{dx}{d\tau} \right| / \left| \frac{d\psi'}{d\tau} \right| \ll 1$$ (5.3.16)

has the same order as *the Duffing parameter* μ. Hence, the *Duffing parameter* μ can be used here as a *hierarchical problem small parameter*. Accomplishing numerical estimations for μ for typical characteristic situations (for instance, taking the device shown in Fig. 5.1.1 as a basis), we can find that, indeed, the Duffing parameter μ might be considered as the small parameter of the problem (really $\mu \sim 10^{-18} \div 10^{-20} \ll 1$).

Apart from that, considering condition (5.3.7) and the fact that $\varepsilon \sim \mu$, we take into account that

$$\left| \mu x^3 \right| \ll x^2.$$

As we did before, we look for the formal solutions of system (5.3.15) in the form of Krylov–Bogolyubov substitutions (see in more detail in Chapter 4, Section 2)

$$a = \bar{a} + \mu u_1\left(\bar{a}, \bar{\psi}'\right) + \mu^2 u_2\left(\bar{a}, \bar{\psi}'\right) + \cdots;$$
$$\psi' = \bar{\psi}' + \mu v_1\left(\bar{a}, \bar{\psi}'\right) + \mu^2 v_2\left(\bar{a}, \bar{\psi}'\right) + \cdots, \qquad (5.3.17)$$

where the averaging variables are determined by the truncated (short-ened) equations (equations for the second hierarchical level)

$$\frac{d\bar{a}}{d\tau} = \mu A_1\left(\bar{a}\right) + \mu^2 A_2\left(\bar{a}\right) + \cdots;$$
$$\frac{d\bar{\psi}'}{d\tau} = \hat{\omega} + \mu B_1\left(\bar{a}, \bar{\psi}'\right) + \mu^2 B_2\left(\bar{a}, \bar{\psi}'\right) + \cdots \qquad (5.3.18)$$

Following with the standard calculational procedures described in Chapter 4, Subsection 2.2 we obtain the set of equations similar to (4.2.6) for determining the unknown functions u_1, v_1, u_2, v_2, ...:

$$\hat{\omega}\frac{\partial u_1}{\partial \bar{\psi}'} = \frac{\bar{a}^3}{4\hat{\omega}}\sin 2\bar{\psi}' + \frac{\bar{a}^3}{8\hat{\omega}}\sin 4\bar{\psi}' - A_1\left(\bar{a}\right);$$

$$\hat{\omega}\frac{\partial v_1}{\partial \bar{\psi}'} = \frac{3\bar{a}}{8\hat{\omega}} + \frac{\bar{a}^2}{2\hat{\omega}}\cos 2\bar{\psi}'$$
$$+ \frac{\bar{a}^2}{8\hat{\omega}}\cos 4\bar{\psi}' - B_1\left(\bar{a}\right);$$

$$\hat{\omega}\frac{\partial u_2}{\partial \bar{\psi}'} = -A_1\left(\bar{a}\right)\frac{\partial u_1}{\partial \bar{a}} - B_1\frac{\partial u_1}{\partial \bar{\psi}'} + \frac{\bar{a}^3 v_1}{2\hat{\omega}}\cos 2\bar{\psi}'$$
$$+ \frac{3\bar{a}^2 u_1}{4\hat{\omega}}\sin 2\bar{\psi}' + \frac{\bar{a}^3 v_1}{2\hat{\omega}}\cos 4\bar{\psi}' + \frac{3\bar{a}^2 u_1}{8\hat{\omega}}\sin 4\bar{\psi}' - A_2\left(\bar{a}\right);$$

$$\hat{\omega}\frac{\partial v_2}{\partial \bar{\psi}'} = -A_1\frac{\partial v_1}{\partial \bar{a}} - B_1\frac{\partial v_1}{\partial \bar{\psi}'} + \frac{3\bar{a} u_1}{4\hat{\omega}} - \frac{\bar{a}^2 v_1}{\hat{\omega}}\sin 2\bar{\psi}' + \frac{\bar{a} u_1}{\hat{\omega}}\cos 2\bar{\psi}'$$
$$- \frac{\bar{a}^2 v_1}{2\hat{\omega}}\sin 4\bar{\psi}' + \frac{\bar{a} u_1}{4\hat{\omega}}\cos 4\bar{\psi}' - B_2\left(\bar{a}\right);$$

$$(5.3.19)$$

. . . .

Solutions of system (5.3.19) can be found using the method that is set forth in Chapter 4, Subsection 2. For solution of the problem of secular terms (see Chapter 3, Subsection 5.2) we should accept the presump-tion about absence of non-periodical terms in right parts of equations (5.3.18). As a result we obtain

$$A_1\left(\bar{a}\right) = 0, \quad B_1\left(\bar{a}\right) = 3\bar{a}^2/8\hat{\omega}. \qquad (5.3.20)$$

Integration of the first equation in system (5.3.19) yields

$$u_1\left(\bar{a},\bar{\psi}'\right) = -\frac{\bar{a}^3}{8\hat{\omega}^2}\cos 2\bar{\psi}' - \frac{\bar{a}^3}{32\hat{\omega}^2}\cos 4\psi';$$

$$v_1\left(\bar{a},\bar{\psi}'\right) = \frac{\bar{a}^2}{4\hat{\omega}^2}\sin 2\bar{\psi}' + \frac{\bar{a}^2}{32\hat{\omega}^2}\sin 4\bar{\psi}'. \tag{5.3.21}$$

The expressions (5.3.20), (5.3.21) determine solutions obtained for the first approximation of the averaging method. In analogous way the solutions for the second approximation can be obtained:

$$A_2\left(\bar{a}\right) = 0,$$

$$u_2\left(\bar{a},\bar{\psi}'\right) = -\frac{41\bar{a}^5}{512\hat{\omega}^4}\cos 2\bar{\psi}' - \frac{\bar{a}^5}{1024\hat{\omega}^4}\cos 4\bar{\psi}'$$
$$- \frac{9\bar{a}^5}{1536\hat{\omega}^4}\cos 6\bar{\psi}' - \frac{\bar{a}^5}{4096\omega^4}\cos 8\bar{\psi}'; \tag{5.3.22}$$

$$B_2\left(\bar{a}\right) = -51\bar{a}^4/256\hat{\omega}^3;$$

$$v_2\left(\bar{a},\bar{\psi}'\right) = -\frac{23\bar{a}^4}{128\hat{\omega}^4}\sin 2\bar{\psi}' - \frac{\bar{a}^4}{512\hat{\omega}^4}\sin 4\bar{\psi}'$$
$$+ \frac{3\bar{a}^4}{384\omega^4}\sin 6\bar{\psi}' + \frac{\bar{a}^4}{2048\hat{\omega}^4}\sin 8\bar{\psi}'. \tag{5.3.23}$$

Therein the analysis shows that for any approximation

$$A_s\left(\bar{a}\right) = 0 \quad (s = 1,\,2,\,3,\,\ldots). \tag{5.3.24}$$

This means that

$$\frac{d\bar{a}}{d\tau} = 0; \quad \bar{a} = \bar{a}\left(\tau = 0\right) = \text{const}, \tag{5.3.25}$$

always, i.e., the *'super-slow' phase of the second hierarchical level* (see definition (5.3.3))

$$\bar{\Theta} = l\bar{a}/c\omega_0 \tag{5.3.26}$$

indeed describes the highest hierarchical level of the considered system (see the diagram in Fig. 5.2.5 and corresponding commentaries to formula (5.3.5)). It is interesting to note that the velocity of changing of the fast averaging phase $\bar{\psi}'$ in this situation is constant:

$$\hat{\omega}^* = d\bar{\psi}'/d\tau = \hat{\omega} + \mu B_1\left(\bar{a}\right) + \mu^2 B_2\left(\bar{a}\right) + \cdots = \text{const}. \qquad (5.3.27)$$

Thus above we have illustrated the represented calculational technology on the example of the single particle theory of the modulation section of the device shown in Fig. 5.1.1. As it was found, such a system can be treated (in general case) as a three level hierarchical oscillatative system (see Fig. 5.2.5). The methodical peculiarity of this system is that the first hierarchical level is formed by some parametrically resonant oscillatative process, when both fast oscillatative phases (5.2.2) are represented by *evident* ones. Whereas, the second hierarchical level is formed as a result of the non-resonant electron oscillations, which are characterized by *the hidden* phase (5.2.33).

4.   MODEL 'AN ELECTRON IN THE FIELD OF THREE ELECTROMAGNETIC WAVES' AS A FOURTH-LEVEL HIERARCHICAL OSCILLATATIVE SYSTEM

4.1   Stimulated Oscillation of a Charged Particle

Analyzing the second design version of the discussed device (see Fig. 5.1.2) we can discover that, in contrast with the above discussed illustration example in Fig. 5.1.1, this device can be described as a *fourth-level hierarchical system*. Characteristic feature of this system is that here the second hierarchical level, as well as the first one, has clearly expressed oscillative-resonant nature. According to the classification performed in Chapter 4, Subsection 1.3, this resonance could be determined as a quasilinear one. However, in spite of the obvious physical novelty, this model does not contain any new, of principle, calculational features, in comparison with the previously discussed illustration example. We will confine ourselves by a short semi-qualitative analysis only below in this Subsection.

So let us discuss the theoretical model that can be considered to be equivalent to the device shown in Fig. 5.1.2. The graphical illustration of this model is shown in Fig. 5.4.1. It can be easily seen that the only difference from the previously discussed model shown in Fig. 5.1.3 is the presence here of the third electromagnetic wave $\left\{\vec{A}, \vec{k}_3\right\}$. This is the transformed wave within the retarding system of the device (see Fig. 5.1.2). Considering influence of this wave we should take into account that in the discussed system both resonance types (parametric

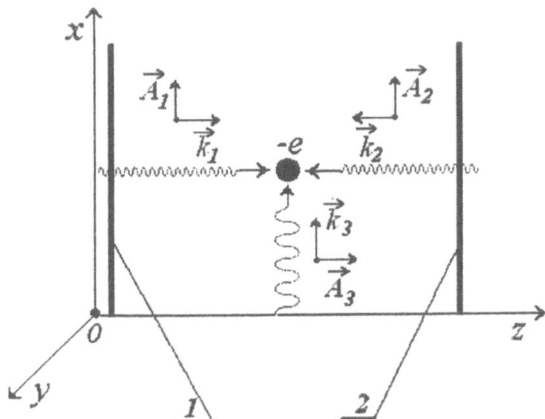

Figure 5.4.1. The illustration of the single-particle model of the device that is shown in Fig. 5.1.2. Here 1 is left boundary of the model, 2 is right boundary of the model, x, y, and z are spatial coordinate axes, $\vec{A}_{1,2}$ are the vector-potentials and $\vec{k}_{1,2}$ are the wave vectors of the opposite directed (transforming and heterodyne, respectively) linearly polarized electromagnetic waves, $\vec{A}_3$ is the vector-potential and $\vec{k}_3$ is the wave vector of the transformed electromagnetic wave, $-e$ is the charged particle (in this specific case it is an electron).

resonance determined by condition (5.1.2) and the quasilinear resonance (5.1.3)) occur simultaneously. As we did before, we will perform the analysis in the moving coordinate system K'. We assume that the transformed (third) electromagnetic wave $\left\{\vec{A}, \vec{k}_3\right\}$ in the coordinate system K' can be approximated with a transverse retarded electromagnetic wave propagating along x axis (see Fig. 5.4.1). We use the method of artificial magneto-dielectric (see this method is described further in Chapter 11, Subsection 2.1, Volume II, and in reference [4]) for description of the retarding property and electrodynamic arrangement of this wave.

The vector-potential of the electromagnetic field acting on electrons can be represented in the following form:

$$\vec{A} = \sum_{j=1}^{2} \vec{A}_j + \vec{A}_3 = \sum_{j=1}^{2} \frac{c\mathcal{E}_j}{\omega_j} \sin(p_j + \varphi_j) \cdot \vec{e}_x + \frac{c\mathcal{E}_3}{\omega_3} \sin(p_3 + \varphi_3)\,\vec{n},$$

$$(5.4.1)$$

where $p_3 = \omega_3 t - k_3 z$, ω_3 and k_3 are the cyclic frequency and wave number of the transformed (third, retarded) electromagnetic wave in the moving coordinate system K'. The Hamilton equations (5.1.4) are chosen as an initial basis.

Then we follow the calculational scheme that was set forth in the previous case (see Subsection 3.2). The difference is only in the following: in the studied case we consider that the phase p_3 is a slowly varying variable (for the zeroth hierarchical level, naturally). This supposition follows from the physical sense of synchronism condition (5.1.3). The fact of satisfying the condition (5.1.3) means that any electron during of its motion along axis z 'sees' practically the same phase of the transformed (third, retarded) electromagnetic wave $\left\{\vec{A}_3, \vec{k}_3\right\}$. Or, more exactly, this phase could slowly change only during many electron oscillations with fast phases $\psi_{1,2}$. This effect is referred to as the *elementary mechanism of the stimulated Cherenkov instability* [4]. It is also put in the basis of operation principle of traveling wave tubes (TWT) [5].

As a result of performed transformations we obtain an equation for nonlinear pendulum similar to (5.2.36), where influence of the transformed (third) electromagnetic wave is taken into account in the right hand side of the *equation of stimulated nonlinear pendulum*

$$\frac{d^2\bar{\theta}}{dt^2} - \omega_0^2 \sin(\bar{\theta} + \bar{\alpha}) = f \cos\left(\bar{p}_3 + \varphi_3\right), \qquad (5.4.2)$$

where f is the amplitude of corresponding normalized 'force' (it is not important to specify its expression for the discussed illustration example), $\bar{p}_3 = \omega_3 t - k_3 \left(d\bar{z}/dt\right)$ is the averaged phase velocity of the transformed wave $\left\{\vec{A}_3, \vec{k}_3\right\}$ in the moving coordinate system K'.

Analogously, taking into consideration expressions (5.3.1), (5.3.2), (5.3.8)–(5.3.11) we could obtain the corresponding so called *'stimulated'* Duffing equation. Passing to the Van der Pol variables (5.3.3) and performing other relevant transformations, we can, in principle, come to the standard system like to (4.3.1) or (5.3.15) (except the right hand side only), obtained before for the case discussed above of 'non-stimulated' nonlinear pendulum (5.2.36).

Let us briefly illustrate the main idea of the procedure of asymptotic solution of the 'stimulated' Duffing equation. We shall do it for the discussed example of a four-level hierarchical oscillatative system. Following the calculational procedures described before in Subsection 3.2 the 'stimulated' Duffing equation can be obtained in the form

$$\frac{dx}{d\tau} + \hat{\omega}^2 x = -\mu x^3 + \hat{f} \cos \bar{p}_3, \qquad (5.4.3)$$

where we presume for the sake of simplicity that $\varphi_3 = 0$. Then, using the Van der Pol variables (5.3.14), we reduce equation (5.3.3) to the standard form similar to (5.3.18)

$$\frac{da}{d\tau} = \mu \frac{a^3}{4\hat{\omega}} \sin 2\psi' + \mu \frac{a^3}{8\hat{\omega}} \sin 4\psi' - \frac{\hat{f}}{2\hat{\omega}} \left[\sin \left(\psi' + \bar{p}_3 \right) + \sin \left(\psi' - \bar{p}_3 \right) \right];$$

$$\frac{d\psi'}{d\tau} = \omega + \mu \frac{3a^2}{8\hat{\omega}} + \mu \frac{a^2}{2\hat{\omega}} \cos 2\psi' + \mu \frac{a^2}{8\hat{\omega}} \cos 4\psi'$$

$$- \frac{\hat{f}}{2a\hat{\omega}} \left[\cos \left(\psi' + \bar{p}_3 \right) + \cos \left(\psi' - \bar{p}_3 \right) \right];$$

$$\frac{d\bar{p}_3}{d\tau} = \hat{\omega}_3 - \hat{\Omega} a \cos \psi' - \hat{\mu} \left(a \cos \psi' \right)^3 + \hat{f} \cos \bar{p}_3,$$

$$(5.4.4)$$

where $\hat{\omega}_3 = l\omega_3/c$, $\hat{\Omega} = k_3 lc\hat{\omega}^2/2\omega\omega_0$, $\mu = \mu k_3 lc/2\omega\omega_0$, $\hat{f} = k_3 lc\hat{f}/2\omega\omega_0$.

All other notations were already given above in this Section.

Comparing systems (5.4.4) and (4.3.1) we can observe that both these expressions possess the similar mathematical structure. This means that here, in principle, two different ways of constructing the solutions system (5.4.4) can be realized. The first is *non-resonant case*, when velocities of the phases ψ' and $\bar{p}_3$ changing are incommensurable values. In this situation the calculational scheme described in Chapter 4, Subsection 3.2 can be used. In the second case, if the resonant condition

$$\left| \frac{d\psi'}{d\tau} \right| \approx \left| \frac{d\bar{p}_3}{d\tau} \right|, \qquad (5.4.5)$$

is satisfied, the *resonant case* (one version of the Cherenkov instability [4], as in is mentioned above already) takes place. The calculational scheme, which has described in Chapter 3, Subsection 4.3, allow us to obtain asymptotic of the solutions of system (5.4.4). So, the 'super-slow'

$$\hat{\Theta} = \psi' - \bar{p}_3 \qquad (5.4.6)$$

and quasi-fast

$$\Psi = \psi' + \bar{p}_3 \qquad (5.4.7)$$

phases can be introduced. In the terms of concept of hierarchical three this means that the following (second) hierarchical level is formed by the slowly varying phase $\bar{\hat{\Theta}} = \bar{\psi}' - \bar{\bar{p}}_3$. In turn, regarding the phase $\bar{\hat{\Theta}} = \bar{\hat{\Theta}} \left(\tau \right)$ as a basis for forming of the next (third) hierarchical level (with some 'super-super-slow' phase $\bar{\bar{\hat{\Theta}}}$ = const that represents the highest level of the system — see below), we can conclude that the discussed model could be treated as a fourth-level hierarchical system.

The hierarchical structure of this system is illustrated in Fig. 5.4.2.

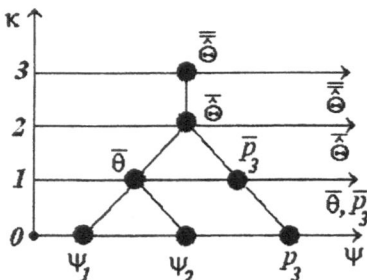

Figure 5.4.2. Graphical illustration of the model with two oppositely and third transversely propagated electromagnetic waves as a fourth-level hierarchical oscillatative system (hierarchical tree). Here κ is the number of a hierarchical level, $\psi_j = p_{1,2}$ and p_3 are the fast phases of the electromagnetic waves, $\bar{\theta} = \bar{p}_1 - \bar{p}_2$ is the averaged slow combinative phase of the first hierarchical level, $\bar{\hat{\Theta}}$ is the averaged 'super-slow' phase of the second hierarchical level, $\bar{\bar{\Theta}}$ is the averaged 'super-super-slow' phase of the third hierarchical level (in the considered case the velocity of changing of this phase equals to zero: $d\bar{\bar{\Theta}}/dt = 0$ — see also Fig. 4.1.1 and corresponding commentaries).

4.2 Stimulated Oscillations of an Electron Ensemble

In the preceding Subsection we considered behavior of a separate particle (electron) within the field of two electromagnetic waves under stimulated influence of the third transversely directed electromagnetic wave (*single-particle models*). But all practical electrodynamics systems of the discussed type (see, in particular, the examples in Figs. 5.1.1, 5.1.2) really have to deal with *particle ensembles*. I.e., such systems always consist of many particles *(multi-particle models)*. From the point of view of general theory of systems (see earlier Chapter 1, Section 5) difference between the single-particle and multi-particle models is in fact that the first one can be treated as a subsystem of the second one. This means that the multi-particle systems (ensembles) should have some '*specific system properties*'. This, in principle, distinguishes it from any other subsystem (e.g., electrons in this case, see Chapter 1, Subsection 5.1 for more details).

Relevant analysis shows that the discussed multi-particle model (see Fig. 5.4.1) really indeed possesses such 'specific system properties'. One of them is a possibility of realization of the so called *grouping mechanisms*. In this case, owing to simultaneous realization of many nonlinear single-particle interactions, the particle ensemble evolves into the form of clearly expressed *particle bunch*. In reference to physical microwave electronics [13, 15] the bunches of this type are quite often treated also as peculiar *electron clusters*. Such bunch (cluster), as a whole, can behave itself as a peculiar *'large particle'*.

As it will be shown below, the grouping mechanism discussed can be achieved in our model, too. Including the case when the electron oscillatative ensemble within two electromagnetic waves (transforming and heterodyne ones, respectively) is exposed to the third (transformed, retarded) electromagnetic wave (see Fig. 5.4.1). Physical features of this process will be discussed somewhat below in more detail. Here we state only that, in spite of obvious difference between the both (single- and multi-particle, respectively) models, results of the single-particle analysis could be effectively applied for analysis of the multi-particle systems. First, for studying of the above mentioned process of the electron cluster forming. Second, for description of dynamics of this cluster as a large particle, i.e., the dynamics of the ensemble, as a whole.

So, let us illustrate some characteristic details of semi-qualitative technique of dynamics analysis for a multi-particle oscillatative particle ensemble. We shall take the above discussed single-particle results as a basis for such analysis. As before, we consider that this single-particle dynamics is described in the terms of theory of hierarchical oscillations. At that, it is assumed for simplicity that the considered multi-particle oscillatative ensemble consists of many electrons that do not interact with each other via the Coulomb repulsion and interact with the electromagnetic waves only.

Let us remind once more that the chosen system is characterized by the presence of two resonances of essentially different physical nature. The first is the parametric resonance (determined by resonant conditions given by the last of formulae (5.2.20)). It can be realized on the initial (zeroth) hierarchical level. Detailed quantitative and qualitative multi-particle studies of this physical mechanism will be performed in Volume II). Therefore, further in this Subsection we will pay our attention mainly to the discussion of the second resonant mechanism. The characteristic feature of this mechanism is, as it mentioned above already, that it, in contrast with the first one, takes place on the first (averaged) hierarchical level (see Fig. 5.4.2).

Thus taking into account what has been said, we can begin our study beginning from the first hierarchical level immediately. According to the previously performed single-particle analysis, we have two main types of oscillatative motion on this level (see Fig. 5.2.1 and corresponding comments). The first is the case that is determined by the condition $H > \omega_0^2$ ($\kappa > 1$). Here the particle moves 'under maximums of the curve $V(\bar{\theta} + \alpha)$' (see Fig. 5.2.1). In the theory of nonlinear pendulum such type of motion is known as the *rotating motion of the nonlinear pendulum*. In the framework of the studying model, accepting the terminology of physical microwave electronics, we classify these electrons as the *passage motion of particles* [17]. The general motion process in this case can be represented as a superposition of translation and oscillatative motions. As analysis shows, the averaged combinative phase $\bar{\theta}$ (which we classified above as the explicit one) characterizes the oscillatative component of this motion [17]. Therein the area of such particle motion is, in principle, unlimited.

The second characteristic type of motion is the case $H < \omega_0^2$ ($\kappa < 1$) (see Fig. 5.2.1). In this situation 'walls' of the Miller–Gaponov potential well (5.2.41) limit the aria of electron motion. In other words, here, in contrast with the first case, the region of the particle motion turns out to be limited. This means that the motion process, in itself, have purely oscillatative nature. Here, these oscillations are determined by the hidden phase $\hat{\psi}$ (or ψ'), which, in turn, is connected with the phase of free oscillation of an electron in the bucket (see definitions (5.3.3), (5.3.14), (5.4.6), (5.4.7) and the corresponding comments). Further we limit ourselves by analysis of the second type model only.

Let us begin our semi-qualitative analysis with the *unperturbed* model, i.e., the model in which the influence of the third (transformed) electromagnetic wave (see. Fig. 5.4.1) is neglected. It is considered that the initial kinetic energy of all electrons is less than the 'height' of the Miller–Gaponov well $U_{MG\,max}$ (see Figs. 5.4.3 and 5.4.4). Kinetic and potential energies of each electron of the ensemble change in mutually opposite directions (see Figs. 5.4.3 and 5.4.4). As it can be easily seen in Fig. 5.4.3, electrons which increase their kinetic energy (Fig. 5.4.3a)) at the same time decrease their potential energy (Fig. 5.4.3b), and vice versa. As a result the total averaged energy $\bar{\mathcal{H}}$ for each electron is conserved in time (the conservation law for the averaged energy — see (5.2.32) and Fig. 5.4.4).

Then we assume that the system is exposed to an external disturbing (third, transformed) electromagnetic wave that is characterized by the frequency ω_3 (which, in turn, is close to ω_0) and the phase $\bar{p}_3 = \omega_3 t - k_3 \bar{x}$. We will refer to such a model as *perturbed*. The state of this model on certain time $t = t_1$ is illustrated in Fig. 5.4.5.

For explanation of the illustration given in Fig. 5.4.5 let us remind once again that different electrons of the ensemble oscillate along $\bar{z}$-axis with the differing initial phases $\hat{\psi}_{0s}$.

Besides that, we remember the widely known fact that any transverse electromagnetic wave consists of crossed electric and magnetic fields. Consequently, the electric and magnetic Lorenz forces (described, for example, by equation (4.1.4)) should act on electrons in the bucket.

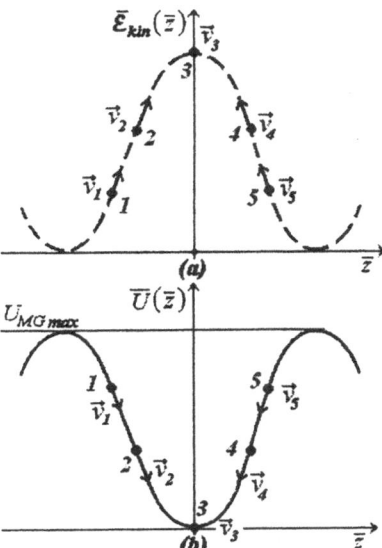

Figure 5.4.3. Illustration of unperturbed oscillations dynamics of electrons in the Miller–Gaponov potential well (the case $H < \omega_0^2$ ($\kappa < 1$) (see Fig. 5.2.1) in some time t. Therein Fig. 5.4.3a describes the dynamics of the averaged kinetic energy of certain five electrons $\bar{\mathcal{E}}_{kin}(\bar{z}) = \bar{\mathcal{H}}(\bar{z}) - \bar{U}(\bar{z}) - mc^2$, and Fig. 5.4.3b shows the dynamics of their averaged potential energy $\bar{U}(\bar{z})$. Here the general dynamics is illustrated on the example of certain five separate electrons with the numbers $s = 1, 2, 3, 4$, and 5; $U_{MG\,max}$ is the height of the potential wells (maximum of the Miller–Gaponov potential $U_{MG}(\bar{z})$); $\vec{v}_s$ are the velocities of the five oscillating electrons, $\bar{z}$ is the averaged longitudinal coordinate connected with the averaged combinative phase $\bar{\theta}$ by relationship (5.2.33).

Simplest analysis show that in the nonrelativistic case we can neglect the influence of the magnetic component of the Lorenz force in comparison with the influence of its electric component (electric Lorentz force)

$$F_{eL} = F_3 = -eE_3 = -eE_{3m}\sin\left(\omega_3 t - k_3\bar{x}\right), \qquad (5.4.8)$$

where E_3 is the electric field intensity of the third field, E_{3m} is the amplitude of the latter. Here we remind that namely non-relativistic case takes place in the discussed model (see above the model description given before in Subsection 1.1). Graphical illustration of the electric Lorentz force (5.4.8) in some cross-section $\overline{YZ}$ with the coordinate $\bar{x} = \bar{x}_1$ is shown in Fig. 5.4.6.

One of the results of an action of the electric Lorentz force is that some electrons (similar to ones that have numbers 1 and 2) move along the positive direction whereas other electrons (numbers 4 and 5) move along its negative direction. In the considered 'perturbed' case, in contrast

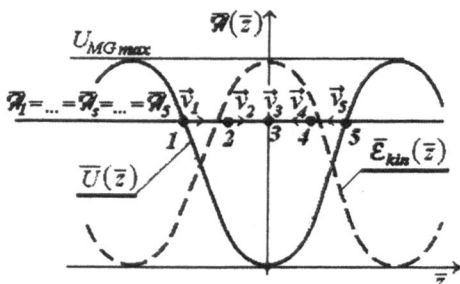

Figure 5.4.4. Illustration of the conservation law for the averaged energy $\bar{\mathcal{H}}_s = \bar{\mathcal{H}}(\bar{z}) = $ const of the unperturbed ensemble of electrons. Here $\bar{\mathcal{E}}_{kin}(\bar{z}) = \bar{\mathcal{H}}(\bar{z}) - \bar{U}(\bar{z}) - mc^2$ is the dependence of the averaged kinetic energy of electrons on the averaged longitudinal coordinate $\bar{z}$; $\bar{U}(\bar{z})$ is the similar dependence but for the averaged potential electron energy. As it can be easily seen, analogous dependence $\bar{\mathcal{H}}_s = \bar{\mathcal{H}}(\bar{z}) = $ const is represented in this case by the horizontal line. Other designations are given in Fig. 5.4.3.

to the situation with 'unperturbed' model (see Figs. 5.4.3 and 5.4.4), this means that in any fixed cross-section $\overline{YZ}$ with coordinate $\bar{x} = \bar{x}_1$ part of electrons always decelerate under the influence of the electric Lorentz force F_3 (electrons number 4,5 — see Fig. 5.4.5b). At the same time, other particles (electrons number 1,2 and 3) accelerate under action of this force. As a result, the electrons of the first group lose their energy, whereas electrons of the second group increase the energy (see Fig. 5.4.5a).

Then we recall that above in our model we have made the supposition that the frequency of free electron oscillations ω_0 is close to the wave frequency ω_3 (i.e., the following resonance condition is achieved: $\omega_0 \approx \omega_3$). This means that all electrons oscillate synchronously collinearly to the vector of Lorentz force $\vec{F}_3$. But, as was mentioned above, this occurs with different (accelerative and decelerative) initial phases. Owing to such synchronous arrangement of the system, each electron 'sees' approximately the same oscillatative phase of the transformed wave during many of its oscillations. In this case it is usually said that the *quasi-stationary action* of oscillatative electric wave field on the (also oscillated) electrons of the ensemble [4] takes place. In vacuum physical electronics such formulation of the resonant condition ($\omega_0 \approx \omega_3$) is traditionally called the *principle of quasi-stationary interaction* [4]. Such treatment turns out to be very useful for analysis of so called electron devices with long-time interaction [4].

However, the question arises: what consequences will the fact of realization of the above mentioned long time mechanism have for general

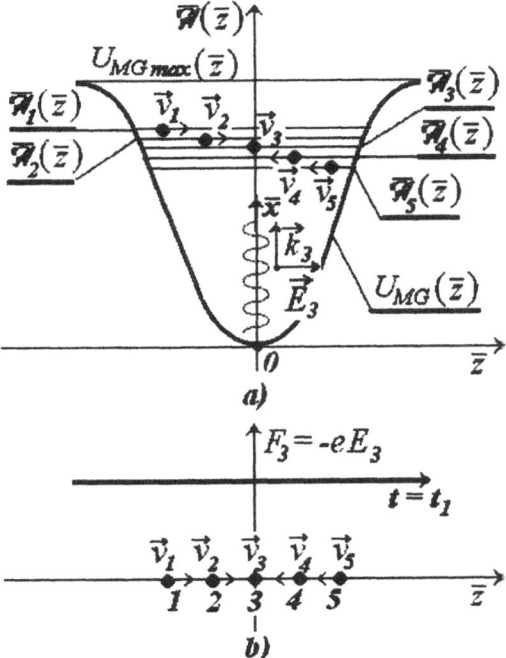

Figure 5.4.5. Initial disturbed state (perturbed model) of the electron ensemble in the bucket; therein, Fig. 5.4.5a illustrates the appearance of the energy modulation of electrons, whereas Fig. 5.4.5b illustrates the nature of a perturbed action of the electric field of third (transformed) electromagnetic wave $\left\{\vec{k}_3, \vec{E}_3\right\}$ on the ensemble. Here $F_3 = -eE_3$ is the electric component of the Lorenz force (see definition (5.4.8)) — this is the force of action of the electric field of the transformed (third) electromagnetic wave $\left\{\vec{E}_3, \vec{k}_3\right\}$ on electrons), $E_3 = \left|\vec{E}_3\right|$, $\vec{E}_3$ and $\vec{k}_3$ are the intensity of electric field and the wave vector (directed along the $\bar{x}$-axis, see Fig. 5.4.5a) of the third electromagnetic wave, respectively; $-e$ is the electron charge. The solid line in Fig. 5.4.5b is the dependence $F_3(z)$ for a time $t = t_1$. Other designations have been explained above in Figs. 5.4.3, 5.4.4

oscillation dynamics of the ensemble? Later we will follow up nonlinear dynamics of the electron oscillatative frequencies to clarify this problem. First, we turn to analysis of the 'unperturbed' frequency of nonlinear electron oscillations ω_b (5.2.53). Let us simplify the expression (5.2.53) accepting that the case of small oscillations is achieved (see condition (5.3.7)). Apart from that we consider that amplitude of the third (perturbed) electromagnetic wave is small, too. In such a situation we can expect that perturbed oscillatative frequency ω_b' should be close to unperturbed one ω_b. Besides, we can expect that both these frequencies are close to the 'linear' frequency ω_0 (5.2.35). Then accomplishing rele-

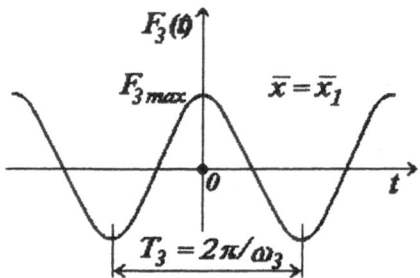

Figure 5.4.6. Dependence of the electric Lorenz force $F_3(t)$ (the third electromagnetic wave) acting on an electron in some cross-section $\overline{YZ}$ with the coordinate $\bar{x} = \bar{x}_1$. Here $F_{3\,max}$ is the amplitude of the force F_3, $T_3 = 2\pi/\omega_3$ is its temporal period. The equivalent dependence $F_3(\bar{z})$ is shown in Fig. 5.4.5b.

vant expansion in a power series in (5.2.53) by the wave amplitude one can find the following expression of rough estimation for the nonlinear frequency

$$\omega_b \cong \omega_0 \left(1 - \frac{\omega^2 \bar{Z}_m^2}{4c^2} \right), \qquad (5.4.9)$$

where $\bar{Z}_m$ is the amplitude of oscillation of the electron in the bucket (potential well) along $\bar{z}$-axis, ω_0 is the linear frequency (see definition (5.2.35)), ω is, as before, the frequency of the standing wave (i.e., the frequencies of the first and second waves in moving coordinate system K' — see corresponding commentaries for definitions (5.1.5)), c is the light velocity in vacuum. Very important thing for further explanations is the clarification of nature of the dependence of the amplitude $\bar{Z}_m$ on the averaged kinetic electron energy $\bar{\mathcal{E}}_{kin}(\bar{z}) = \bar{\mathcal{H}}(\bar{z}) - \bar{U}(\bar{z}) - mc^2$. For this we use the solutions (5.2.58)–(5.2.61) and set of simplifying suppositions accepted above. After accomplishing relevant calculations for corresponding approximate expression the required dependence could be obtained in the following form:

$$\bar{Z}_m \cong \sqrt[3]{\frac{4c^2}{\omega^2 \omega_0}} \sqrt{\frac{2\bar{\mathcal{E}}_{kin}}{m}}, \qquad (5.4.10)$$

where m is the electron rest mass. Substitution (5.4.10) into (5.4.9) yields:

$$\omega_b \cong \omega_0 - \sqrt[3]{\frac{\omega_0 \omega^2}{2mc^2} \bar{\mathcal{E}}_{kin}} = \omega_0 - \Delta\left(\bar{\mathcal{E}}_{kin}\right). \qquad (5.4.11)$$

Let us turn the reader attention to the two following characteristic features of the dependence (5.4.11). The first is that the nonlinear frequency ω_b is close to the linear frequency ω_0, i.e., $\Delta\left(\bar{\mathcal{E}}_{kin}\right) \ll \omega_0$. The second is inverse dependence of nonlinear frequency ω_b on the averaged electron energy $\bar{\mathcal{E}}_{kin}$. Namely, the bigger is the energy, the smaller is the frequency. Or, in other words, the electrons with bigger energy are characterized with a bigger oscillatative period.

Let us analyze shortly the dynamics of oscillatative phases of electrons, taking into consideration the obtained results (5.4.9)–(5.4.11), and accepted above supposition about small oscillations (see (5.3.17)). First, we consider the model with unperturbed oscillations. Using solutions (5.3.17) after not complex (but rather cumbersome) transformation we can write the approximate expression

$$\hat{\psi}_s \cong \omega_b t + \hat{\psi}_{0s} = \omega_0 t - \Delta\left(\bar{\mathcal{E}}_{kins}\right) \ t + \hat{\psi}_{0s}, \qquad (5.4.12)$$

where $\hat{\psi}_s$ is the phase of oscillations of the s-th electron of the ensemble, $\hat{\psi}_{s0} = \hat{\psi}_s\left(t = t_1\right)$ is the initial phase, $\bar{\mathcal{E}}_{kins}$ is the averaged kinetic energy of the s-th electron. Above we have accepted that the initial energies of electrons of the unperturbed ensemble are equal. This means, from the point of view of relationships (5.4.11) and (5.4.12), that the time dependencies $\hat{\psi}_s\left(t\right)$ for electrons can be represented with approximately straight lines with equal inclination (see Fig. 5.4.7a). Then we discuss the analogous dependencies for the perturbed model. In this case perturbed nonlinear frequency is also approximately equal to unperturbed one

$$\omega_b' \cong \omega_b\left[\bar{\mathcal{E}}_{kin}\left(t\right)\right]. \qquad (5.4.13)$$

The difference is only in the following: in this case the averaged kinetic energies of different electrons are found to be different. Apart from that, they adiabatically slowly change in time. The electrons that are additionally accelerated by electric field of the third (perturbed) wave have bigger energies, and the decelerated electrons have smaller energies (this situation is clearly illustrated above in Fig. 5.4.5a. The result of this difference is somehow unexpected. Because the electrons with bigger energies oscillate more slowly than the decelerated electrons which drive oscillatative phases of the second ones to phases of the first ones. In more detail this situation is shown in Fig. 5.4.7b. It can be easily seen that the ultimate time comes when the slowly and fast oscillating electrons meet in the vicinity of the same point, where they form the *phase bunch* (phase cluster — see in Fig. 5.4.7b). During some time this cluster

remains relatively compact. This means that in this case it behaves as
a large particle.

Thus the *process of self-organization* (see above Chapter 1, Sections 5,
and 6) occurs in the studied perturbed electron ensemble owing to the
nonlinearity of the system. This example obviously illustrates the discussed above (in Chapter 1, Sections 5, and 6) so difficult to understand '*specific system property*'. Indeed, collective nonlinear evolution
of the electron ensemble demonstrates the properties that not each of
separate electrons do not possess (in this case this is the property of
self-organization). Here, some initially non-organized electron ensemble
eventually organizes into an electron cluster.

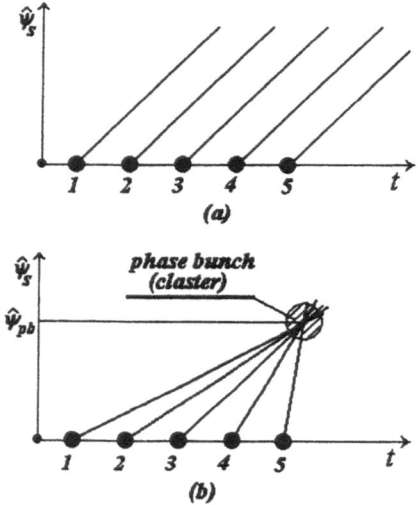

Figure 5.4.7. Illustration of the process of forming the phase bunch (cluster) in the
electron ensemble. E.g., Fig. 5.4.7a illustrates the dynamics of electron oscillatative
phases in the unperturbed model, and Fig. 5.4.7b shows a similar process in the
perturbed model. Here $\hat{\psi}_{pb}$ is the cluster oscillation phase as a whole.

The occurrence of the discussed process, self-organization, from the
practical point of view means that the ensemble as a whole obtain a
possibility to interact with the third (perturbing) electromagnetic wave
coherently, in spite of the fact that the initial system was not coherent
(see Fig. 5.4.5). The physical phenomena of such kind are called in vacuum physical electronics the *mechanisms of phase grouping* [1–4]. They
play very important role in work of many electronic devices such as gy-
roresonant systems (gyrotrons, cyclotron masers), free electron lasers,
traveling wave tubes, etc. (for more detail see Chapters 7 and 8, Volume II).

At last, let us point out especially that the discussed mechanisms of phase grouping, as the rigorous analysis shows, is not basic work effect for the system shown in Fig. 1.1.2, although it could be realizes here beyond doubt. However, as we had possibility be convinced, it is very interesting, in itself. It is interesting as well from mathematical point of view, and the physical one also. Beside that it can have independent practical applications [17].

References

[1] L.A. Vainstein, V.A. Solntzev. *Lectures on microwave electronics*. Sov. Radio, Moscow, 1973.

[2] V.I. Gaiduk, K.I. Palatov, D.M. Petrov. *Principles of microwave physical electronics*. Sov. Radio, Moscow, 1971.

[3] A.N. Kondratenko, V.M. Kuklin. *Principles of plasma electronics*. Energoatomizdat, Moscow, 1988.

[4] V.V. Kulish. *Methods of averaging in non-linear problems of relativistic electrodynamics*. World Federation Publishers, Atlanta, 1998.

[5] V.V. Kulish. On the theory of devices with difference-frequency signal separation in an electron beam. *Sov. Microwave Electronics*, 4:25–38, 1978. Super-high Frequency Electronics.

[6] T.C. Marshall. *Free electron laser*. Mac Millan, New York, London, 1985.

[7] C. Brau. *Free electron laser*. Academic Press, Boston, 1990.

[8] P. Luchini, U. Motz. *Undulators and free electron lasers*. Clarendon Press, Oxford, 1990.

[9] Ju.I. Klymenko, V.V. Kulish, P.V. Birulin. Transformer of frequency of an optical radiation. Patent of USSR No. 520857 (Cl.H 03D 7/00). Priority of 01 April 1974.

[10] V.V. Kulish, Ju.I. Klymenko. Transformer of frequency of an optical radiation. Patent of USSR No. 599719 (Cl.H 03D 7/00). Priority of 04 February 1976.

[11] V.V. Kulish. Aspects of the theory of charged particle interaction with electromagnetic wave fields. Ph.D. Diss. Theses, Kiev State University, Kiev, 1978.

[12] V.V. Kulish, Ju.M. Jermolaev. Demodulator. Patent of USSR No. 563882 (Cl.H 03D 9/00). Priority of 20 September 1974.

[13] V.V. Kulish, N.Ja. Kotsarenko. Optical frequency transformer. Patent of USSR No. 615799 (Cl.H 03D 7/00). Priority of 04 June 1976.

[14] V.V. Kulish, N.Ja. Kotsarenko. Free electron laser. Patent of USSR (Cl. H 01 J 25/00; H 01 S 3/00). Priority 05 October 1978.

[15] A.M. Kalmykov, N.Ja. Kotsarenko, V.V. Kulish. Transformation of laser radiation frequency in electron beams. *Pis'ma Zh. Tecn. Fiz. (Sov. Tech. Phys. Lett.)*, 14:820–824, 1978.

[16] L.P. Landau, E.M. Liftshitz. *Field theory*, volume 2. Nauka, Moscow, 1974.

[17] V.V. Kulish, A.V. Lysenko. Nonlinear selfconsistent theory of free electron lasers. stimulated radiation of electrons oscillating in buckets. *Ukrainian Physical Journal*, 37(5):651–659, 1992.

[18] A.V. Gaponov, M.A. Miller. On the acceleration of charged particles in moving high-frequency potential well potential wells. *Zh. Eksp. Fiz. (Sov.Phys. – JETP)*, 34(3):751–755, 1958.

[19] G.M. Zaslavsky, R.Z. Sagdeev. *Introduction in nonlinear physics*. Nauka, Moscow, 1988.

Chapter 6

HIERARCHICAL SYSTEMS WITH PARTIAL DERIVATIVES. METHOD OF AVERAGED CHARACTERISTICS

We have mentioned that nonlinear self-consistent resonant wave problems in electrodynamics challenge of many theorists over the last hundred years (i.e., from Maxwell and Heavyside up to now) [1–10]. In the general case mathematical description requires taking into account, at the same time, nonlinearity in the right-hand parts of the corresponding equations, partial derivatives in both their parts, the wave nature of initial and boundary conditions, etc.. Up to now the number of effective methods the solving problems of such a type was limited. These methods include *the method of nonlinear dispersion equation* [2–6,8–11], the *slowly varying amplitude* method (see below Chapter 8) [4,5,12] and some less widespread approaches. In studying the nonlinear mechanisms of higher orders (cubic and higher with respect to amplitudes of oscillations) all these methods are either too inaccurate (the method of nonlinear dispersion equation, for instance) or limited in their application (method of slow varying amplitudes and others).

Thus the problem of developing some new and more conventional approaches and improving the traditional ones is rather topical for the general theory of nonlinear oscillations and waves in electrodynamic systems. This Chapter (as well as the two following) is dedicated to the presentation of such a kind of new calculational technology, which, in turn, is based on the discussed earlier in the book hierarchical ideology. The totality of these general ideas and calculational technologies, in combination with the well known *method of characteristics*, are called the *method of averaged characteristics* [13].

The method mentioned and its some particular realizations (such as methods of averaging kinetic [14], quasi-hydrodynamic [15,16], current-density [15] equations), are described below in this Chapter. The ex-

amples of its practical application are given in the following Chapter
[17]. The plasma-like model of a two-beam relativistic electron system
is chosen there as a convenient illustration example.

As practice shows, some traditional methods (method of slowly varying amplitudes, for instance) in the case of plasma-like systems should be
also upgraded because such a type of objects studied are characterized
by a number of specific features. Chapter 8 is devoted to a discussion of
similar traditional and modernized versions of well known calculational
schemes. Except for this, there are some other relatively new analogous
versions, which are not now too widespread in modern electrodynamic
practice. They are described there too. The examples of practical application of these calculational technologies are also given in Chapter 8.
This is done in the framework of nonlinear theory of two-stream instability.

1.   SOME PRELIMINARY INFORMATION
1.1   Motion Equations

Let us consider the model of a plasma-like wave electrodynamic system, for instance, an intense relativistic charged particle beam. We
consider that within the interaction volume of the plasmas both possible types of electromagnetic fields (as well intrinsic (including proper),
and the external ones) may exist.

It is well known that the *Euler approach* is preferable (see Chapter 1,
Subsection 4.9) for describing of plasma-like systems similar high intensity plasma beams. In contrast to the Lagrange approach, it allows us
to observe the kinematic processes in an arbitrary spatial point that is
characterized by the position vector $\vec{r}$. Therein magnitudes of the instantaneous velocity $\vec{v}_\alpha$ and the canonical momentum $\vec{\mathcal{P}}_\alpha$ of the particle,
passing through this point, are fixed. Therein each point of plasma beam
in the chosen model corresponds to some point in the *velocity space*, given
by a velocity field $\vec{v}_\alpha(\vec{r}, t)$, or in the *momentum space* $\vec{\mathcal{P}}_\alpha(\vec{r}, t)$. So the
plasmas could be regarded as a stochastic complex system (about the
stochastic systems see in Chapter 3, Section 2) in the considered model.
The so called *Boltzmann kinetic equation* in canonical form is chosen
(see also (3.2.4)) for description of motion of a partial beam containing
of α-sort particles:

$$\left\{ \frac{\partial}{\partial t} + \frac{\partial \mathcal{H}}{\partial \vec{\mathcal{P}}} \frac{\partial}{\partial \vec{r}} - \frac{\partial \mathcal{H}}{\partial \vec{r}} \frac{\partial}{\partial \vec{\mathcal{P}}} \right\} f_\alpha \left(t, \vec{r}, \vec{\mathcal{P}} \right) = \zeta_\varepsilon^{-1} I_{\alpha\beta}, \qquad (6.1.1)$$

where $f_\alpha \left(t, \vec{r}, \vec{\mathcal{P}} \right)$ is the distribution function of particles α and $\zeta_{\alpha\beta}^{-1} I_{\alpha\beta}$ is
the collision integral which describes interactions of the α-sort particles

with some β-sort particles, $\zeta_{\alpha\beta} \gg 1$ is corresponding scale parameter. The kinetic approach (including the so called *one-particle* case (6.1.1) [3]) is rather general description method for plasma-like (including electron beam relativistic) systems.

Thus the model discussed can be treated in framework of the considered approach as an obvious example of a hierarchical stochastic oscillatory ensemble (see Chapter 3, Section 2). It is important to note that all particle motions in this case are described by the same single-particle dynamical equations (4.1.3)) or (4.1.4). The motion of each particle is characterized by random initial conditions. In our case thermal nature of initial particle velocities and random electromagnetic interactions between the particles (which are interpreted here as particle 'collisions') cause its stochastic nature.

In general, we can formulate the following affirmation: the considered plasma-like system can be represented as a stochastic particle ensemble described by the distribution function $f_\alpha\left(t, \vec{r}, \vec{\mathcal{P}}\right)$. It will be shown below that equations like (6.1.1) can be reduced to standard hierarchical form like (4.1.5). The mathematical treatment of the beam motion process could be possible in the framework of kinetic description by means of the hierarchical method. However, another description method for the beam motion problem can be realized, too. But let us discuss this topic in more detail.

It might seem that the kinetic approach is the only appropriate method for studying of any electrodynamic plasma-like problems. However, it is not correct always. The *quasi-hydrodynamic approach* can be used also in the cases when certain conditions are held [2, 3, 6]. This approach is based on the *quasi-hydrodynamic equation* as an initial one. Motion of the plasma beam can be described in this case using model of a flow of a two-component charged fluid [3]. A few versions of quasi-hydrodynamic description are known in modern electrodynamics. Below we will comment shortly some simplest way of obtaining the quasi-hydrodynamic equation.

Let us accept some simplifying assumptions. In particular, we assume that the pressure difference between output and input of a chosen beam element can be neglected

$$\Delta P \approx 0, \tag{6.1.2}$$

transverse characteristic dimensions a of the beam (for instance, transverse size) are much greater than the *Debye radius* (see [2, 3, 6])

$$a \gg r_D, \tag{6.1.3}$$

and the number of particles inside the sphere of radius r_D is much greater than unity, i.e.

$$r_D^3 n_0 \gg 1. \qquad (6.1.4)$$

Other appropriate conditions are held to maintain the above mentioned simplest two-fluid hydrodynamic model [3, 6] (here n_0 is the averaged (unperturbed) particle density of the plasma beam).

Numerical analysis of collisions between particles in traditional vacuum electron devices and accelerators shows that their role more often is inessential there. Having this in mind, we will consider below that the collision processes in the studied model is governed by a qualitative interest mainly. Employing the Bhatnahar–Gross–Krook collision model [3, 6] allows to realize such a semi-qualitative description. In this way we obtain the simplest form of relativistic quasi-hydrodynamic equation (see for more details in [1, 2])

$$\left(\frac{\partial}{\partial t} + \vec{u}_\alpha \frac{\partial}{\partial \vec{r}} + \frac{\nu}{\gamma^2}\right)\vec{u}_\alpha = \frac{q_\alpha}{m_\alpha \gamma_\alpha}\left\{\vec{E} + \frac{1}{c}\left[\vec{u}_\alpha \vec{B}\right] - \frac{\vec{u}_\alpha}{c^2}\left(\vec{u}_\alpha \vec{E}\right)\right\}$$
$$- \frac{v_T}{n_\alpha \gamma_\alpha}\left[\frac{\partial n_\alpha}{\partial \vec{r}} - \frac{\vec{u}_\alpha}{c^2}\left(\vec{u}_\alpha \frac{\partial}{\partial \vec{r}}\right)n_\alpha\right], \quad (6.1.5)$$

where $\vec{u}_\alpha$ is the velocity of the beam of α-particles as a whole, ν is the effective particle collision frequency, $\gamma_\alpha = \left(1 - \vec{u}_\alpha^2/c^2\right)^{-1/2}$ is the *beam relativistic factor* of particles α; q_α is their charge, m_α is their rest mass, n_α is the particle density and v_T is the root-mean-square particle thermal velocity.

1.2   Field Equations

It is necessary to add the Maxwell equations (1.4.23) for self-consistent description of the beam motion:

$$\left[\vec{\nabla}\vec{E}\right] = -\frac{1}{c}\frac{\partial \vec{B}}{\partial t}; \qquad \left[\vec{\nabla}\vec{H}\right] = \frac{1}{c}\frac{\partial \vec{D}}{\partial t} + \frac{4\pi}{c}\left(\vec{j}_0 + \vec{j}\right);$$
$$\left(\vec{\nabla}\vec{D}\right) = 4\pi\left(\rho_0 + \rho\right); \qquad \left(\vec{\nabla}\vec{B}\right) = 0, \qquad (6.1.6)$$

where $\vec{E}$, $\vec{D}$, $\vec{B}$ and $\vec{H}$ are the standard designations for electric and magnetic field vectors (see Chapter 1, Subsection 4.8 for more details); $\vec{j}_0$ and $\vec{j}$ are the current density vectors arising from external and internal sources of the fields, ρ_0 and ρ are the space charge densities caused by similar sources, and $\vec{\nabla}$ is the nabla operator.

In the case of kinetic approach the induced densities $\vec{j}_\alpha$ and ρ_α are related to the distribution function $f_\alpha\left(t,\vec{r},\vec{\mathcal{P}}\right)$ through the following relations:

$$\vec{j}_\alpha = q_\alpha n_{0\alpha} \int_{-\infty}^{\infty} \vec{v} f_\alpha\left(t,\vec{r},\vec{\mathcal{P}}\right) d^3\mathcal{P}; \qquad (6.1.7)$$

$$\rho_\alpha = q_\alpha n_{0\alpha} \int_{-\infty}^{\infty} f_\alpha\left(t,\vec{r},\vec{\mathcal{P}}\right) d^3\mathcal{P}; \qquad (6.1.8)$$

$$\vec{j} = \sum_\alpha \vec{j}_\alpha; \quad \rho = \sum_\alpha \rho_\alpha. \qquad (6.1.9)$$

Using (6.1.6), one can obtain the *continuity equation* in the form:

$$\frac{\partial \rho}{\partial t} + \left(\vec{\nabla}\vec{j}\right) = 0. \qquad (6.1.10)$$

In the case of quasi-hydrodynamic approach the densities $\vec{j}_\alpha$ and ρ_α can be expressed as [8, 18, 19]:

$$\vec{j}_\alpha = q_\alpha n_\alpha \vec{u}_\alpha, \quad \rho = q_\alpha n_\alpha. \qquad (6.1.11)$$

Here $n_{0\alpha}$ and n_α are unperturbed and induced densities of corresponding particles.

The joint integration of the Maxwell equations (6.1.6) with motion equations (6.1.1) and (6.1.5) enables solution of the complete self-consistent problem of motion of plasma-like beam.

1.3 Some General Information about Equations with Partial Derivatives

Let us recall some general concepts and definitions, which are usually studied in standard course of differential equations with partial derivatives.

Single differential equation. The following equation of general type usually is replaced with the differential equation with partial derivatives of a n-th order:

$$F\left(z_1, \ldots, z_n, u, \frac{\partial u}{\partial z_1}, \ldots, \frac{\partial u}{\partial z_n}, \frac{\partial^2 u}{\partial z_1 \partial z_1}, \frac{\partial^2 u}{\partial z_1 \partial z_2}, \right.$$
$$\left. \ldots, \frac{\partial^2 u}{\partial z_n \partial z_n}, \ldots, \frac{\partial^n u}{\partial z_n \ldots \partial z_n} \right) = 0. \quad (6.1.12)$$

This equation contains at least one n-th order partial derivative of a unknown function $u(z_1, \ldots, z_n)$, z_j $(j = 1, \ldots, n)$ are independent variables.

It is necessary to find a function $u(z_1, \ldots, z_n)$ such that the function F becomes equal to zero after substitution of $u(z_1, \ldots, z_n)$ into (6.1.12). Such a function is called the solution or the *integral of a differential equation with partial derivatives*. All *partial solutions* (excepting *additional solutions*) can be obtained from the general integral.

It is well known that any general solution of exact differential equation (equation with momentum derivatives) can be found with the accuracy of some arbitrary constants. The specific feature of the equation with partial derivatives is that instead of the arbitrary constants we obtain some *arbitrary functions*. Use some initial or boundary conditions for the function u and its derivatives can define the latter. In the general case the number of these arbitrary functions is equal to the order of the differential equation. If we have n independent variables then the arbitrary functions are the functions of $n-1$ variables.

System of Equations with Partial Derivatives . The systems of equations for a few unknown functions $u_1(z_1, \ldots, z_n)$, $\ldots, u_k(z_1, \ldots, z_n)$ can be written in the form

$$F_i\left(z_j, u_1, \ldots, u_k, \frac{\partial u_1}{\partial z_j}, \ldots, \frac{\partial u_k}{\partial z_j}, \ldots, \frac{\partial^n u}{\partial z_j \ldots \partial z_l}\right) = 0, \quad (6.1.13)$$

where $i = 1, \ldots, h$; $j, l = 1, \ldots, n$; k is the current number of the unknown functions, h is the total number of the system equations.

In the case $k = h$ the system (6.1.13) is called *determined* one, and in the cases $k < h$ or $k > h$ we have to deal with *overdetermined* or *underdetermined systems*.

Each differential equation of type (6.1.12) or system of equations with higher derivatives like (6.1.13) can be reduced to the system with only first derivatives (system *of the first order*).

Conditions of compatibility. The system of differential equations (6.1.13) allows obtaining the solutions $u_1(z_1, \ldots z_n)$, $u_2(z_1, \ldots, z_n)$ only in the case when the given functions F_i and their derivatives satisfy the *conditions of compatibility*. These conditions guarantee that the differentiation of two or more equations (6.1.13) leads to coincidence of the highest derivatives of the sought functions u_k. The conditions of compatibility can be obtained by excluding of the functions u_k and their derivatives from some sequence of the equations. In turn, this sequence could be obtained by differentiation of equations (6.1.13).

2. METHOD OF AVERAGED CHARACTERISTICS

Let us start discussion of the hierarchical methods of asymptotic integration of differential equations with partial derivatives with the *method of averaged characteristics* [13].

2.1 Concept of the Standard Form

As analysis shows, the method of averaged characteristics is suitable for asymptotic integration of the systems like (6.1.13) that could be described using so called *standard form* (or, the same, *standard system with partial derivatives* — see also equation (3.3.1) and corresponding commentaries)

$$A' (U,z,t) \frac{\partial U}{\partial t} + \left(Z' (U,z,t) \times \frac{\partial}{\partial z} \right) U + C'(z,t)U = R'(U,z,t), \quad (6.2.1)$$

where A', Z', C', R', Z, R are square matrices of size $l \times l$, $U = U(z,t)$ is some vector-function in Euclidean n-*dimensional* space R^n with coordinates $\{z_1, z_2, ..., z_n\}$, i.e. $\forall z \in R^n z = (z_1, z_2, ..., z_n)^T$, $\forall z_i \in (-\infty, +\infty)$, $i \in (1, 2, ..., n)$, $R(...)$ is a given weakly nonlinear periodical (m-fold, in the general case) vector function, t is some scalar variable, for instance, the laboratory time. Therein it is considered that some hierarchy of the dynamical values (in time or spatial coordinates) could be determined.

Then let us turn to the terminology of the theory of method of characteristics (see also the following Sections for more details). It is not difficult to prove that the standard form (6.2.1) can be transformed into so called *quasilinear* (with respect to the derivatives!) *homogeneous equation*

$$\frac{\partial U}{\partial t} + Z(U,z,t) \frac{\partial U}{\partial z} = R(U,z,t). \quad (6.2.2)$$

Further in this Chapter we shall be oriented on the studying the standard equations (and systems of equations) of form (6.2.2).

2.2   General Scheme of the Method

The calculational procedure of the method of averaged characteristic includes three specific steps. The *first step* is the *straight transformation* of an initial system of equations with partial derivatives to some κ-th higher level. The main task of the *second step* is the *solving of this equation.* Finally, the transformation of the obtained solution from this κ-th hierarchical level to the initial (zero) level (*back transformations*) is accomplished during the *third step.* In turn, each of the mentioned steps contains a few smaller substeps (sub-stages). All these steps and substeps are illustrated graphically in Fig. 6.2.1.

We begin with discussing the first step of the considered calculational algorithm. It, as it can be easily seen in Fig. 6.2.1), includes two more particular substeps. The main idea of the first substep is based on some known properties of *characteristics of initial differential equation* (6.2.2). Owing to these properties (see the following Subsection for more details) we have a possibility of reducing the initial equation with partial derivatives (6.2.2) to some system with momentum derivatives (exact differential equations — see Fig. 6.2.1):

$$\frac{dz}{dt} = Z\left(U, z, t\right);$$
$$\frac{dU}{dt} = R\left(U, z, t\right), \tag{6.2.3}$$

where all designations are self-evident. Corresponding essential simplification of the initial problem (6.2.2) can be attained as a result of the transformations performed. This simplifying could be explained using the fact that the integration procedure for the equations with momentum derivatives like (6.2.3) usually is essentially simpler than the procedure of finding solutions of corresponding equations with partial derivatives (like (6.2.2)).

Unfortunately, the achieved simplification very often is insufficient for successive solving of (6.2.2). The point is that problem (6.2.3), especially in the case of multi-periodical nonlinear oscillations, could be far from simple, from calculational point of view (this circumstance has been clearly illustrated previously in Chapters 4 and 5). So the goal of the second substep (see Fig. 6.2.1) is the accomplishing the asymptotic integration of equations (6.2.3). The *hierarchical calculational technology, which has been developed previously in Chapters 4 and 5, is proposed as a methodical basis for this part of the discussed general calculational procedure.* So, equations (6.2.3), as before, can be regarded as another form of the initial equation of the zeroth hierarchical level (see Fig. 6.2.1).

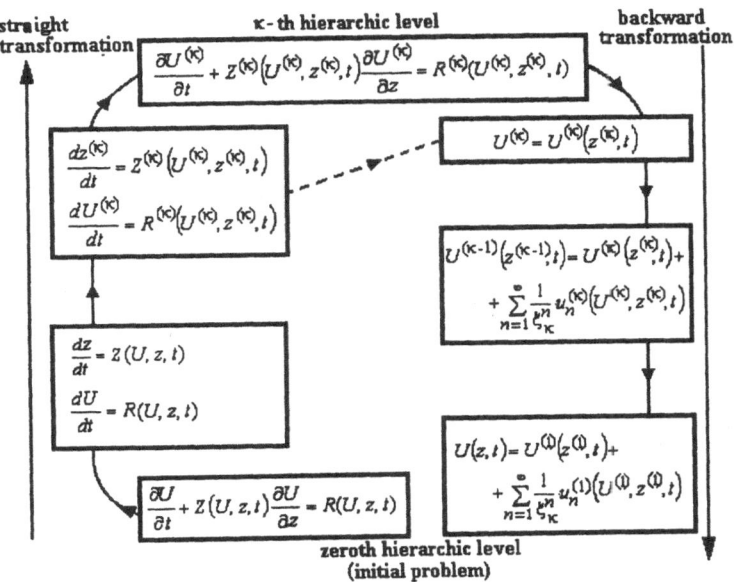

Figure 6.2.1. General calculational scheme of the method of averaged characteristic

Below we perform the procedure of introducing the fast rotating phases (see Chapter 4 for more details) as parameters:

$$\frac{dz}{dt} = Z\left(U, z, \psi, t\right);$$
$$\frac{dU}{dt} = R\left(U, z, \psi, t\right); \qquad (6.2.4)$$
$$\frac{d\psi}{dt} = \Omega\left(z, U\right) + Y\left(U, z, \psi, t\right),$$

where essence of all designations is obvious from what was set forth before in Chapter 4. Then according to the general procedure (see Chapter 4) we separate the complete set of problem large parameters and construct the hierarchical series

$$\xi_1 \gg \xi_2 \gg \cdots \gg \xi_\kappa \gg \xi_m \gg 1, \qquad (6.2.5)$$

where all large parameters are determined by the standard method (see (4.1.7) and corresponding commentaries). The asymptotic solutions of the system (6.2.4) can be represented in the form of hierarchical sequence of the Krylov–Bogolyubov substitutions (see Chapter 4, Subsection 1.4).

In particular, in the case of two-level hierarchical system the asymptotic solutions for system (6.3.2) can be represented in the following form:

$$U = \bar{U} + \sum_{n=1}^{\infty} \frac{1}{\xi^n} u_u^{(n)}\left(\bar{z}, \bar{U}, \bar{\psi}\right);$$

$$z = \bar{z} + \sum_{n=1}^{\infty} \frac{1}{\xi^n} u_z^{(n)}\left(\bar{z}, \bar{U}, \bar{\psi}\right); \qquad (6.2.6)$$

$$\psi = \bar{\psi} + \sum_{n=1}^{\infty} \frac{1}{\xi^n} v^{(n)}\left(\bar{z}, \bar{U}, \bar{\psi}\right),$$

where all averaged values can be found from the equations for the first hierarchical level:

$$\frac{d\bar{U}}{dt} = \sum_{n=1}^{\infty} \frac{1}{\xi^n} A_u^{(n)}\left(\bar{z}, \bar{U}\right);$$

$$\frac{d\bar{z}}{dt} = \sum_{n=1}^{\infty} \frac{1}{\xi^n} A_z^{(n)}\left(\bar{z}, \bar{U}\right); \qquad (6.2.7)$$

$$\frac{d\bar{\psi}}{dt} = \Omega\left(\bar{z}, \bar{U}\right) + \sum_{n=1}^{\infty} \frac{1}{\xi^n} B^{(n)}\left(\bar{z}, \bar{U}\right),$$

$\xi \equiv \xi_1$. All auxiliary functions could be taken into account by use of the algorithms described in Chapter 4. For instance, result of such calculations for the first (by the small parameter $1/\xi$) approximation could be represented in the form

$$A_u^{(1)} = \frac{1}{2\pi} \int_0^{2\pi} R\left(\bar{z}, \bar{U}, \bar{\psi}\right) d\bar{\psi}; \qquad (6.2.8)$$

$$u_u^{(1)} = \frac{1}{\Omega} \int_0^{\psi} \left(R\left(\bar{z}, \bar{U}, \bar{\psi}\right) - A_u^{(1)}\right) d\bar{\psi}; \qquad (6.2.9)$$

$$A_z^{(1)} = \frac{1}{2\pi} \int_0^{2\pi} Z\left(\bar{z}, \bar{U}, \bar{\psi}\right) d\bar{\psi}; \qquad (6.2.10)$$

$$u_z^{(1)} = \frac{1}{\Omega} \int \left(Z\left(\bar{z}, \bar{U}, \bar{\psi}\right) - A_z^{(1)}\right) d\bar{\psi}; \qquad (6.2.11)$$

$$B^{(1)} = \frac{1}{2\pi} \int_0^{2\pi} \left(\frac{\partial \Omega}{\partial U} u_u^{(1)} + \frac{\partial \Omega}{\partial \bar{z}} u_z^{(1)} + Y\left(\bar{z}, \bar{U}, \bar{\psi}\right) \right) d\bar{\psi}; \qquad (6.2.12)$$

$$v^{(1)} = \frac{1}{\bar{\Omega}} \int_0^{2\pi} \left(\frac{\partial \Omega}{\partial U} u_u^{(1)} + \frac{\partial \Omega}{\partial \bar{z}} u_z^{(1)} + Y\left(\bar{z}, \bar{U}, \bar{\psi}\right) - B^{(1)} \right) d\bar{\psi}. \qquad (6.2.13)$$

Taking into account the parametrical nature of the variable ψ (see Chapter 4), equations for some κ-th hierarchical level in the general case can be written as (see Fig. 6.2.1):

$$\begin{aligned} \frac{dz^{(\kappa)}}{dt} &= Z^{(\kappa)}\left(U^{(\kappa)}, z^{(\kappa)}\right); \\ \frac{dU^{(\kappa)}}{dt} &= R^{(\kappa)}\left(U^{(\kappa)}, z^{(\kappa)}\right), \end{aligned} \qquad (6.2.14)$$

where $R^{(\kappa)}\left(U^{(\kappa)}, z^{(\kappa)}\right)$ is some nonlinear functional matrix, the upper index κ denotes that the corresponding values belong to the κ-th hierarchical level. The values $z^{(\kappa)}$ and t in (6.2.14) are, in principle, dependent variables (because Lagrange formulation of the problem). However, they are considered here (that is very important!) as *parameters of the problem*. According to what was set forth in Chapter 4, equations for any κ-th hierarchical level (4.2.14) by themselves are the κ-fold averaged equations (6.2.3) (equations for the κ-th hierarchical level).

Then, keeping the parametrical form of the equations (6.2.14), a special hypothesis is put in the basis of the considered step of the discussed algorithm. Namely, it is supposed that equations (6.2.14) can be treated as characteristics of some other equation with partial derivatives:

$$\frac{\partial U^{(\kappa)}}{\partial t} + Z^{(\kappa)}\left(U^{(\kappa)}, z^{(\kappa)}, t\right) \frac{\partial U^{(\kappa)}}{\partial z} = R^{(\kappa)}(U^{(\kappa)}, z^{(\kappa)}, t). \qquad (6.2.15)$$

It should be specially pointed out that, in contrast to (6.2.14), the variables $z^{(\kappa)}$ and t are the independent ones (because of the Euler form of representation).

Comparing (6.2.15) and the initial equation (6.2.2) we find that both these equations have analogous in principle mathematical structure, i.e. hierarchical resemblance principle remains here. Because of their special form, we call equation (6.2.15) the *averaged (truncated) quasilinear*

equation of the κ-th hierarchical level. Besides that, the second hierarchical principle is also satisfied (hierarchical compression principle — see
Chapter 2, Section 2 for more details). According to this principle the
initial hierarchical system (6.2.2) (zeroth hierarchical level) reproduces
itself on each hierarchical level in its general dynamical structure. So it
turns out to grow simpler as the hierarchy level grows higher. The other
two hierarchical principles are satisfied here, too.

The passage from the equation with partial derivatives (6.2.2) to equations of characteristics (6.2.3) (exact differential equations) is similar to
the passage from the Euler form of writing to the Lagrange form (about
the Lagrange and Euler formulations of the problem see in Chapter 1,
Subsection 4.9). Correspondingly, the passage from (6.2.14) to (6.2.15)
describes the passage like the passage from the Lagrange form to the
Euler one. However, it should be mentioned that we are talking about
some resemblance between both procedures only in the general case.
Exact equivalence takes place only in some particular cases (see below
Subsections 5–7). Mutual resemblance of these passages here is kept due
to the use of parametrical (with respect to the variables $z^{(\kappa)}$ and t) form
of representation in corresponding differential equations.

It can easily be seen that equation (6.2.15) could be treated as initial equation (6.2.2) (zeroth hierarchical level), which is transformed to
the κ-th hierarchical level by means of the *straight transformations* (see
Fig. 6.2.1). So, taking into consideration the said above, we can affirm that eventual result of the first step of the discussed calculational
algorithm is the *obtaining of a κ-fold averaged equation with partial
derivatives* like (6.2.15) (i.e., the equation of κ-th hierarchical level).
The totality of calculational procedures of the first step we treat as the
straight transformation of the initial equation (6.2.2) into hierarchically
equivalent equation (6.2.15) for some κ-th level (see Fig. 6.2.1).

The main task of the *second step* of the discussed algorithm is the
solving of the equation (equations) (6.2.15). Analogously to the situation that takes place in the framework of the hierarchical methods
for exact differential equations (see Chapters 4 and 5), the transformed
equation for any κ-th hierarchical level (6.2.15) is also simpler than the
initial equation (6.2.2). Owing to this, the general procedure of solving
κ-fold averaged equation (6.2.15) turns out to be simpler than the analogous procedure for equation (6.2.2). In principle, maximal simplification
could be attained in the case when the straight transformation of initial equation (6.2.2) is accomplished up to the highest hierarchical level
$\kappa = m$. But it should be mentioned especially that quite often there is
no strong necessity for performing of such 'complete' hierarchical transformations. The point is that sufficiently simplified (because they are

averaged), solutions could be found on some of the lower hierarchical levels $\kappa < m$.

In what follows, it should be noted that any known analytical or numerical calculational method can be applied for the solution of equation (6.2.15). However, very interesting situation occurs in the case of application for this purpose of the method of characteristics. As it will be shown below in the next Section 3, this method has one specific feature. Namely, the possibility of constructing solutions of a corresponding equation with partial derivatives, basing on the solutions of the equations for characteristics. The characteristics, as was mentioned above, are exact equations (equations with momentum derivatives). This means that in that case, if the characteristics method is used, we, strictly speaking, have no necessity for constructing of κ-fold averaged equation (6.2.15) (see the dotted line in Fig. 6.2.1). Their corresponding solutions can be constructed immediately on the basis of solutions of equations for characteristic (6.2.14) (see the following Sections for more details). But, in any case, independently with the solution method used, we eventually obtain the solutions of equation (6.2.15) in the form

$$U^{(\kappa)} = U^{(\kappa)}\left(z^{(\kappa)}, t\right). \tag{6.2.16}$$

Constructing solution like (6.2.16) completes the main task of the second step of the discussed calculational algorithm.

Analyzing the form of solution (6.2.16) and possibilities of constructing corresponding solutions of initial problem (6.2.2), we can observe a new peculiar calculational situation. It realizes owing to the use of the hierarchical procedure, which has been developed earlier for solving of the exact equations (see Chapters 4, 5). However, let us discuss this problem in more detail.

First of all we turn the reader's attention to the following circumstance. In fact, the problems discussed earlier in Chapters 4 and 5 can be classified as one-dimensional, because all dynamical variables change in this case with the temporal coordinate t only. (Here let us point out that speaking further in this Chapter about one-, two- and more dimensional problems, we will mean the total quantity of variables (the spatial coordinates and the temporal one). In contrast, in other Chapters in analogous situations we will talk about the *spatial dimensionality* only). On the other hand, we used the same time t for description of functional dependencies on any κ-th hierarchical level, i.e., $z^{(\kappa)} = z^{(\kappa)}(t)$, $z^{(\kappa-1)} = z^{(\kappa-1)}(t)$, and so on. Spatial coordinates, vice versa, are different for different hierarchical levels. However, in the case of exact equations this does not lead to any difficulties because we have the temporal dependencies like $z^{(\kappa)} = z^{(\kappa)}(t)$ on any hierarchical level. Owing to this we obtained a chance of constructing solutions for any lower hierarchical level, substituting the solutions for higher hierarchical levels into the Krylov–Bogolyubov substitutions (4.2.2)

$$x^{(\kappa-1)}(t) = x^{(\kappa)}(t) + \sum_{n=1}^{\infty} \frac{1}{\xi_{\kappa}^{n}} u_{x\kappa}^{(n)}\left(x^{(\kappa)}(t), \psi^{(\kappa)}(t)\right);$$

$$\psi^{(\kappa-1)}(t) = \psi^{(\kappa)}(t) + \sum_{n=1}^{\infty} \frac{1}{\xi_{\kappa}^{n}} v_{\kappa}^{(n)}\left(x^{(\kappa)}(t), \psi^{(\kappa)}(t)\right),$$

$$(6.2.17)$$

where all designations were given earlier in Chapter 4, and, therefore, they are self-evident. Thus the relationships like (6.2.17) play double role in the discussed situation. Namely, using them we have accomplished the *straight as well as the back transformations* in the framework of method described in Chapters 4, 5.

We have a different, in principle, situation in the case of differential equations with partial derivatives. Here, besides the temporal functional dependencies, we have also dependencies on spatial coordinates (see the form of standard equations (6.2.1) and (6.2.2)). In contrast to the preceding case, variables of both these types are independent from each other (the Euler description — see Chapter 1, Subsection 4.9). But as was mentioned above, the spatial coordinates are different for different hierarchical levels. This means that corresponding functions $U^{(\kappa)}$ depend — on each hierarchical level — on a proper set of variables for this level. As before, the Krylov–Bogolyubov substitutions give us a connection between different neighboring hierarchical levels:

$$U^{(\kappa-1)}\left(z^{(\kappa-1)}, t\right) = U^{(\kappa)}\left(z^{(\kappa)}, t\right) + \sum_{n=1}^{\infty} \frac{1}{\xi_{\kappa}^{n}} u_{U\kappa}^{(n)}\left(U^{(\kappa)}, z^{(\kappa)}, t\right), \quad (6.2.18)$$

i.e., it could be used for the straight transformations in this case also. But in contrary with the above discussed situation with exact differential equations, here we can not immediately use the relationships like (6.2.8) for constructing solutions of the initial problem (6.2.2) (i.e., for performing the back transformations). For this we additionally should have some special correlation between proper coordinates of the neighboring hierarchical level like $z^{(\kappa)} = z^{(\kappa)}\left(z^{(\kappa-1)}\right)$. The correlation of such type, as will be shown below, can be constructed really. Modernized in such manner calculational procedures embody the main task of the *third step* of the considered algorithm. Totality of all procedures, which allow to accomplish the transformation of solutions from any κ-th hierarchical level to the zeroth hierarchical level (see Fig. 6.2.1), we call *back transformations*.

The main idea of the back transformations is based on the following observation. If we look at the Krylov–Bogolyubov substitutions (6.2.18) we can see that difference between any variables of two neighboring hierarchical levels $((\kappa - 1)$-th and κ-th, for example) is proportional to $1/\xi_\kappa \ll 1$:

$$U^{(\kappa-1)}\left(z^{(\kappa-1)}, \psi^{(\kappa-1)}, t\right) - U^{(\kappa)}\left(z^{(\kappa)}, \psi^{(\kappa)}, t\right)$$

$$= \sum_{n=1}^{\infty} \frac{1}{\xi_\kappa^n} u_{U\kappa}^{(n)}\left(U^{(\kappa)}, z^{(\kappa)}, \psi^{(\kappa)}, t\right), \quad (6.2.19)$$

and so on.

Taking this into account, let us expound the function $U^{(\kappa)}(z^{(\kappa)}, \psi^{(\kappa)}, t)$ in the Taylor series in the vicinity of equivalent spatial point of $(\kappa-1)$-th hierarchical level.

Substituting then all differences like (6.2.9) into this series we obtain the required set of approximate correlation between any two neighboring hierarchical levels. For instance, we can do it with respect to the difference between the first (averaged) and zero (initial) levels, respectively:

$$U\left(z, t\right) = U\left(\bar{z}, \bar{\psi}\right)\bigg|_{\substack{\bar{z}=z \\ \bar{\psi}=\psi}}$$

$$+ \left\{\frac{\partial \bar{U}\left(\bar{z}, \bar{\psi}\right)}{\partial \bar{z}}\left(\sum_{n=1}^{\infty} \frac{1}{\xi^n} u_z^{(n)}\right) + \frac{\partial \bar{U}\left(\bar{z}, \bar{\psi}\right)}{\partial \bar{\psi}}\left(\sum_{n=1}^{\infty} \frac{1}{\xi^n} v^{(n)}\right)\right\}\bigg|_{\substack{\bar{z}=z \\ \bar{\psi}=\psi}} + \cdots$$

$$(6.2.20)$$

and so on.

Below we accomplish the chain of successive analogous back transformations between each pair of neighboring hierarchical levels. As a result, the back transformation procedure sought could be constructed eventually. It allows to accomplish the back transformations from any higher hierarchical level (including the highest one $\kappa = m$) to initial (zero) hierarchical level ($\kappa = 0$).

Thus we can realize that key point of the discussed above calculational technology is a special combination of the concept of characteristic and, at the same time, the hierarchical algorithm for asymptotic integration of exact differential equations described above in Chapters 4 and 5. In other words, the *method of characteristics* is important part of the discussed *method of averaged characteristics*. That is why further we shall pay special attention to the discussion of specific properties of the characteristics, including the method of characteristics itself.

3.   CHARACTERISTICS AND THE METHOD OF CHARACTERISTICS

3.1   Method of Characteristics. The Scalar Case

We begin with the scalar differential equation of the first order because it is the most convenient for the following:

$$F\left(z_1, ..., z_n, \frac{\partial u}{\partial z_1}, ..., \frac{\partial u}{\partial z_n}\right) = 0, \qquad (6.3.1)$$

where F is a given function of its arguments, u is the unknown function on arguments $z_1, ..., z_n$. In this Subsection let us restrict ourselves with study of the specific case of equation (6.3.1) which can be described by the quasilinear differential equation of the type

$$\sum_{i=1}^{n} a_i\left(z_1, ..., z_n, u\right) \frac{\partial u}{\partial z_i} = R\left(z_1, ..., z_n, u\right), \qquad (6.3.2)$$

where a_i, R are some given functions on $z_1, ..., z_n$, u. It should be mentioned that equations of such type are rather widespread in electrodynamics. Besides that, they are also typical for nonlinear problems of hydrodynamics, thermodynamics, quantum physics, etc.. So, our choice, apart from merely methodical meaning, could have some practical significance, too.

A few calculational schemes are known for solving equation (6.3.2) using the method of characteristics [20–25]. Let us consider two of them.

The first scheme of constructing the general solution.   The first scheme of constructing the general solution of equation (6.3.2) we realize in the form of the following four steps:

1) the system of equations that determines the characteristics of system (6.3.2) is constructed:

$$\frac{dz_1}{a_1} = \cdots = \frac{dz_n}{a_n} = \frac{du}{R}, \qquad (6.3.3)$$

2) n independent integrals of the system (6.3.3) are found:

$$\psi_1 (z_1, ..., z_n, u) = c_1;$$
$$\cdots\cdots\cdots\cdots\cdots\cdots\cdots \quad ; \qquad (6.3.4)$$
$$\psi_n (z_1, ..., z_n, u) = c_n,$$

3) some arbitrary function Φ of these integrals is constructed; this function is set equal to zero (that is implicit solution of equation (6.3.2)):

$$\Phi \left(\psi_1 (z_1, ...z_n, u), ...\psi_n (z_1, ...z_n, u) \right) = 0, \qquad (6.3.5)$$

4) further, the general solution of equation (6.3.2) should be obtained by the solving of equation (6.3.5) with respect to the function u.

The second scheme of constructing the general solution. The second scheme of constructing the general solution of equation (6.3.2) differs with the first one only with the scheme of constructing of the characteristics:

$$\frac{dz_1}{dt} = a_1 (z_1, ..., z_n, u) ; ...; \frac{dz_n}{dt} = a_n (z_1, .. , z_n, u) ;$$
$$\frac{du}{dt} = R (z_1, ..., z_n, u) . \qquad (6.3.6)$$

Then, steps 2–4 of the preceding scheme are performed.

The complete solution of the considered problem could be found only in the case when initial and boundary conditions are given, i.e., the Cauchy problem should be solved.

The Cauchy problem. The essence of the Cauchy problem is finding the solution $u = u(z)$ that satisfies the initial condition

$$u|_{z \in \gamma} = \varphi(z) \qquad (6.3.7)$$

on some surface γ. The latter is given by the equation

$$f (z_1, ..., z_n) = 0. \qquad (6.3.8)$$

This means that amongst interminable number of functions (6.3.5) there should be chosen functions that satisfy the initial condition (6.3.7) in case when condition (6.3.8) is satisfied.

A few algorithms of the Cauchy problem solution are known [20–25]. Now let us comment briefly two of them.

The first algorithm. For the sake of finding the solution $u = u(z)$ for equation (6.3.2) (for given initial conditions (6.3.7) or (6.3.8)) it is necessary:

1) to construct the system for characteristic (6.3.3) or (6.3.6);

2) to find all independents integrals (6.3.4) for the characteristics;

3) to construct the algebraic system for the founded integrals and for the initial conditions:

$$\psi_1 (z_1, ..., z_n, u) = c_1;$$

$$\dotfill$$

$$\psi_n (z_1, ..., z_n, u) = c_n;$$

$$f (z_1, ..., z_n) = 0;$$

$$u_0 = u|_{z \in \gamma} = \varphi (z_1, ..., z_n), \qquad (6.3.9)$$

4) to exclude all variables z and u to obtain the equation of the type

$$F (c_1, c_2, ...c_n) = 0, \qquad (6.3.10)$$

5) substitution the left sides of equations (6.3.4) into (6.3.10) to obtain the equation

$$F (\psi_1 (z_1, ..., z_n, u), ..., \psi_n (z_1, ..., z_n, u)) = 0, \qquad (6.3.11)$$

6) solving (6.3.11) to find solution sought $u = u (z_1, ..., z_n)$ for initial condition (6.3.7), (6.3.8).

The second algorithm. For the finding the solution of equation (6.3.3)

$$u = \varphi (z_1, ..., z_{n-1}) \qquad (6.3.12)$$

with the given initial condition

$$z_n = z_n^{(0)} \qquad (6.3.13)$$

it is necessary:

1) to construct the system of characteristics (6.3.3) or (6.3.6);

2) to find all independent integrals (6.3.4);

3) to replace the independent variable z_n in integrals (6.3.4) with its initial value $z_n^{(0)}$:

$$\psi_1 \left(z_1, ..., z_{n-1}, z_n^{(0)}, u \right) = c_1';$$

$$\dotfill \qquad (6.3.14)$$

$$\psi_n \left(z_1, ..., z_{n-1}, z_n^{(0)}, u \right) = c_n',$$

4) to solve equation (6.3.14) with respect to $z_1, ..., z_{n-1}, u$ to obtain the system:

$$z_1 = \omega_1\left(c_1', ..., c_n'\right);$$

$$\cdots\cdots\cdots\cdots\cdots\cdots\cdots$$

$$z_{n-1} = \omega_{n-1}\left(c_1', ..., c_n'\right);$$

$$u = \omega\left(c_1', ..., c_n'\right),$$

(6.3.15)

5) to construct the relationship:

$$\omega\left(c_1, ..., c_n\right) - \varphi\left[\omega_1\left(c_1, ..., c_n\right), ..., \omega_{n-1}\left(c_1, ..., c_n\right)\right] = 0, \qquad (6.3.16)$$

which gives the sought solution of the Cauchy problem in an implicit form,

6) to solve (6.3.16) with respect to u to find sought solution $u = u\left(z_1, ..., z_n\right)$ for initial conditions (6.3.13).

We will widely use below in this Chapter the described algorithms for constructing the calculational schemes by the method of averaged characteristics. But considering the important role of the characteristic method, let us give two simple examples for illustrating these algorithms.

The illustration example 1. The first scheme of constructing of general solution and the second algorithm of solving of the Cauchy problem. Let us find a solution of the equation

$$\left(1 + \sqrt{z - x - y}\right)\frac{\partial z}{\partial x} + \frac{\partial z}{\partial y} = 2. \qquad (6.3.17)$$

The initial condition is chosen in the form

$$z\left(x, y\right) = 2x \quad \text{if} \quad y = 0. \qquad (6.3.18)$$

Then, we follow the above-described scheme and algorithm:

1) the system for characteristics (6.3.3) is constructed:

$$\frac{dx}{1 + \sqrt{z - x - y}} = \frac{dy}{1} = \frac{dz}{2}; \qquad (6.3.19)$$

2) the system integrals (6.3.4) are found:

$$c_1 = z - 2y, \qquad c_2 = 2\sqrt{z - x - y} + y; \qquad (6.3.20)$$

3) substituting the initial conditions in integrals (6.3.20) we find:

$$c_1 = z, \qquad c_2 = 2\sqrt{z - x}; \qquad (6.3.21)$$

4) solving this system with respect to x, z we get:

$$z = c_1, \qquad x = c_1 - \frac{c_2}{4}; \qquad (6.3.22)$$

5) relation (6.3.16) is constructed

$$c_1 - 2\left(c_1 - \frac{c_2}{4}\right) = 0, \quad 2c_1 - c_2^2 = 0; \tag{6.3.23}$$

6) accomplishing the replacement of the found constants $c_{1,2}$ in (6.3.23) (6.3.21) we eventually have the solution of the problem in the form

$$2z - 4y - \left(2\sqrt{z - x - y} + y\right)^2 = 0. \tag{6.3.24}$$

The illustration example: . The second scheme of constructing general solution and the first algorithm of solving the Cauchy problem. We shall look for solution of the equation

$$x\frac{\partial u}{\partial x} + y\frac{\partial u}{\partial y} = 1 \tag{6.3.25}$$

with the initial condition

$$u(x, y) = x \quad \text{if} \quad y = 1. \tag{6.3.26}$$

Further we follow the scheme described above and algorithm:
1) the system for characteristic (6.3.6) is constructed:

$$\frac{dx}{dt} = x; \quad \frac{dy}{dt} = y; \quad \frac{du}{dt} = 1, \tag{6.3.27}$$

2) the system integrals are found:

$$c_1 = xe^{-t}; \quad c_2 = ye^{-t}; \quad c_3 = u - t, \tag{6.3.28}$$

3) substituting the initial condition (6.3.26) in (6.3.28) we obtain:

$$c_1' = xe^{-t}; \quad c_2' = e^{-t}; \quad c_3' = x - t, \tag{6.3.29}$$

4) it could be found after simple transformations of (6.3.29) that

$$c_3' - \frac{c_1'}{c_2'} - \ln\left(c_2'\right) = 0, \tag{6.3.30}$$

5) substituting the expression for constants $c_{1,2,3}$ (6.3.28) in (6.3.30) we get the problem solution:

$$u - t - \frac{xe^{-t}}{ye^{-t}} - \ln\left(ye^{-t}\right) = 0, \tag{6.3.31}$$

6) which can be rewritten in more convenient form:

$$u = \frac{x}{y} + \ln y. \tag{6.3.32}$$

3.2   Method of Characteristics. The Vector Case

Above we discussed the simplest version of the method of characteristics when the function sought u is represented with a scalar. Now we shall generalize these results for the case of a vector-functional variable u. Therefore we start with the studying of the one-dimension problem, which is described with a system with two independent variables z, t.

One-dimension problem. Let us write down the system of scalar equations that describes some two-dimension problem in the following form:

$$A_{11}(z,t)\frac{\partial u_1}{\partial t} + A_{12}(z,t)\frac{\partial u_2}{\partial t} + B_{11}(z,t)\frac{\partial u_1}{\partial z} + B_{12}(z,t)\frac{\partial u_2}{\partial z}$$
$$= f_1(z,t,u);$$

$$A_{21}(z,t)\frac{\partial u_1}{\partial t} + A_{22}(z,t)\frac{\partial u_2}{\partial t} + B_{21}(z,t)\frac{\partial u_1}{\partial z} + B_{22}(z,t)\frac{\partial u_2}{\partial z}$$
$$= f_2(z,t,u), \tag{6.3.33}$$

where A_{ij}, B_{ij} and f_i $(i,j = 1,2)$ are some given functions.

The system (6.3.33) could be rewritten also in the matrix form:

$$A\frac{\partial u}{\partial t} + B\frac{\partial u}{\partial z} = f, \tag{6.3.34}$$

where

$$A = \begin{vmatrix} A_{11} & A_{12} \\ A_{21} & A_{22} \end{vmatrix}, \; B = \begin{vmatrix} B_{11} & B_{12} \\ B_{21} & B_{22} \end{vmatrix}, \; u = \begin{pmatrix} u_1 \\ u_2 \end{pmatrix}, \; f = \begin{pmatrix} f_1 \\ f_2 \end{pmatrix} \tag{6.3.35}$$

are the corresponding vectors and vector-functions. It should be mentioned that the matrix form (6.3.34) is interesting also because it allows simple generalization the discussed scheme on the case of more then two (for instance, n) equations (6.3.33). In this case the matrices A and B have size $n \times n$, and the matrices u and f are determined as n-size column matrices.

Characteristics of the equations (6.3.33), (6.3.34) can be defined as lines (which are given by the differentials of shift dz, dt) along which the following condition is satisfied:

$$\det \begin{vmatrix} A & B \\ E \cdot dt & E \cdot dz \end{vmatrix} = 0, \tag{6.3.36}$$

where E, in the general case, is the unity matrix of size $n \times n$. Therein the vector du is not arbitrary. It should satisfy the relationship

$$\operatorname{rank} \begin{vmatrix} A & B & f \\ dt \cdot E & dz \cdot E & du \end{vmatrix} = \operatorname{rank} \begin{vmatrix} A & B \\ dt \cdot E & dz \cdot E \end{vmatrix}. \tag{6.3.37}$$

The condition (6.3.37) is referred to as the *relationship of characteristics*.

Canonical form. The particular case of equation (6.3.34)

$$\frac{\partial u}{dt} + C\left(z, t\right) \frac{\partial u}{\partial z} = g\left(z, t, u\right) \tag{6.3.38}$$

is called the *canonical form*. Here $C = A^{-1}B$, $g = A^{-1}f$. The characteristics are the lines that are described with the equations

$$\frac{dz}{dt} = k_i\left(z, t\right), \tag{6.3.39}$$

where the characteristic inclinations k_i are calculated as *characteristic roots* of the matrix C

$$\det\left(C - k \cdot E\right) = 0. \tag{6.3.40}$$

In the case we have to deal with multiple and real roots only, the considered system is related with the *parabolic type* of equations with partial derivatives. Analogously, the system could be classified as a *hyperbolic* one in the case when all roots are complex.

Now let us make one more remark. The systems of the type

$$A\left(z, t, u\right) \frac{\partial u}{\partial t} + B\left(z, t, u\right) \frac{\partial u}{\partial z} = f\left(z, t, u\right) \tag{6.3.41}$$

form the class of so called *quasi-hydrodynamic* (hydrodynamic) equations (see, for instance, equation (6.1.5)). Comparing (6.3.41) with (6.3.33), (6.3.34) we can conclude that matrices A and B in the quasi-hydrodynamic case are found to be dependent on the sought function u. Nevertheless, as well known [20–25], in spite of this difference such system can also be solved using the method of characteristics. Specific feature of this calculational scheme is that the characteristics depend on the solution considered. Or, in other words, the characteristics depend in this case on the solution that is chosen for analysis, i.e., the line that is a characteristic for certain solution is not the characteristic for another one.

Two-dimensional problem. Later we shall discuss the calculational situation when the number of variables is three (x, y and t, respectively). Let us write down the system of three equations with three unknown functions

$$A\left(x,y,t\right)\frac{\partial u}{\partial t} + B\left(x,y,t\right)\frac{\partial u}{\partial x} + C\left(x,y,t\right)\frac{\partial u}{\partial y} = f\left(x,y,t,u\right). \quad (6.3.42)$$

In this case the characteristics of the system (6.3.42) are the surfaces $S\left\{\varphi\left(x,y,t\right)=0\right\}$, for which the conditions

$$\det\left|\frac{\partial\varphi}{\partial t}A + \frac{\partial\varphi}{\partial x}B + \frac{\partial\varphi}{\partial y}C\right| = 0 \quad (6.3.43)$$

or

$$\det\left|\tau A + \xi B + \mu C\right| = 0, \quad (6.3.44)$$

are satisfied. Here (τ,ξ,μ) is the vector of the normal to the surface S. Solutions for the characteristics could be obtained if the solutions of equations (6.3.43) or (6.3.44) are substituted into the initial equation (6.3.42). However, it should be mentioned that, in the general case, calculational technology in this situation becomes somewhat complicated. It becomes even more complicated in the four- and greater dimensional cases.

The illustration example for the vector case. Let us consider the problem of propagation of the sound wave within some physical medium (see the corresponding qualitative picture in Section 3, Chapter 1). The equation system that describes this process can be written in the form

$$\begin{aligned} \frac{\partial u}{\partial t} + \frac{1}{\rho_0}\frac{\partial p}{\partial z} &= 0, \\ \frac{\partial p}{\partial t} + \rho_0 c_0^2\frac{\partial u}{\partial z} &= 0, \end{aligned} \quad (6.3.45)$$

where ρ_0 and c_0 are some constants. System (6.3.45) in the matrix form can be written as

$$\begin{pmatrix} 1 & 0 \\ 0 & 1 \end{pmatrix}\frac{\partial}{\partial t}\begin{pmatrix} u \\ p \end{pmatrix} + \begin{pmatrix} 0 & \frac{1}{\rho_0} \\ \rho_0 c_0^2 & 0 \end{pmatrix}\frac{\partial}{\partial z}\begin{pmatrix} u \\ p \end{pmatrix} = \begin{pmatrix} 0 \\ 0 \end{pmatrix}. \quad (6.3.46)$$

The equation for characteristics (6.3.36) is obtained in the form

$$\det \begin{pmatrix} 1 & 0 & 0 & \frac{1}{\rho_0} \\ 0 & 1 & \rho_0 c_0^2 & 0 \\ dt & 0 & dz & 0 \\ 0 & dt & 0 & dz \end{pmatrix} = dz^2 - c_0^2 dt^2 = 0. \qquad (6.3.47)$$

Thus we obtain the characteristics of the system (6.3.45) as lines that are described by the equation

$$\frac{dz}{dt} = \pm c_0. \qquad (6.3.48)$$

It can easily be seen that they are straight lines:

$$z \mp c_0 t = \text{const}. \qquad (6.3.49)$$

In what follows, we find the relationship for the characteristics like in (6.3.37). Analyzing the rank of the matrix in view of the relationship (6.3.37)

$$\begin{pmatrix} 1 & 0 & 0 & \frac{1}{\rho_0} & 0 \\ 0 & 1 & \rho_0 c_0^2 & 0 & 0 \\ dt & 0 & dz & 0 & du \\ 0 & dt & 0 & dz & dp \end{pmatrix} \qquad (6.3.50)$$

we eventually have the condition

$$\det \begin{pmatrix} 1 & 0 & 0 & 0 \\ 0 & 1 & \rho_0 c_0^2 & 0 \\ dt & 0 & dz & du \\ 0 & dt & 0 & dp \end{pmatrix} = dz dp + \rho_0 c_0^2 du dt = 0. \qquad (6.3.51)$$

Substituting the characteristics (6.3.48) into (6.3.51) one can obtain the following equations:

$$\begin{aligned} \frac{\partial u}{\partial t} + c_0 \frac{\partial u}{\partial z} &= 0; \\ \frac{\partial p}{\partial t} - c_0 \frac{\partial p}{\partial z} &= 0, \end{aligned} \qquad (6.3.52)$$

which could be easily integrated by the method described above:

$$\begin{aligned} u(z,t) &= \frac{f(z - c_0 t) + g(z + c_0 t)}{2}, \\ g(z,t) &= \rho_0 c_0 \frac{f(z - c_0 t) - g(z + c_0 t)}{2}, \end{aligned} \qquad (6.3.53)$$

where the arbitrary functions f and g are determined by the corresponding initial conditions.

4. EXAMPLE: APPLICATION OF THE METHOD OF AVERAGED CHARACTERISTICS FOR A SIMPLEST SYSTEM WITH OSCILLATATIVE RIGHT PARTS

As we can see, arrangement and calculational technology of the method of averaged characteristics are far from simple. A practice shows that their comprehension sometimes requires some additional training. Therefore, this Section and the following Chapter are dedicated to discussion of illustrative examples which allow better understanding of the main calculational peculiarities of the method.

The choice of the examples was done according to the following principle. The first example (that is set forth below in this Section) have purely abstract nature. Here a simplest system of two equations with oscillatative right parts is chosen as a basic illustration object. The goal of this example is to demonstrate the practical calculational technology of the method using simplest means. The second example [17] (which is set forth below in Chapter 7) has some other purpose. Namely, specific features of the use of considered method in a real electrodynamic problem are shown there. This includes as well the problem statement (i.e., reducing the initial electrodynamic equations to the standard form) and the solving of the obtained standard equations. The nonlinear theory of two-stream instability in two-velocity relativistic electron beam is chosen as a convenient illustration object for this.

4.1 Initial Equations

Let us write down the considered system of equations in the following form:

$$
\begin{aligned}
\frac{\partial u_1(z,t)}{\partial t} + \frac{1}{u_0}\frac{\partial u_2(z,t)}{\partial z} &= a_1 \cos(\omega t - kz); \\
\frac{\partial u_2(z,t)}{\partial t} + u_0 c^2 \frac{\partial u_1(z,t)}{\partial z} &= a_2 \cos(\omega t - kz),
\end{aligned}
\tag{6.4.1}
$$

where u_1 and u_2 are some unknown functions sought, u_0, c, a_1, a_2 are given parameters of the problem.

4.2   Characteristics

According to the general procedure of the method of averaged characteristics, we should construct on the first step the equations for characteristics of equation (6.4.1) (see Fig. 6.2.1). For this we find the determinant (6.3.36)

$$
\det \begin{pmatrix}
1 & 0 & 0 & \frac{1}{u_0} \\
0 & 1 & u_0 c^2 & 0 \\
dt & 0 & dz & 0 \\
0 & dt & 0 & dz
\end{pmatrix} = dz^2 - c^2 dt = 0. \qquad (6.4.2)
$$

Thus we find for the system characteristics:

$$
\frac{dz}{dt} = \pm c. \qquad (6.4.3)
$$

Using well known definition for momentum derivative

$$
\frac{du_i}{dt} = \frac{\partial u_i}{\partial t} + \frac{\partial u_i}{\partial z}\frac{dz}{dt}, \qquad (6.4.4)
$$

we can get (from equation (6.4.1)) the two following systems:

$$
\begin{aligned}
\frac{dz}{dt} &= c; \\
\frac{du_1}{dt} &= a_1 \cos(\omega t - kz),
\end{aligned} \qquad (6.4.5)
$$

and

$$
\begin{aligned}
\frac{dz}{dt} &= c; \\
\frac{du_2}{dt} &= a_2 \cos(\omega t - kz).
\end{aligned} \qquad (6.4.6)
$$

respectively.

4.3   Passage to the First Hierarchical Level

Below, regarding the system (6.4.5), (6.4.6) as equations for the zeroth hierarchical level, we perform the passage to the first (one-fold averaged) hierarchical level. For this we transform them into the standard form like (6.3.2). The first stage of this transformation procedure is introducing the new dependent variable (oscillation phase ψ)

$$
\psi = \omega t - kz. \qquad (6.4.7)
$$

The following stage is the subdividing of all variables into slow and fast ones. Taking into consideration the specific mathematical structure of systems (6.4.5), (6.4.6), we can assume that the variables u_j and z can be treated as the slow variables. (Here let us recall that slow variables are the ones whose magnitudes do not essentially change during the system period — see Chapter 1, Section 2). The phase ψ, correspondingly, plays the role of the fast variable (because it changes greatly during the system period). Then the large parameter of the system (hierarchical large parameter) can be determined by the standard method (see Chapters 3–5):

$$\xi \sim \left| \frac{d\psi}{dt} \right| \bigg/ \left| \frac{dx_i}{dt} \right| \gg 1, \tag{6.4.8}$$

where $x = \{x_i\} = \{x_1, x_2\}$, $x_1 \equiv u_j$; $x_2 \equiv z$, $j = 1,2$.

Differentiating the definition (6.4.7) by t, it is not difficult to obtain the differential equation for the phase ψ. As a result of the performed transformation we can reconstruct equations (6.4.5), (6.4.6) into the two parametrical standard forms like (6.3.2):

$$\frac{dz}{dt} = c;$$
$$\frac{du_1}{dt} = a_1 \cos(\psi);$$
$$\frac{d\psi}{dt} = \omega + kc,$$

and

$$\frac{dz}{dt} = c;$$
$$\frac{du_2}{dt} = a_2 \cos(\psi); \tag{6.4.9}$$
$$\frac{d\psi}{dt} = \omega + kc,$$

Below we will accomplish the transformation of systems (6.4.9) into the first hierarchical level. The formal solutions systems (6.4.9) can be represented in the form of Krylov–Bogolyubov substitutions:

$$u_1 = \bar{u}_1 + \sum_{n=1}^{\infty} \frac{1}{\xi^n} U_{u1}^{(n)} \left(\bar{z}, \bar{\psi} \right);$$

$$u_2 = \bar{u}_2 + \sum_{n=1}^{\infty} \frac{1}{\xi^n} U_{u2}^{(n)} \left(\bar{z}, \bar{\psi} \right);$$

$$z = \bar{z} + \sum_{n=1}^{\infty} \frac{1}{\xi^n} U_{z}^{(n)} \left(\bar{z}, \bar{\psi} \right); \tag{6.4.10}$$

$$\psi = \bar{\psi} + \sum_{n=1}^{\infty} \frac{1}{\xi^n} V^{(n)} \left(\bar{z}, \bar{\psi} \right),$$

where the unknown functions $U_{u,z}^{(n)}$ and $V^{(n)}$ can be found by the use of corresponding procedures of the Bogolyubov–Zubarev method described in Chapter 4. The corresponding equations for the first hierarchical level could be obtained analogously. Regarding this equations as characteristics of corresponding averaged equations in partial derivatives (i.e., equations for the first hierarchical level — see Fig. 6.2.1), we can construct them in the form

$$\left(\frac{\partial}{\partial t} + c \frac{\partial}{\partial \bar{z}} \right) \bar{u}_1 \left(\bar{z}, t \right) = 0;$$

$$\left(\frac{\partial}{\partial t} - c \frac{\partial}{\partial \bar{z}} \right) \bar{u}_2 \left(\bar{z}, t \right) = 0. \tag{6.4.11}$$

The solutions of truncated system (6.4.11) can easily be obtained:

$$\bar{u}_1 \left(\bar{z}, t \right) = \bar{u}_{01} \left(\bar{z} - ct \right);$$

$$\bar{u}_2 \left(\bar{z}, t \right) = \bar{u}_{02} \left(\bar{z} + ct \right), \tag{6.4.12}$$

where the initial (or boundary) conditions for the arbitrary functions $\bar{u}_{01}$, $\bar{u}_{02}$ (solutions!) coincide with the analogous conditions for equivalent functions for equations (6.4.1). According to the described in Section 2, the back transformations should be performed (see Fig. 6.2.1) for obtaining of the complete asymptotic solutions for the zeroth hierarchical level.

4.4 Back Transformations

It is not difficult to reconstruct the first two Krylov–Bogolyubov substitutions (6.4.9) into the form (first approximation on the small parameter $1/\xi$:

$$u_{1,2}\left(\bar{z}, \bar{\psi}, t\right) \cong \bar{u}_{1,2}(\bar{z}, t) + \frac{a_{1,2}}{\omega \pm kc} \sin\left(\bar{\psi}\right). \qquad (6.4.13)$$

Then we use the calculational scheme of back transformations described previously in Section 2 (see (6.2.19), (6.2.20) and corresponding commentaries). In particular, we note that difference between any variables of the first and zeroth hierarchical levels is proportional to $1/\xi_\kappa \ll 1$:

$$u_j - \bar{u}_j = \sum_{n=1}^{\infty} \frac{1}{\xi^n} u_{ui}^{(n)}\left(\bar{z}, \bar{u}_i, \bar{\psi}\right);$$

$$z - \bar{z} = \sum_{n=1}^{\infty} \frac{1}{\xi^n} u_z^{(n)}\left(\bar{z}, \bar{u}_i, \bar{\psi}\right); \qquad (6.4.14)$$

$$\psi - \bar{\psi} = \sum_{n=1}^{\infty} \frac{1}{\xi^n} v^{(n)}\left(\bar{z}, \bar{u}_i, \bar{\psi}\right).$$

Using this observation we can accomplish corresponding expansions in the Taylor series in (6.4.13). As a result, we obtain the transformation formulae like (6.2.20):

$$u_j\left(z, t\right) = u_j\left(\bar{z}, \bar{\psi}\right)\Big|_{\substack{\bar{z}=z \\ \bar{\psi}=\psi}}$$

$$+ \left\{ \frac{\partial u_j\left(\bar{z}, \bar{\psi}\right)}{\partial \bar{z}} \left(\sum_{n=1}^{\infty} \frac{1}{\xi^n} u_z^{(n)}\right) + \frac{\partial u_j\left(\bar{z}, \bar{\psi}\right)}{\partial \bar{\psi}} \left(\sum_{n=1}^{\infty} \frac{1}{\xi^n} v^{(n)}\right) \right\} \Bigg|_{\substack{\bar{z}=z \\ \bar{\psi}=\psi}}$$

$$+ \cdots \qquad (6.4.15)$$

and so on.

After not very difficult calculations we have the complete approximate solutions of the considered problem (here we restrict ourselves by keeping only the terms of order $1/\xi$ in (6.4.15):

$$u_1\left(z, t\right) = \bar{u}_{01}\left(z - ct\right) + \frac{a_1}{\omega + kc} \sin\left(\omega t - kz\right); \qquad (6.4.16)$$

$$u_2\left(z, t\right) = \bar{u}_{02}\left(z + ct\right) + \frac{a_2}{\omega - kc} \sin\left(\omega t - kz\right). \qquad (6.4.17)$$

Thus the initial problem (6.4.1) is solved analytically by using the method of averaged characteristics.

5.   HIERARCHICAL METHOD OF AVERAGED QUASI-HYDRODYNAMIC EQUATION

As has been illustrated above, one of the key points of the method of averaged characteristics is the transformation of an equation with partial derivatives into some equivalent system of exact differential equations (see Fig. 6.2.1 and corresponding commentaries). However, it should be mentioned that such a type of straight transformation technology is not the only one. Moreover, historically the first time the analogous, in principle, idea (that is, however, based on the use of definition for momentum derivative (3.2.3), (6.4.4)) was when it was proposed for nonlinear wave resonant electrodynamic problems [14, 16, 26, 27]. This hierarchical approach is found to be suitable for solving motion problems mostly. The Boltzmann kinetic equation (3.2.4), (6.1.1) has been used there as a basis (see Chapter 3, Subsection 2.2). That is why the developed method is called the *method of averaged kinetic* equation (see Section 7 below in this Chapter).

Unfortunately, the method of averaged kinetic equation turns out to be inconvenient in some calculational situations. First, this takes place quite often in the case of self-consistent problems. The matter is that the field part of this problem is solved in three-dimensional space (because of the Maxwell equations), whereas Boltzmann's equation (3.2.4), (6.1.1) (the motion part of problem) is written for six-dimensional space. The connection relationships (6.1.7)–(6.1.9) are used for the spatial coordination of the field and motion problems. As a result the general calculational scheme sometimes looks somewhat complicated.

Second, the taking into account of kinetic properties of the considered plasma-like system quite often turns out not necessary (for instance, in the theory high-current electron and ion beams in accelerators or relativistic electron devices [1, 2, 28–30] — see Volume II). So the kinetic approach, which is put in the basis of the method of averaging kinetic equation, possesses superfluous possibilities in such situations. But the cost of this superfluous generalization is the unjustified complication of calculational procedures.

Some other version of the same idea had been proposed in [16] and developed in [15]. The quasi-hydrodynamic equation (6.1.5), (6.3.41) is used here as an initial basis. Because of this the calculational procedure proposed is called the *method of averaged quasi-hydrodynamic equation.* Inasmuch as the quasi-hydrodynamic equation is three-dimensional, the above mentioned problem concerned the spatial coordination does not arise there. Besides that, the kinetic plasma processes are not considered directly in the framework of the quasi-hydrodynamic description (the semi-qualitative estimation of influence of the kinetic effects was done there only). Due to these peculiarities, the method of averaged quasi-hydrodynamic equation in such cases provides essentially less labour consuming nature of corresponding calculational procedures.

In what follows, let us note the following. In spite of their external differences, the method of averaged characteristics, as well as the methods of averaged kinetic and quasi-hydrodynamic equations, actually have very close general calculational ideology. The method of averaged characteristics looks more general than the other two. However, this advantage can be really attained in the case only when number of equations in the considered system does not exceed three. At the same time, the methods of averaged kinetic and quasi-hydrodynamic equations have no such kind limitations. Therefore, all these method have the 'right to live' in practice as well as optimal areas of their application.

Considering the above mentioned resemblance of the general calculational ideologies, we classify the methods of averaged kinetic and quasi-hydrodynamic equations as some special modification of the method of averaged characteristics.

5.1 Averaged Quasi-Hydrodynamic Equation

Let us choose quasi-hydrodynamic equation (6.1.5) as a basic equation for description of an electron beam motion. So, as in the case of the method of averaged characteristics, the main task is the representing of the initial problem (6.1.5) in the hierarchical standard form like (4.1.5). Comparing (4.1.5) and (6.1.5), we see clearly that, as before, formally the problem actually consists in reducing the equation in partial derivatives (6.1.5) to system of exact differential equations (4.1.5).

In contrast to the method of averaged characteristics, this time we solve the straight and back transformation problems by means of the passage from the Euler description to the Lagrange one and then in the back direction. Here, the definition for momentum derivative (3.2.3), (6.4.4) is put in the basis of these transformations. Below we express $\vec{u}_\alpha(\vec{r}, t)$ from (6.1.5) in terms of Lagrange rotating phases ψ_κ. It is noteworthy that the transformation of variables drives us into the space whose *each point oscillates with all the frequencies of particle beam oscillations*. It is obvious that this oscillation coordinate system is not inertial because it moves with an acceleration.

Thus at the first stage of calculational we replace time coordinate t with particle oscillation phases p_μ that are considered as parameters. Then we pass to the momentum derivative using (3.2.3), (6.4.4). This way we can replace quasi-hydrodynamic equation (6.1.5) with the new set of exact differential equations:

$$\frac{d\vec{u}_\alpha\left(\vec{r}, p_\mu\right)}{dt} = \vec{U}_\alpha\left(\vec{u}_\alpha, \vec{r}, p_\mu\right);$$

$$\frac{d\vec{r}}{dt} = \vec{u}_\alpha\left(\vec{r}, p_\mu\right); \qquad (6.5.1)$$

$$\frac{dp_\mu}{dt} = \chi_\mu\left(\vec{u}_\alpha, \vec{r}, p_\mu\right);$$

where sense of the functions $\vec{U}_\alpha$ and χ_μ is obvious. For instance, $\vec{U}_\alpha$ can be determined as a sum of electromagnetic forces acting on the particle beam (including electric and magnetic Lorentz forces (see (4.1.4)), and the forces of electromagnetic 'friction' and 'viscosity', respectively):

$$\vec{U}_\alpha = \frac{q_\alpha}{m_\alpha \gamma_\alpha} \left\{ \vec{E} + \frac{1}{c}\left[\vec{u}_\alpha \vec{B}\right] - \frac{\vec{u}_\alpha}{c^2}\left(\vec{u}_\alpha \vec{E}\right) \right\}$$
$$- \nu \frac{\vec{u}_\alpha}{\gamma_\alpha^2} - \frac{v_T}{n_\alpha \gamma_\alpha}\left[\frac{\partial n_\alpha}{\partial \vec{r}} - \frac{\vec{u}_\alpha}{c^2}\left(\vec{u}_\alpha \frac{\partial}{\partial \vec{r}}\right)n_\alpha\right]. \quad (6.5.2)$$

In the case that some or all phases of p_μ are wave, use of (4.1.2) yields

$$\chi_q\left(\vec{u}_\alpha, \vec{r}\right) = \omega_q - k_q \vec{u}_\alpha, \qquad (6.5.3)$$

For hidden phases we have somewhat more complex definitions (see Chapter 4, Subsection 1.3), and so on.

The quantities $\vec{E}$, $\vec{B}$, n_α are functions of $\vec{r}$ and t obeying Maxwell's equations (6.1.6) and the particle/plasma continuity equation (6.1.10).

Now we classify the variables as slow and fast ones. In similar way with 'purely' single-particle case (see, for instance, the example considered in Chapter 5) we take beam velocity $\vec{u}_\alpha$ and position vector $\vec{r}$ as slow variables. Moreover, the combination phases $\theta_{\nu g}$ (see (4.1.9)) are slowly varying quantities (this allows to choose the type of resonance in the system — see Chapter 1, Subsection 2.3 and Chapter 4, Subsection 1.3), too. Analogously, amongst other (non-resonant) phases of the set $\{p_\mu\}$ not included in combinations (4.1.9), we also find slowly varying variables. The rest of the phases of such type and the combination phases $\psi_{\nu g}$ (see (4.1.9)) are the fast ones.

Below we use the calculational hierarchical scheme described in Chapter 4. After corresponding calculations the set (6.5.1) can be reduced to the form coinciding (with respect to its general mathematics structure) with standard hierarchical form (4.1.5). In this case, dynamical operators (2.2.7) for the first hierarchical level can be written as the Krylov–Bogolyubov substitutions (see (6.2.6)):

$$\vec{u}_\alpha = \bar{\vec{u}}_\alpha + \sum_{n=1}^{\infty} \frac{1}{\xi_1^n} \vec{u}_u^{(n)} \left(\bar{x}, \bar{\psi}_1\right) ; \qquad (6.5.4)$$

$$\vec{r} = \bar{\vec{r}} + \sum_{n=1}^{\infty} \frac{1}{\xi_1^n} \vec{u}_r^{(n)} \left(\bar{x}, \bar{\psi}_1\right) , \qquad (6.5.5)$$

and so on.

For equations of the first hierarchical level we have (see (6.2.7)):

$$\frac{d\bar{\vec{u}}_\alpha}{dt} = \sum_{n=1}^{\infty} \frac{1}{\xi_1^n} \vec{A}_n^{(n)} (x) ; \qquad (6.5.6)$$

$$\frac{d\bar{\vec{r}}}{dt} = \sum_{n=1}^{\infty} \frac{1}{\xi_1^n} \vec{A}_r^{(n)} (x) , \qquad (6.5.7)$$

and so on, where $\bar{\vec{u}}_\alpha = \bar{\vec{u}}_\alpha (\bar{r}, t)$, functions $\vec{u}_n^{(n)}$, $\vec{u}_r^{(n)}$, $\vec{u}_n^{(n)}$, $\vec{A}_r^{(n)}$, and other similar functions are calculated using written above formulae of averaging method (see Chapter 4).

Furthermore, with (3.2.3) we come back from the momentum derivative (with respect to time t) to a partial one (i.e., with respect to t and Euler's vector-coordinates $\vec{r}$) taking into account the equation (6.5.7). As a result we obtain the new equation:

$$\frac{\partial \bar{\vec{u}}_\alpha}{\partial t} + \sum_{n=1}^{\infty} \frac{1}{\xi_1^n} \vec{A}_r^{(n)} (\bar{x}) \frac{\partial \bar{\vec{u}}_\alpha}{\partial \bar{r}} = \sum_{n=1}^{\infty} \frac{1}{\xi_1^n} \vec{A}_U^{(n)} (\bar{x}) . \qquad (6.5.8)$$

Comparing (6.5.8) and initial one (6.1.5) we find that both these equations have analogous mathematical structure, i.e. *hierarchical resemblance principle holds* in the problem wave considered. Because of special form of the latter we call equation (6.5.8) the *averaged quasi-hydrodynamic equation* of the first hierarchical level. Besides that, the second hierarchical principle is also satisfied (hierarchical *compression principle*). This means that the original hierarchical system on each hierarchical level is reproduced in its general dynamical structure, but essentially simpler. The other two hierarchical principles are satisfied here, too.

In the simplest case the system contains only two hierarchical levels — the zeroth and the first ones. So, in order to obtain complete solution of initial non-averaged equation (6.1.5) we find solutions of truncated equation (6.5.8) and substitute them into transformation formulae (6.5.4), (6.5.5).

5.2   Back Transformations

Let us recall once more (see (6.2.19), (6.2.20)) that the main idea of the backward transformation is based on the fact that difference between of solutions of two neighboring levels is (see also formulae (6.2.18) — (6.2.20) and corresponding commentaries)

$$\vec{u}_\alpha - \bar{\vec{u}}_\alpha = \sum_{n=1}^{\infty} \frac{1}{\xi_1^n} \vec{u}_U^{(n)}; \quad \vec{r} - \bar{\vec{r}} = \sum_{n=1}^{\infty} \frac{1}{\xi_1^n} \vec{u}_r^{(n)};$$

$$\psi_1 - \bar{\psi}_1 = \sum_{n=1}^{\infty} \frac{1}{\xi_1^n} \vec{u}_\psi^{(n)}, \tag{6.5.9}$$

and so on, i.e., inversely proportional to the power of the scale parameter $1/\xi_1^n$. Consequently, we can obtain approximate solutions of required accuracy using corresponding expansions:

$$\vec{u}_\alpha\left(\vec{r}, t\right) = \bar{\vec{u}}_\alpha\left(\bar{\vec{r}}, \bar{\theta}\left(\bar{\vec{r}}, t\right), \bar{\psi}_1\left(\bar{\vec{r}}, t\right)\right)\Bigg|_{\bar{\vec{r}}=\vec{r}} + \left[\frac{\partial \bar{\vec{u}}_\alpha}{\partial \bar{\vec{r}}} \sum_{n=1}^{\infty} \frac{1}{\xi_1^n} \vec{u}_r^{(n)}\right]\Bigg|_{\bar{\vec{r}}=\vec{r}}$$

$$+ \frac{1}{2!}\left[\frac{\partial^2 \bar{\vec{u}}}{\partial \bar{\vec{r}}^2}\left(\sum_{n=1}^{\infty} \frac{1}{\xi_1^n} \vec{u}_r^{(n)}\right)^2\right]\Bigg|_{\bar{\vec{r}}=\vec{r}} + \cdots \tag{6.5.10}$$

Here, in eventual results, they can be expressed in terms of the original phases using expressions (4.1.9) and (6.5.9). If more than one hierarchical level exists in our system, asymptotic integration procedure is used according to (6.5.8) as a new base for calculational. Or, in other words, we again reduce (6.5.8) to (6.5.1), separate the next scale parameter ξ_2 (in a hierarchical series (4.1.6)) and the next vector of fast phases ψ_2, and so on, until we arrive at the last hierarchical level m. Then we solve corresponding *m-fold* averaged quasi-hydrodynamic equation and perform back transformations using the above scheme.

6.   THE METHOD OF AVERAGED CURRENT–DENSITY EQUATION

Below, let us discuss hierarchical method for calculational of nonlinear current density in plasma-like wave systems. We call the discussed calculational method the *method of averaged current–density equation*.

6.1   Averaged Current–Density Equation

We use relativistic quasi-hydrodynamic equation (6.1.5) and relations (6.1.11) as initial set of equations. After trivial transformation, we obtain the working form of current–density equation for the zeroth hierarchical level:

$$
\frac{\partial \vec{j}}{\partial t} + \frac{\vec{j}}{\rho}\frac{\partial \vec{j}}{\partial \vec{r}} = -\nu \frac{\vec{j}}{\gamma^2 \left(\vec{j}\right)} + \frac{q}{m\gamma\left(\vec{j}\right)}\left\{\rho\vec{E} + \frac{1}{c}\left[\vec{j}\vec{B}\right] - \frac{1}{c^2\rho^2}\vec{j}\left(\vec{j}\vec{E}\right)\right\}
$$

$$
+ \left(\frac{\vec{j}^2}{\rho^2} - \frac{v_T^2}{\gamma\left(\vec{j}\right)}\right)\frac{\partial \rho}{\partial \vec{r}} + \frac{v_T^2}{c^2\rho^2\gamma\left(\vec{j}\right)}\vec{j}\left(\vec{j}\frac{\partial \rho}{\partial \vec{r}}\right) + \frac{\vec{j}}{\rho}\frac{\partial \rho}{\partial t}, \quad (6.6.1)
$$

where subscript α is omitted for simplicity, the quantities $\rho, \vec{E}, \vec{B}$ are considered as predetermined functions of time t and coordinates $\vec{r}$; the relativistic factor γ in (6.6.1) is represented in the form

$$
\gamma\left(\vec{j}\right) = \left(1 - \frac{\vec{j}^2}{\rho^2}\right)^{-1/2}. \quad (6.6.2)
$$

The other parameters were defined earlier already. Then, using similar (to the case of quasi-hydrodynamic equation method) calculational scheme we write down the dynamical operators and corresponding equations for the first level of hierarchy:

$$
\vec{j}\left(\vec{r}, t\right) = \bar{\vec{j}}\left(\vec{r}, \bar{\theta}_1\right) + \sum_{n=1}^{\infty}\frac{1}{\xi_1^n}\vec{u}_{j1}^{(n)}\left(\bar{\vec{j}}, \bar{\vec{r}}, \bar{\theta}_1, \bar{\psi}_1\right) = \bar{\vec{j}}\left(\vec{r}, \bar{\theta}_1\right) + \tilde{\vec{j}}\left(\bar{\vec{j}}, \bar{\vec{r}}, \bar{\theta}_1, \bar{\psi}_1\right);
$$

$$
(6.6.3)
$$

$$
\frac{d\bar{\vec{j}}}{dt} = \sum_{n=1}^{\infty}\frac{1}{\xi_1^n}\vec{A}_{j1}^{(n)}\left(\bar{\vec{j}}, \bar{\vec{r}}, \bar{\theta}_1\right); \quad (6.6.4)
$$

$$
\frac{d\bar{\vec{r}}}{dt} = \sum_{n=1}^{\infty}\frac{1}{\xi_1^n}\vec{A}_r^{(n)}\left(\bar{\vec{j}}, \bar{\vec{r}}, \bar{\theta}_1\right); \quad (6.6.5)
$$

$$
\frac{d\bar{\theta}_1}{dt} = \sum_{n=1}^{\infty}\frac{1}{\xi_1^n}\vec{A}_\theta^{(n)}\left(\bar{\vec{j}}, \bar{\vec{r}}, \bar{\theta}_1\right); \quad (6.6.6)
$$

$$
\frac{d\bar{\psi}_1}{dt} = \omega_1\left(\bar{\vec{j}}, \bar{\vec{r}}, \bar{\theta}_1\right)\xi_1 + \sum_{n=1}^{\infty}B_1^{(n)}\left(\bar{\vec{j}}, \bar{\vec{r}}, \bar{\theta}_1\right) \quad (6.6.7)
$$

where $\bar{\theta}$ and $\bar{\psi}$ are the slowly and fast varying averaged vector-phases, which include as components the ordinary phases as well as combination ones (4.1.9); $A_x^{(n)}$, $B_x^{(n)}$, ω_x, $u_x^{(n)}$ are functions, whose calculational procedures are given in Chapter 4.

Further we construct the hierarchical series of large parameters. In the case we restrict ourselves with taking into consideration the first hierarchical level only (characterized by the large parameter $\xi\ 1¿¿1$), we can transform (6.6.5) into the Euler form (i.e., the form with partial derivatives):

$$\frac{\partial \vec{\bar{j}}}{\partial t} + \sum_{n=1}^{\infty} \frac{1}{\xi_1^n} \vec{A}_{r1}^{(n)} \left(\vec{\bar{j}}, \vec{r}, \bar{\theta}_1 \right) \frac{\partial \vec{\bar{j}}}{\partial \vec{r}} = \sum_{n=1}^{\infty} \frac{1}{\xi_1^n} \vec{A}_{j1}^{(n)} \left(\vec{\bar{j}}, \vec{r}, \bar{\theta}_1 \right). \qquad (6.6.8)$$

Here all phases $\bar{\theta}_1 = \bar{\theta}_1 \left(\vec{r}, t \right)$, $\bar{\psi}_1 = \bar{\psi}_1 \left(\vec{r}, t \right)$ are considered as parameters. The equation (6.6.8) is called the *averaged current–density equation*.

In what follows we solve (6.6.8) considering the averaged current $\vec{\bar{j}}$ as function of $\vec{r}$, $\bar{\theta}_1$, and $\bar{\psi}_1$. It is obvious that equation (6.6.8) is essentially simpler than initial equation (6.6.1) because it does not comprise dependence on fast phases ψ_1 or $\bar{\psi}_1$. During the next step of the discussed calculational scheme we perform backward transformation from the $\vec{r}$-averaged space back to $\vec{r}$-non-averaged space. There we keep in mind that functions $\bar{\theta}_1 \left(\vec{r} \right)$ and $\bar{\psi}_1 \left(\vec{r} \right)$ can be found according to the scheme (6.5.9), (6.5.10).

In opposite case when the number of hierarchical levels that we take into account exceeds two, we must turn to use multi-level version of hierarchical scheme. Namely, we take equation (6.6.5) as a new basis for calculations and separate out from the slowly varying vector phases $\bar{\theta}_1$ the fast phase of the second hierarchy $\bar{\psi}_2$. The latter is characterized by the next hierarchical scale parameter $\xi_2 \gg 1$ in hierarchical series (4.1.6). Then equation (6.6.8) is reduced to standard form for the second hierarchical level. Accomplishing an analogous transformation we obtain the Krylov–Bogolyubov substitutions connecting the first and the second hierarchical levels:

$$\vec{\bar{j}} = \vec{\bar{\bar{j}}} \left(\vec{r}, \bar{\bar{\theta}}_2 \right) + \sum_{n=1}^{\infty} \frac{1}{\xi_2^n} \vec{u}_{j2}^{(n)} \left(\vec{\bar{\bar{j}}}, \vec{r}, \bar{\bar{\theta}}_2, \bar{\psi}_2 \right) = \vec{\bar{\bar{j}}} + \vec{\tilde{j}}. \qquad (6.6.9)$$

The two-fold averaged values are calculated by using the equation for the second hierarchical level:

$$\frac{d\bar{\bar{\vec{j}}}}{dt} = \sum_{n=1}^{\infty} \frac{1}{\xi_2^n} \vec{A}_{j2}^{(n)} \left(\bar{\bar{\vec{j}}}, \bar{\bar{\vec{r}}}, \bar{\theta}_2 \right); \qquad (6.6.10)$$

$$\frac{d\bar{\bar{\vec{r}}}}{dt} = \sum_{n=1}^{\infty} \frac{1}{\xi_2^n} \vec{A}_{r2}^{(n)} \left(\bar{\bar{\vec{j}}}, \bar{\bar{\vec{r}}}, \bar{\theta}_2 \right); \qquad (6.6.11)$$

$$\frac{d\bar{\bar{\theta}}}{dt} = \sum_{n=1}^{\infty} \frac{1}{\xi_2^n} \vec{A}_{\theta 2}^{(n)} \left(\bar{\bar{\vec{j}}}, \bar{\bar{\vec{r}}}, \bar{\theta}_2 \right); \qquad (6.6.12)$$

$$\dots\dots\dots\dots\dots\dots\dots\dots\dots\dots\dots\dots\dots\dots\dots\dots,$$

and so on.

Here $\bar{\theta}_2$ is the averaged part of the vector $\bar{\theta}_1$ from which new fast phase $\bar{\psi}_2$ was separated out. Considering the above reasoning, the meaning of other functions is evident.

Following successive calculational scheme eventually we can reach the ultimate (i.e. *m-fold*) hierarchical level. It is important that *m-fold* averaged current density function does not depend on any oscillating phases:

$$\vec{j}^{(m-1)} = \vec{j}^{(m)} + \sum_{n=1}^{\infty} \frac{1}{\xi_m^n} \vec{u}_{jm}^{(n)} \left(\vec{j}^{(m)}, \vec{r}^{(m)}, \bar{\psi}_m \right); \qquad (6.6.13)$$

$$\frac{\partial \vec{j}^{(m)}}{\partial t} + \sum_{n=1}^{\infty} \frac{1}{\xi_m^n} \vec{A}_{rm}^{(n)} \left(\vec{j}^{(m)}, \vec{r}^{(m)} \right) \frac{\partial \vec{j}^{(m)}}{\partial \vec{r}^{(m)}} = \sum_{n=1}^{\infty} \frac{1}{\xi_m^n} \vec{A}_{jm}^{(n)} \left(\vec{j}^{(m)}, \vec{r}^{(m)} \right);$$

$$(6.6.14)$$

$$\dots\dots\dots\dots\dots\dots\dots\dots\dots\dots\dots\dots\dots\dots\dots\dots,$$

and so forth.

Here, as before, subscript and superscript m denotes the order of multiplication of performed averaged procedure.

The obvious advantage of equation (6.6.14) lies in the fact that the latter is simpler mathematically than similar equations of lower hierarchical levels, including the lowest (zero) level (6.6.1). We employ some calculational method (numerical or analytical) to find solutions of m-averaged (m-fold averaded) current density equation (6.6.14). Then performing back transformations we obtain non-averaged function $\vec{j}(\vec{r}, t)$, i.e., the solution determining the dynamics of the zeroth hierarchical level equation.

6.2   Back Transformations

The equation (6.6.14) can be solved by means of the successive approximation method or the characteristics method. Then, we substitute the solutions, which are obtained for $\vec{j}^{(m)}$, into Krylov–Bogolyubov substitution (6.6.13). This allows to construct corresponding solution for $\vec{j}^{(m-1)}$. In turn, we substitute the latter into similar expression for $\vec{j}^{(m-2)}$, and so on. In the final stage of successive substitution procedure, we have an analytical solution for non-averaged current density (6.6.3) depending on corresponding averaged values of quantities involved. We take into account algorithms analogous to (6.5.9) and (6.5.10). For this, we rewrite (6.6.3) as

$$\vec{j} - \bar{\vec{j}} = \sum_{n=1}^{\infty} \frac{1}{\xi_1^n} \vec{u}_{j1}^{(n)} \left[\vec{r}, \bar{\vec{j}} \left(\vec{r}, t \right), \bar{\theta}_1 \left(\vec{r}, t \right), \bar{\psi}_1 \left(\vec{r}, t \right) \right]; \qquad (6.6.15)$$

from which the difference

$$\left| \vec{j} - \bar{\vec{j}} \right| \approx \frac{1}{\xi_1} << 1. \qquad (6.6.16)$$

Consequently, we can write down the solution for non-averaged current density $\vec{j}\left(\vec{r}, t \right)$ as the expansion:

$$\vec{j} \left(\vec{r}, t \right) \cong \bar{\vec{j}} \left(\vec{r}, \bar{\theta}_1 \right) \Big|_{\bar{\vec{r}} = \vec{r}} + \left[\frac{\partial \bar{\vec{j}} \left(\vec{r}, \bar{\theta}_1, t \right)}{\partial \bar{\vec{r}}} \sum_{n=1}^{\infty} \frac{1}{\xi_1^n} \vec{u}_r^{(n)} \left(\vec{r}, \bar{\theta}_1, \bar{\psi}_1 \right) \right] \Big|_{\bar{\vec{r}} = \vec{r}} + \dots$$
$$(6.6.17)$$

where $\bar{\theta}_1 = \bar{\theta}_1 \left(\bar{\vec{r}} = \vec{r} \right)$ and $\bar{\psi}_1 = \bar{\psi}_1 \left(\bar{\vec{r}} = \vec{r} \right)$ are initial values of these functions (see definitions (4.1.9) and corresponding comments).

7.   HIERARCHICAL METHOD OF THE AVERAGED KINETIC EQUATION

7.1   Averaged Kinetic Equation

Similar calculational scheme can be constructed for the Boltzmann kinetic equation (6.1.1) chosen as an initial equation. Let us discuss main peculiarities and advantages of such calculational procedure.

First of all we note that characteristics of equations (6.1.1) coincide with the single-particle Hamiltonian equations (4.1.3). So the method of averaged kinetic equation, as well as the method of quasi-hydrodynamic equation, can actually be considered as specific case of the method of averaged characteristics.

By virtue of this or using relationship (3.2.3), we can reduce the general calculational procedure to the form analogous to presented above in the method of quasi-hydrodynamic equation. For this we express the functions $\partial \mathcal{H}/\partial \vec{\mathcal{P}}$ and $-\partial \mathcal{H}/\partial \vec{r}$ (in the left-hand side of (6.1.1)) in terms of Lagrange rotating electron phases used further as parameters. Then using Liouville's theorem and passing to momentum derivatives we replace the kinetic equation (6.1.1) with the new set

$$\frac{d}{dt} f\left(\vec{r}, \vec{\mathcal{P}}, p_\mu\right) = \zeta^{-1} I_{st}; \tag{6.7.1}$$

$$\frac{dp_\mu}{dt} = \chi_\mu\left(\vec{r}, \vec{\mathcal{P}}, p_\mu\right); \tag{6.7.2}$$

$$\frac{d\vec{r}}{dt} = -\frac{\partial}{\partial \vec{\mathcal{P}}} \mathcal{H}\left(\vec{r}, \vec{\mathcal{P}}, p_\mu\right); \tag{6.7.3}$$

$$\frac{d\vec{\mathcal{P}}}{dt} = \frac{\partial}{\partial \vec{r}} \mathcal{H}\left(\vec{r}, \vec{\mathcal{P}}, p_\mu\right). \tag{6.7.4}$$

The set (6.7.1)–(6.7.4) is similar to the set of quasi-hydrodynamic equations (6.7.6), (6.7.7), with respect to general mathematical structure. Therein we employ further the same calculational scheme. That is, vectors of slow and fast variables can be formed, the largest scale parameter ξ_κ (see the definitions (4.1.7), (4.1.10)) is separated out, and so on. We obtain functional operator $\hat{U}_1$ of the first hierarchical order (formal solutions of the problem) in 'usual' form of the Krylov–Bogolyubov substitutions:

$$f = \bar{f} + \sum_{n=1}^{\infty} \frac{1}{\xi_1^n} u_f^{(n)}\left(\bar{x}, \bar{\psi}_1\right);$$

$$\vec{r} = \vec{\bar{r}} + \sum_{n=1}^{\infty} \frac{1}{\xi_1^n} \vec{u}_r^{(n)}\left(\bar{x}, \bar{\psi}_1\right); \tag{6.7.5}$$

$$\vec{\mathcal{P}} = \vec{\bar{\mathcal{P}}} + \sum_{n=1}^{\infty} \frac{1}{\xi_1^n} \vec{u}_{\mathcal{P}}^{(n)}\left(\bar{x}, \bar{\psi}_1\right),$$

and so forth, where corresponding averaged values are found from corresponding reduced (truncated) equations for the first hierarchical level:

$$\frac{d\bar{f}}{dt} = \sum_{n=1}^{\infty} \frac{1}{\xi_1^n} A_f^{(n)}(\bar{x}); \tag{6.7.6}$$

$$\frac{d\vec{\bar{r}}}{dt} = \sum_{n=1}^{\infty} \frac{1}{\xi_1^n} \vec{A}_r^{(n)}(\bar{x}); \tag{6.7.7}$$

$$\frac{d\bar{\vec{P}}}{dt} = \sum_{n=1}^{\infty} \frac{1}{\xi_1^n} \vec{A}_{\mathcal{P}}^{(n)}(\bar{x});$$

(6.7.8)

Using (6.7.6) together with (6.7.7) and (6.7.8), we obtain *averaged kinetic equation* of the first hierarchy:

$$\left\{ \frac{\partial}{\partial t} + \left[\sum_{n=1}^{\infty} \frac{1}{\xi_1^n} \vec{A}_r^{(n)} \right] \frac{\partial}{\partial \bar{\vec{r}}} + \left[\sum_{n=1}^{\infty} \frac{1}{\xi_1^n} \vec{A}_{\mathcal{P}}^{(n)} \right] \frac{\partial}{\partial \bar{\vec{P}}} \right\} \bar{f} \left(t, \bar{\vec{r}}, \bar{\vec{P}} \right) = \bar{I}_{st};$$

(6.7.9)

where *new the collision integral* is

$$\bar{I}_{st} = \sum_{n=1}^{\infty} \frac{1}{\xi_1^n} A_f^{(n)}.$$

(6.7.10)

The corresponding relations for the second hierarchical level are found in similar way. The next large parameter ξ_2 of hierarchical series (4.1.6) serves as required scale basis. In a similar way we deal with sequence of levels until hierarchical series (4.1.6) is exhausted. The corresponding back transformations are accomplished according to the scheme described above (see (6.7.9), (6.7.10), and other).

It is interesting to compare both the latter discussed hierarchical approaches (quasi-hydrodynamic and kinetic) in the beam motion problem. One can be sure that essential point is in the choice of specific basis metrics of the problem space. In the quasi-hydrodynamic case only three-dimensional space coordinates are involved. In the kinetic case six-dimensional coordinate space is required. Therein the kinetic approach seems more complicated for use. On the other hand, this approach has a wider area of application, inasmuch as it is more general. Their common characteristic is that both cases contain some *m-fold-averaged* set of equations much simpler mathematically than the initial ones on final stage of calculations.

References

[1] A.A. Ruhadze, L.S. Bogdankevich, S.E. Rosinkii, V.G. Ruhlin. *Physics of high-current relativistic beams.* Atomizdat, Moscow, 1980.

[2] R.C. Davidson. *Theory of nonlinear plasmas.* Mass: Benjamin, Reading, 1974.

[3] A.G. Sitenko and V.M. Malnev. *Principles of the plasma theory.* Naukova Dumka, Kiev, 1994.

[4] L.A. Vainstein, V.A. Solnzev. *Lectures on Microwave electronics*. Sov. Radio, Moscow, 1973.

[5] V.I. Gaiduk, K.I. Palatov, D.M. Petrov. *Principles of microwave physical electronics*. Sov. Radio, Moscow, 1971.

[6] A.F. Alexandrov, L.S. Bogdankevich, A.A. Ruhadze. *Principles of plasma electrodynamics*. Vyschja Shkola, Moscow, 1978.

[7] A.N. Kondratenko, V.M. Kuklin. *Principles of plasma electronics*. Energoatomizdat, Moscow, 1988.

[8] B.E. Zshelezovskii. *Electron beam parametric microwave amplifiers*. Nauka, Moscow, 1971.

[9] I.V. Savel'jev. *Principles of theoretical physics*. Nauka, Moscow, 1977.

[10] E.M. Liftshitz, L.P. Pitayevskiy. *Physical kinetics*. Nauka, Moscow, 1979.

[11] I.V. Dzedolik, V.V. Kulish. To the nonlinear theory of parametrical resonance of electromagnetic waves in plasmas of a high-current relativistic electron flux. *Ukrainian Physical Journal*, 32(11):1672–1677, 1987.

[12] J. Weiland, H. Wilhelmsson. *Coherent nonlinear interactions of waves in plasmas*. Pergamon Press, Oxford, 1977.

[13] V.V. Kulish, V.I. Savchenko. Method of averaged characteristics in nonlinear electrodynamic problems. *Gerald of Sumy State University, ser. Physics and Mathematics*, 2:5–12, 2002.

[14] V.V. Kulish, A.V. Lysenko. Method of averaged kinetic equation averaged kinetic equation and its use in the nonlinear problems of plasma electrodynamics. *Fizika Plasmy (sov. Plasma Physics)*, 19(2):216–227, 1993.

[15] V.V. Kulish, P.B. Kosel, A.G. Kailyuk. New acceleration principle of charged particles for electronics applications. the general hierarchical approach. *International Journal of Infrared and Millimeter Waves*, 19(1):3–93, 1993.

[16] V.V. Kulish. *Methods of averaging in nonlinear problems of relativistic electrodynamics*. World Federation Publishers, Atlanta, 1998.

[17] V.V. Kulish, V.I. Savchenko. Method of averaged characteristics in the nonlinear theory of two-stream instability two-stream instability. *Gerald of Sumy State University, ser. Physics and Mathematics*, 2:13–18, 2002.

[18] A.I. Olemskoi, A.Ya. Flat. Application of the factual concept in the condensed-matter physics. *Physics-Uspekhy*, 163(12):101–104, 1993.

[19] R. Rammal, G. Toulouse, M.A. Virasoro. The gnats and gnus document preparation system. *Reviews of Modern Physics*, 50(3):765–788, 1986.

[20] F.G. Tricomi. Lezioni sulle equzioni a derivative partziali. Editrice Gheroni Torino via Carlo Alberto, 1954.

[21] G.A. Korn, T.W. Korn. Mathematical handbook for scientists and engineers. NY: McGraw Hill, 1961.

[22] A.M. Samoylenko, S.A. Kryvoshyja, N.A. Perestyuk. *Differential equations: examples and exercises.* Vysshaja Shkola, Moscow, 1989.

[23] A.N. Tichonov, A.B. Vasil'jeva, A.G. Sveshnikov. *Differential equations.* Nauka, Moscow, 1980.

[24] N.M. Matvejev. *Integration methods integration of differential equations.* Vysshaja Shkola, Moscow, 1963.

[25] S.K. Godunov. *Equations of mathematical physics.* Nauka, Moscow, 1971.

[26] V.V. Kulish. Nonlinear self-consistent theory of free electron lasers. method of investigation. *Ukrainian Physical Journal*, 36(9):1318–1325, 1991.

[27] V.V. Kulish, S.A. Kuleshov, A.V. Lysenko. Nonlinear self-consistent theory of superheterodyne and free electron lasers. *The International journal of infrared and millimeter waves*, 14(3):451–568, 1993.

[28] T.C. Marshall. *Free electron laser.* Mac Millan, New York, London, 1985.

[29] C. Brau. *Free electron laser.* Academic Press, Boston, 1990.

[30] P. Luchini, U. Motz. *Undulators and free electron lasers.* Clarendon Press, Oxford, 1990.

Chapter 7

EXAMPLE: APPLICATION OF THE METHOD OF AVERAGED CHARACTERISTICS IN NONLINEAR THEORY OF THE TWO-STREAM INSTABILITY

Again about instabilities and resonances. It is well known that *two-stream instability* in electron beams (as well as its specific case — the *plasma-beam instability*) is one of the most complex objects for studying in electrodynamics [1–9]. Two main causes determine such situation.

The first cause is the uniqueness of a physical nature of the two-stream instability. It should be mentioned that the concept of '*instability*' in electrodynamics is often confused erroneously with the concept of the '*resonance*' (about these concepts see also in Chapter 1, Section 2). Such situation could be explained by the following: most of the real oscillation–wave systems, which are traditionally studied in electrodynamics, are characterized by the presence of instabilities as well as resonances, at the same time. On the other hand, according to the standard definition, the resonance is usually treated *as any steep increase of the oscillation amplitude*, but only in the case if the so called *resonant condition* (like in (1.2.13), for instant) is satisfied (see Chapter 1, Section 2 for more details). This condition, in turn, is satisfied if the frequency of an *external* (stimulated) perturbed force coincides with one of the *proper* system frequencies (see, for instance, the definitions (1.2.13)). Or in the other words, the simultaneous presence of external (stimulated) and proper oscillations is necessary condition for realization of the resonance.

The weakly nonlinear models with *harmonic* (i.e., *one-harmonic*) external stimulated oscillations (or waves) are studied in practice most often. The presence of such harmonic stimulated oscillation (wave) 'imposes' an evolving scenario for the resonant process. One of specific features of the resonant interactions is a relatively narrow *resonant frequency band* around of the *resonant point* vicinity, where this phenomenon appears remarkably (see Fig. 5.2.1b and corresponding commentaries, and Chapter 1, Section 2). Roughly speaking, we can treat such an interaction process as a quasi-harmonic one because the first harmonic here is explicitly predominated in relevant Fourier spectrum. One easily may make certain of this expanding corresponding solutions, that describe such process, in the Fourier–Taylor series. As a result, it could be found that such types of series, as a rule, are characterized by a *good convergence* with respect to the harmonic amplitudes. Generalizing, we can state that the discussed behavior of the Fourier–Taylor series is typical calculational feature of the resonant problems.

The two-stream, as well as the plasma-beam instabilities, represents some other kind of examples. Namely, they characterize the systems where *instability* takes place, but the resonant conditions are not satisfied [4]. Or, in other words, these physical mechanisms *can be treated as instabilities, but not resonances.* In contrast to the resonant systems, any external stimulated oscillations are absent there. The instability evolves only owing to the interaction between the system proper waves. Hence, we cannot connect it with any resonant conditions like (1.2.13). This means, in particular, that no external (stimulated) influences can be 'imposed' at this time, and, consequently, no 'external quasi-harmonic scenario' can occur in the system of the discussed type. As a result of the absence of this 'imposing', in contrast to the resonant systems, the wave process carry multiharmonical, of principle, nature. *A large number of harmonics might be characterized by amplitudes of approximately equal order* (see further Figs. 8.7.1–8.7.3 and corresponding discussion). Moreover, the situations when one of the higher harmonics of beam waves becomes maximal could be realized, too (see further Fig. 8.7.3). As a result, *poor convergence* of the Fourier–Taylor series which is constructed as a factorization of this process is typical calculational feature of such kind of instability problems.

Thus summarizing, we can state that *resonance* is a specific particular case of instability. Or, that is the same, the concept of instability includes the resonant as a specific particular case.

Then let us briefly discuss the second type of the above mentioned causes, which determine the difficulties that arose in study of the two-stream instability. These difficulties are connected with mathematical side of the problem. The matter is that most widespread traditional mathematical methods, such as the method of nonlinear dispersion equation [1–9], the method of slowly varying amplitudes (see below Chapter 8) etc., in this case turns out to be ineffective very often. Overcoming of such problems requires creation of a new methodical ideas and calculational schemes. One of such new approaches was proposed previously (see Chapter 6) in the form of the method of averaged characteristics.

The demonstration of effectiveness of application of this method to the two-stream instability nonlinear theory [10–12] is performed below in this Chapter.

Two-Stream Superheterodyne Free Electron Lasers (TSFELs). In what follows, let us discuss briefly the applied significance of the two-stream instability, as a potentially useful practical phenomenon. We have in view the possibility of constructing of the so called *Two-Stream Superheterodyne Free Electron Lasers* (TSFELs) on its basis [13–25]. The TSFELs possess a number of unique properties that make them very promising for various practical applications. *The klystron-TSFEL amplifiers* [18, 20, 21, 23, 24] occupy a special place amongst the devices of this type.

The simplest variant of scheme of the klystron TSFEL amplifier is shown in Fig. 7.0.1.

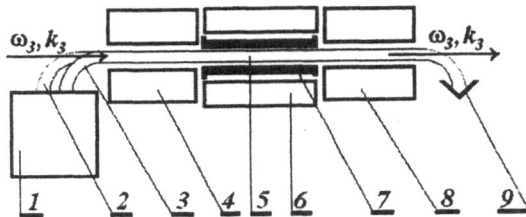

Figure 7.0.1. The simplest version of the scheme of klystron TSFEL amplifier [18, 20, 21]. Here 1 is the two-beam electron accelerator, 2 is the first electron beam, 3 is the second electron beam, 4 is the first pumping system, 5 is the two-velocity electron beam, 6 is the transit section, 7 is the absorber system, 8 is the second pumping system, 9 is the electron collector, ω_3, k_3 are the characteristics of amplified electromagnetic wave signal (frequency and wave vector, respectively).

The device works in the following way. Two-beam electron accelerator 1 generates two relativistic electron beams 2 and 3, respectively. Both these beams are directed on the same input of first pumping system 4. Later they form joint two-velocity electron beam 5. Electromagnetic signal ω_3, k_3 (in the form of an amplified electromagnetic wave) is directed on the same input of first pumping system 4. Two-velocity electron beam 5 is found to be weakly density modulated as a result of the nonlinear parametric interaction electromagnetic wave ω_3, k_3 with its plasmas. So electron beam 5 then (within transit section 6) moves being modulated by some (in the general case) frequency ω_3'. For simplicity the following is accepted: $\omega_3' = \omega_3$. This means that initial (input) signal ω_3, k_3 in the first pumping system transforms from the electromagnetic form into the electron wave one.

The two-stream instability is evolving in electron beam 5 within transit section 6. Maximum amplification of the electron waves in this case is achieved for the waves with the optimal frequency ω_{opt} [15–25] (here we note especially that this maximum does not have a resonant nature). The key point of the discussed design is that the modulation frequency ω_3 is approximately equal to the optimal frequency ω_{opt}. In this case amplitudes of the electron waves with the frequency ω_3 increase because of the two-stream instability. So the input signal $\omega_3, /, k_3$ exists in transit section 6 in the form of electron waves with frequency ω_3. As a result the depth of the beam modulation strongly increases. At the same time the initial (input) electromagnetic signal ω_3, k_3 is absorbed by absorber system 7. Thus only strongly modulated electron beam is directed on the input of second pumping system 8.

The generation (restoration) of the output electromagnetic signal ω_3, k_3 occurs within the second pumping system 8. This takes place due to nonlinear interaction of strongly modulated electron beam 5 with corresponding pumping field of the second pumping system 8. This means that the back transformation of the amplifying signal (from the electron wave form into the electromagnetic one) realizes within the working bulk of the second pumping system 8.

The worked-out electron beam is collected by the electron collector 9. The amplified electromagnetic signal goes from the system output.

Thus the key part of the discussed design of klystron TSFEL amplifier is transit section 6, where the main signal amplification occurs. Therefore, below (in this and the following Chapters) we will analyze the nonlinear dynamics of the amplification process in this section in more detail. We will do this in the framework of nonlinear theory of the two-stream instability. So, let us turn to the calculational aspects of the two-stream instability nonlinear theory.

1. PROBLEM OF MOTION OF A TWO-VELOCITY ELECTRON BEAM IN GIVEN ELECTROMAGNETIC FIELDS

Following the tradition, we divide the initial general self-consistent problem into two formally independent parts. The first is the problem about motion of a two-velocity relativistic electron beam in some given (i.e., known) electromagnetic fields (it is the *motion problem*). The second is the problem of generation (excitation) of those fields on a given motion of the beam (*field problem*). Self-coordination of both these parts gives the complete self-consistent solution of the initial problem. Let us begin our study with the motion problem.

1.1   Statement of the Motion Problem

Let us consider the system of two-velocity relativistic intensive electron beam as the object of study. The beam model is assumed to be a transversely unrestricted system. Velocities of the both beam components (partial velocities) u_1 and u_2, are considered relativistic. It is supposed that the immobile ion background compensates the averaged spatial charge of the electron beam, as a whole. Systems parameters are chosen so that all required conditions for achievement of the two-stream instability are satisfied [7].

We choose relativistic quasi-hydrodynamic equation (6.1.5) as the basic motion equation for the electron beam that consists of two partial beams:

$$\left\{ \frac{\partial}{\partial t} + \left(\vec{u}_\alpha \frac{\partial}{\partial \vec{r}} \right) + \nu \right\} \vec{u}_\alpha = \frac{q_\alpha}{m_\alpha \gamma_\alpha} \left\{ \vec{E} + \frac{1}{c} \left[\vec{u}_\alpha \vec{B} \right] - \frac{\vec{u}_\alpha \left(\vec{u}_\alpha \vec{E} \right)}{c^2} \right\}$$

$$- \frac{v_T^2}{3} \left\{ \frac{1}{n_\alpha} \frac{\partial n_\alpha}{\partial \vec{r}} + \frac{\gamma_\alpha^2}{2c^2} \frac{\partial \left(\vec{u}_\alpha \vec{u}_\alpha \right)}{\partial \vec{r}} \right\}. \quad (7.1.1)$$

where $\vec{u}_\alpha$ is the velocity vector for the α-th partial beam component ($\alpha = 1, 2$), v is the effective frequency of electron collisions, n_α is the electron density for α-th partial beam (here we suppose, for simplicity, that $n_1 = n_2$), $\gamma_\alpha = \gamma_\alpha \left(\vec{r}, t \right) = \left(1 - \vec{u}_\alpha^2 / c^2 \right)^{-1/2}$ is the relativistic factor for electrons of α-th partial beam, v_T is root-mean-square particle thermal velocity, $\vec{r}$ is the Euler's (spatial) coordinate of an observation spatial point, c is the light velocity in vacuum, m_α is its rest mass of particles of the α-sort, q_α is their charge, $\vec{E}$ is the intensity of electric and $\vec{B}$ is the induction of magnetic fields acting on electrons of the α-th beam. We consider that these fields are the same for both partial components of the two-velocity beam.

Then we take into consideration the solely methodical purpose of this illustration example. Let us somewhat simplify the considered model. Namely, we will assume that the electron beam is 'cold' (i.e., $v = 0$, $v_T = 0$), external magnetic fields as well as internal are absent ($\vec{B} = 0$) and electric fields are represented by the beam electron waves and quasi-stationary electric field only. They can be generated in the system as a result of nonlinear interactions [6–8].

We take into account the supposition made above about the transverse unrestrictedness of the model considered. This automatically means that it is spatially one-dimensional. A simplest analysis shows that in this case we may assume that all electron motion processes occur along z-axis only:

$$\frac{\partial \vec{u}_\alpha}{\partial x} = \frac{\partial \vec{u}_\alpha}{\partial y} = 0; \quad u_x = u_y = 0. \tag{7.1.2}$$

As a result we obtain the fact that the electric field $\vec{E}$ in the system is longitudinal only. The intensity of the wave part of this longitudinal field can be represented in the form of Fourier series:

$$E_3\left(z, t\right) = E_{30}\left(z, t\right) + \frac{1}{2} \sum_{m=1}^{N} \left(E_{3m}\left(z, t\right) e^{imp_3(z,t)} + \text{c.c.}\right), \tag{7.1.3}$$

where m are the numbers of *space charge wave* (SCW) harmonics, N is the number of harmonics that are taken into account, E_{3m} are slowly varying (on time t as well as on the spatial coordinate z) amplitudes, $p_3 = \omega_3 t - k_3 z$ is the wave phase of the SCW, ω_3 is the cyclic frequency, and k_3 is the wane number of the SCW.

Apart from the wave part (7.1.3) of the common electric field $\vec{E}$, the quasi-stationary electric field E_{30}, which can be generated in the system as a result nonlinear wave interactions [6–8], exists here too. But a corresponding analysis shows that we could neglect the influence of this field on the accepted further precision of solving the motion problem.

Making all the suppositions accepted above we can rewrite motion equation (7.1.1) in the simplified form:

$$\left(\frac{\partial}{\partial t} + u_\alpha \frac{\partial}{\partial z}\right) u_\alpha = \frac{q_\alpha E_3\left(z, t\right)}{m_\alpha \gamma_\alpha^3\left(z, t\right)}. \tag{7.1.4}$$

The boundary condition is the following:

$$z = 0, \quad u_\alpha\left(0, t\right) = u_{0\alpha}\left(t\right). \tag{7.1.5}$$

Thus the motion problem is stated, and then we can proceed with the constructing procedure of asymptotic integration of equation (7.1.4).

1.2   Averaged Characteristics for the Motion Problem

Let us use the method of averaged characteristics (see Chapter 6) for this purpose. The constructing of characteristics for the equation (7.1.4) is the first step on this way (see Fig. 6.2.1). Accomplishing the passage to the parametric form of writing (where the phase p_3 is the parameter), we obtain the characteristics of the equation (7.1.4) in the form of standard system like (6.2.4):

$$\frac{du_\alpha}{dt} = \frac{q_\alpha}{m_\alpha \gamma_\alpha^3} \left(E_{30}\left(z, t\right) + \frac{1}{2} \sum_{m=1}^{N} \left\{ E_{3m}\left(z, t\right) e^{imp_3} + \text{c.c.} \right\} \right),$$

$$\frac{dz}{dt} = u_\alpha, \tag{7.1.6}$$

$$\frac{dp_3}{dt} = \omega_3 - k_3 u_\alpha.$$

Then we introduce the concepts of slow time

$$\tau = \frac{1}{\xi} t, \tag{7.1.7}$$

and slow coordinate

$$\chi = \frac{1}{\xi} z, \tag{7.1.8}$$

respectively. Than, accomplishing the following obvious transformations

$$E_{3m} = E_{3m}\left(\tau, \chi\right),$$

$$\frac{\partial E_{3m}}{\partial t} = \frac{\partial E_{3m}}{\partial \tau} \frac{d\tau}{dt} = \frac{1}{\xi} \frac{\partial E_{3m}}{\partial \tau},$$

$$\frac{\partial E_{3m}}{\partial z} = \frac{\partial E_{3m}}{\partial \chi} \frac{d\xi}{dz} = \frac{1}{\xi} \frac{\partial E_{3m}}{\partial z} \tag{7.1.9}$$

it is not difficult to get the standard equation system:

$$\frac{du_\alpha}{dt} = \frac{q_\alpha}{m_\alpha \gamma^3} E_3\left(\tau, \chi, p_3\right) = \frac{1}{\xi} X^u\left(u_\alpha, \chi, \tau, p_3\right)$$

$$\frac{d\chi}{dt} = \frac{1}{\xi} u_\alpha = \frac{1}{\xi} X^\chi\left(u_\alpha, \chi, \tau, p_3\right)$$

$$\frac{d\tau}{dt} = \frac{1}{\xi} = \frac{1}{\xi} X^\tau\left(u_\alpha, \chi, \tau, p_3\right) \tag{7.1.10}$$

$$\frac{dp_3}{dt} = \omega_3 - k_3 u_\alpha = \Omega\left(u_\alpha\right) + \frac{1}{\xi} Y\left(u_\alpha, \chi, \tau, p_3\right)$$

where the function $\frac{1}{\xi} Y\left(u_\alpha, \chi, \tau, p_3\right) = 0$ (see also the detail discussion of analogous situation given previously in Chapter 5).

Then we perform the classification of variables into the fast and slow ones, respectively. For this we should construct the hierarchical series for the problem large parameters like (6.2.5). It can be easily seen that in the discussed case only three large parameters of the problem can be separated:

$$[\xi_i] = [\xi_1, \xi_2, \xi_3],$$ (7.1.11)

where

$$\xi_1 \sim \chi_u \left|\frac{dp_3}{dt}\right| \Bigg/ \left|\frac{du_\alpha}{dt}\right| \gg 1, \quad \xi_2 \sim \chi_\chi \left|\frac{dp_3}{dt}\right| \Bigg/ \left|\frac{d\chi}{dt}\right| \gg 1,$$

$$\xi_3 \sim \chi_\tau \left|\frac{dp_3}{dt}\right| \Bigg/ \left|\frac{d\tau}{dt}\right| \gg 1,$$ (7.1.12)

where $\chi_u, \chi_\chi, \chi_\tau$ are some normalizing parameters. However, corresponding analysis shows that both these large parameters are characterized by approximately equal order (i.e., $\xi_1 \sim \xi_2 \sim \xi_3 \sim \xi$). So, further we could use only one of them. Let us choose the one that is smaller:

$$\xi = \min(\xi_1, \xi_2).$$ (7.1.13)

The passage onto the first hierarchical level is the following step of our calculational procedure. The Krylov–Bogolyubov substitutions are used for performing these hierarchical transformations:

$$u_\alpha = \bar{u}_\alpha + \sum_{n=1}^{\infty} \frac{1}{\xi^n} u_u^{(n)}\left(\bar{\chi}, \bar{\tau}, \bar{u}_\alpha, \bar{p}_3\right)$$

$$\chi = \bar{\chi} + \sum_{n=1}^{\infty} \frac{1}{\xi^n} u_\chi^{(n)}\left(\bar{\chi}, \bar{\tau}, \bar{u}_\alpha, \bar{p}_3\right)$$

$$\tau = \bar{\tau} + \sum_{n=1}^{\infty} \frac{1}{\xi^n} u_\tau^{(n)}\left(\bar{\chi}, \bar{\tau}, \bar{u}_\alpha, \bar{p}_3\right)$$ (7.1.14)

$$p_3 = \bar{p}_3 + \sum_{n=1}^{\infty} \frac{1}{\xi^n} v^{(n)}\left(\bar{\chi}, \bar{\tau}, \bar{u}_\alpha, \bar{p}_3\right),$$

where the averaged variables are determined by the equations of the first hierarchical level:

$$\frac{d\bar{u}_\alpha}{dt} = \sum_{n=1}^{\infty} \frac{1}{\xi^n} A_u^{(n)} \left(\bar{\chi}, \bar{\tau}, \bar{u}_\alpha, \bar{p}_3\right)$$

$$\frac{d\bar{\chi}}{dt} = \sum_{n=1}^{\infty} \frac{1}{\xi^n} A_\chi^{(n)} \left(\bar{\chi}, \bar{\tau}, \bar{u}_\alpha, \bar{p}_3\right)$$

$$\frac{d\bar{\tau}}{dt} = \sum_{n=1}^{\infty} \frac{1}{\xi^n} A_\tau^{(n)} \left(\bar{\chi}, \bar{\tau}, \bar{u}_\alpha, \bar{p}_3\right) \tag{7.1.15}$$

$$\frac{d\bar{p}_3}{dt} = \bar{\Omega}\left(\bar{u}_\alpha\right) + \sum_{n=1}^{\infty} \frac{1}{\xi^n} B^{(n)} \left(\bar{\chi}, \bar{\tau}, \bar{u}_\alpha, \bar{p}_3\right),$$

where $\bar{\Omega}\left(\bar{u}_\alpha\right) = \omega_3 - k_3 \bar{u}_\alpha$. All other auxiliary functions are found using corresponding calculational procedures described earlier in Chapter 4. Let us over again from discussion of the calculations accomplished in the first approximation on $1/\xi$:

$$A_u^{(1)} = \frac{1}{2\pi} \int_0^{2\pi} X^u \left(\bar{\chi}, \bar{\tau}, \bar{u}_\alpha, \bar{p}_3\right) d\bar{p}_3 = \frac{q_\alpha}{m_\alpha \bar{\gamma}^3} E_{30}; \tag{7.1.16}$$

$$u_u^{(1)} = \frac{1}{\bar{\Omega}} \int \left(X^u - A_u^{(1)}\right) d\bar{p}_3 = \frac{q_\alpha}{2\bar{\Omega} m_\alpha \bar{\gamma}_\alpha^3} \left(\sum_{m=1}^{N} \frac{1}{im} E_{3m} e^{im\bar{p}_3} + \text{c. c.}\right); \tag{7.1.17}$$

$$A_\chi^{(1)} = \frac{1}{2\pi} \int_0^{2\pi} X^\chi \left(\bar{\chi}, \bar{\tau}, \bar{u}_\alpha, \bar{p}_3\right) d\bar{p}_3 = \frac{1}{\xi} \bar{u}_\alpha, \tag{7.1.18}$$

$$u_\chi^{(1)} = \frac{1}{\bar{\Omega}} \int \left(X^\chi - A_\chi^{(1)}\right) d\bar{p}_3 = 0; \tag{7.1.19}$$

$$A_\tau^{(1)} = \frac{1}{2\pi} \int_0^{2\pi} X^t \left(\bar{\chi}, \bar{\tau}, \bar{u}_\alpha, \bar{p}_3\right) d\bar{p}_3 = \frac{1}{\xi}, \tag{7.1.20}$$

$$u_\tau^{(1)} = \frac{1}{\bar{\Omega}} \int \left(X^\tau - A_\tau^{(1)}\right) d\bar{p}_3 = 0; \tag{7.1.21}$$

$$B^{(1)} = \frac{1}{2\pi} \int_0^{2\pi} \left(\frac{\partial \bar{\Omega}}{\partial \bar{u}_\alpha} u_u^{(1)} + \frac{\partial \bar{\Omega}}{\partial \bar{\chi}} u_\chi^{(1)} + Y\right) d\bar{p}_3 = 0; \tag{7.1.22}$$

$$v^{(1)} = \frac{1}{\bar{\Omega}} \int \left(\frac{\partial \bar{\Omega}}{\partial \bar{u}_\alpha} u_u^{(1)} + \frac{\partial \bar{\Omega}}{\partial \bar{\chi}} u_\chi^{(1)} + Y - B^{(1)} \right) d\bar{p}_3$$

$$= \frac{k_3 q_\alpha}{2\bar{\Omega}^2 m_\alpha \bar{\gamma}_\alpha^3} \left(\sum_{m=1}^{N} \frac{1}{m^2} E_{3m} e^{im\bar{p}_3} + \text{c. c.} \right), \quad (7.1.23)$$

where $\bar{\gamma}_\alpha = \left(1 - \bar{u}_\alpha^2/c^2 \right)^{-1/2}$ is the averaged relativistic factor.

Then we turn the reader's attention at the following specific calculational peculiarity. The choice of the approximation number at a practice is determined usually by the physical features of a model studied. Let us explain this thought in more detail using the here considered two-stream model. The calculations performed in the first approximation $1/\xi$ allow us to describe, as relevant analysis shows, the beginning of nonlinear stage of the two-stream instability. Or, more exactly, we can 'see' in this case the beginning only of a first distinctions from the simplest quasilinear exponential-like law. The latter are characteristic for wave amplitude dependencies obtained in the framework of simplest quasilinear theory (see, for instance, [13–25]). However, the most characteristic nonlinear phenomenon of the two-stream instability (the so called *percussive excitation of highest harmonics*), as our analysis shows, can not be described in the first approximation. Accomplishing of corresponding calculations in the second approximation $(1/\xi^2)$ allows to solve this problem effectively:

$$A_u^{(2)} = 0; \quad (7.1.24)$$

$$u_u^{(2)} = \frac{k_3 q_\alpha}{2\bar{\Omega}^2 m_\alpha \bar{\gamma}_\alpha^3} \left\{ \frac{q_\alpha}{4 m_\alpha \bar{\gamma}_\alpha^3} \left(\frac{k_3}{\bar{\Omega}} - \frac{3 \bar{u}_\alpha \bar{\gamma}^2}{c^2} \right) \right.$$

$$\times \left(\sum_{m=1}^{N} \left(\frac{E_{3m}}{im} e^{im\bar{p}_3} + \text{c. c.} \right) \right)^2$$

$$+ \frac{k_3 q_\alpha}{\bar{\Omega} m_\alpha \bar{\gamma}_\alpha^3} \left(E_{30} + \frac{1}{2} \sum_{m=1}^{N} \left(E_{3m} e^{im\bar{p}_3} + \text{c. c.} \right) \right)$$

$$\times \left(\sum_{m=1}^{N} \left(\frac{E_{3m}}{m^2} e^{im\bar{p}_3} + \text{c. c.} \right) \right)$$

$$+ \sum_{m=1}^{N} \left(\frac{1}{m^2} \left(\bar{u}_\alpha \frac{1}{\xi} \frac{\partial E_{3m}}{\partial \bar{\chi}} + \frac{1}{\xi} \frac{\partial E_{3m}}{\partial \tau} \right) e^{im\bar{p}_3} + \text{c. c.} \right) \right\} \quad (7.1.25)$$

$$A_\chi^{(2)} = 0; \tag{7.1.26}$$

$$u_\chi^{(2)} = -\frac{q_\alpha}{2\bar{\Omega}^2 m_\alpha \bar{\gamma}_\alpha^3} \left(\sum_{m=1}^{N} \frac{E_{3m}}{m^2} e^{im\bar{p}_3} + \text{c. c.} \right); \tag{7.1.27}$$

$$A_\tau^{(2)} = 0; \tag{7.1.28}$$

$$u_\tau^{(2)} = 0; \tag{7.1.29}$$

$$B^{(2)} = 0. \tag{7.1.30}$$

Substituting expressions (7.1.16), (7.1.18), (7.1.20) and (7.1.22) in (7.1.15) we obtain the equations for the *first hierarchical level* in the following form:

$$d\bar{u}_\alpha = \frac{q_\alpha}{m_\alpha \bar{\gamma}^3} E_{30}$$

$$\frac{d\bar{\chi}}{dt} = \bar{u}_\alpha$$

$$\frac{d\bar{\tau}}{dt} = 1$$

$$\frac{d\bar{p}_3}{dt} = \omega_3 - k_3 \bar{u}_\alpha. \tag{7.1.31}$$

Regarding the first two equations (7.1.31) as characteristics of some averaged quasilinear equation like (6.2.15), we can obtain this equations in the form (see (6.2.15))

$$\left(\frac{\partial}{\partial t} + \bar{u}_\alpha \frac{\partial}{\partial \bar{z}} \right) \bar{u}_\alpha = \frac{q_\alpha}{m_\alpha \bar{\gamma}^3} E_{30}. \tag{7.1.32}$$

Now we should recall that the averaged parameter $\bar{p}_3$ (see the third of equations (7.1.31)) here plays the role of a parameter. Besides that, it is interesting to point out the following. It is easy to be convinced that, in spite of equation (7.1.32) being formally obtained in the first approximation $1/\xi$ only, it turns out to be actually valid for the second approximation $1/\xi^2$, too. This can be explained by that the zeroth results (7.1.14), (7.1.26), (7.1.28), and (7.1.30) are obtained in the second approximation. It is obvious that the mentioned conclusion does not concern the calculational results for oscillation parts of relevant expressions.

Comparing (7.1.32) and (7.1.4), we see that the first of those equations has essentially simpler mathematical structure (because it contains

no oscillations in the right part). Solving (7.1.32) and performing corresponding back transformation (see, for instant, (6.2.19), (6.2.20)) we can
obtain the complete solution of the initial problem (7.1.4). Let us begin
describing these calculational procedures with the back transformations.

1.3   Back Transformations

Substituting (7.1.17), (7.1.21) and (7.1.23) in (7.1.14) we obtain the
formal solutions of the initial problem (7.1.4):

$$u_\alpha\left(\bar{\chi},\bar{\tau},\bar{p}_3\right) = \bar{u}_\alpha\left(\bar{\chi},\bar{\tau}\right) + \frac{1}{\xi}u_u^{(1)}\left(\bar{\chi},\bar{\tau},\bar{u}_\alpha,\bar{p}_3\right) + \frac{1}{\xi^2}u_u^{(2)}\left(\bar{\chi},\bar{\tau},\bar{u}_\alpha,\bar{p}_3\right).$$

$$(7.1.33)$$

Relationships like (6.2.19) in the considered case can be written in the
form:

$$u_\alpha - \bar{u}_\alpha = \sum_{n=1}^{\infty}\frac{1}{\xi^n}u_u^{(n)}\left(\bar{\chi},\bar{\tau},\bar{u}_\alpha,\bar{p}_3\right)$$

$$\chi - \bar{\chi} = \sum_{n=1}^{\infty}\frac{1}{\xi^n}u_\chi^{(n)}\left(\bar{\chi},\bar{\tau},\bar{u}_\alpha,\bar{p}_3\right)$$

$$\tau - \bar{\tau} = \sum_{n=1}^{\infty}\frac{1}{\xi^n}u_\tau^{(n)}\left(\bar{\chi},\bar{\tau},\bar{u}_\alpha,\bar{p}_3\right)$$ $$(7.1.34)$$

$$p_3 - \bar{p}_3 = \sum_{n=1}^{\infty}\frac{1}{\xi^n}v^{(n)}\left(\bar{\chi},\bar{\tau},\bar{u}_\alpha,\bar{p}_3\right).$$

After accomplishing in (7.1.34) corresponding expansion in the Taylor
series (see (6.2.20)) solutions (7.1.33) could be rewritten as:

$$u_\alpha\left(\chi,\tau,p_3\right) = u_\alpha\left(\bar{\chi},\bar{\tau}\right)\Big|_{\substack{\bar{\chi}=\chi \\ \bar{p}_3=p_3}}$$

$$+\left\{\frac{\partial u_\alpha\left(\bar{u}_\alpha,\bar{\chi},\bar{\tau},\bar{p}_3\right)}{\partial\bar{\chi}}\left(\sum_{n=1}^{\infty}\frac{1}{\xi^n}u_\chi^{(n)}\right) + \frac{\partial u_\alpha\left(\bar{u}_\alpha,\bar{\chi},\bar{\tau},\bar{p}_3\right)}{\partial\bar{\tau}}\left(\sum_{n=1}^{\infty}\frac{1}{\xi^n}u_\tau^{(n)}\right)\right.$$

$$\left.+\frac{\partial u_\alpha\left(\bar{u}_\alpha,\bar{\chi},\bar{\tau},\bar{p}_3\right)}{\partial\bar{p}_3}\left(\sum_{n=1}^{\infty}\frac{1}{\xi^n}v^{(n)}\right)\right\}\Bigg|_{\substack{\bar{\chi}=\chi \\ \bar{p}_3=p_3}} + O\left(\frac{1}{\xi^3}\right),   (7.1.35)$$

or, after corresponding calculational, in the form

$$
u_\alpha\left(\chi,\tau,p_3\right) = \left.\left(\bar{u}_\alpha + \frac{1}{\xi}u_u^{(1)} + \frac{1}{\xi^2}u_u^{(2)} - \frac{1}{\xi^2}\frac{\partial u_u^{(1)}}{\partial \bar{p}_3}v^{(1)}\right)\right|_{\substack{\bar{\chi}=\chi,\bar{\tau}=\tau \\ \bar{p}_3=p_3}}
$$

$$
= \bar{u}_\alpha + \frac{q_\alpha}{2\bar{\Omega}m_\alpha\bar{\gamma}_\alpha^3}\left(\sum_{m=1}^{N}\frac{1}{im}E_{3m}e^{imp_3} + \text{c. c.}\right) + \frac{q_\alpha}{2\bar{\Omega}^2 m_\alpha\bar{\gamma}_\alpha^3}
$$

$$
\times \left\{\frac{q_\alpha}{4m_\alpha\bar{\gamma}_\alpha^3}\left(\frac{k_3}{\bar{\Omega}} - \frac{3\bar{u}_\alpha\bar{\gamma}^2}{c^2}\right)\left(\sum_{m=1}^{N}\left(\frac{E_{3m}}{im}e^{imp_3} + \text{c. c.}\right)\right)^2\right.
$$

$$
\left.+ \frac{k_3 q_\alpha E_{30}}{\bar{\Omega}m_\alpha\bar{\gamma}_\alpha^3}\sum_{m=1}^{N}\left(\frac{1}{m^2}\left(E_{3m} + \bar{u}_\alpha\frac{\partial E_{3m}}{\partial z} + \frac{\partial E_{3m}}{\partial t}\right)e^{imp_3} + \text{c. c.}\right)\right\},
$$

$$
(7.1.36)
$$

where $\bar{\Omega} = \omega_3 - k_3\bar{u}_\alpha$, $\bar{\gamma}_\alpha = \left(1 - \bar{u}_\alpha^2/c^2\right)^{-1/2}$ is the averaged relativistic factor as a function of the averaged velocity $\bar{u}_\alpha\left(z,t\right)$.

As can easily be seen, the sought complete solution of the initial problem (7.1.4) could be obtained after the determination of a specific form of the function $\bar{u}_\alpha\left(z,t\right)$. As was mentioned above, such expressions could be obtained by solving of the averaged quasilinear equation (7.1.32).

1.4 Integration of the Averaged Quasi-Linear Equation for the Beam Velocity

Now let us find solutions of equation (7.1.32) ($E_{30} = 0$)

$$
\left(\frac{\partial}{\partial t} + \bar{u}_\alpha\frac{\partial}{\partial \bar{z}}\right)\bar{u}_\alpha = 0 \tag{7.1.37}
$$

for the boundary conditions:

$$
\bar{z} = 0, \quad \bar{u}_\alpha = \bar{u}_{0\alpha}\left(t\right) = u_{0\alpha}\left(t\right), \tag{7.1.38}
$$

where the function $u_{0\alpha}\left(t\right)$ is determined in (7.1.5).

We will use the method of characteristics (see Chapter 6, Section 3) for integration of the equation (7.1.37). Following the general calculational scheme, we should construct characteristics for equation (7.1.37). The peculiarity of the discussed situation lies in fact that these characteristics are already known (see first two equations of (7.1.31)):

$$
\frac{d\bar{u}_\alpha}{dt} = 0, \quad \frac{d\bar{z}}{dt} = \bar{u}_\alpha \tag{7.1.39}
$$

The integrals of the system (7.1.39) should be found at the next step of the used calculational procedure. They can be found easily:

$$C_1 = \bar{u}_\alpha, \quad C_2 = \bar{z} - \bar{u}_\alpha t. \tag{7.1.40}$$

Substitution of boundary condition (7.1.38) in (7.1.40) yields

$$C_1' = \bar{u}_{0\alpha}(t), \quad C_2' = -C_1' t \Rightarrow t = -\frac{C_2'}{C_1'} \tag{7.1.41}$$

Excluding all varying values from (7.1.41), we obtain the functional dependency, which binds together the constants C_1', C_2':

$$C_1' = \bar{u}_{0\alpha}\left(-\frac{C_2'}{C_1'}\right). \tag{7.1.42}$$

Then we replace the constants C_1', C_2' in (7.1.42) with the expressions (7.1.40). As a result we obtain the sought solution of problem (7.1.37), (7.1.38):

$$\bar{u}_\alpha = \bar{u}_{0\alpha}\left(-\frac{\bar{z} - \bar{u}_\alpha t}{\bar{u}_\alpha}\right) = \bar{u}_{0\alpha}\left(t - \frac{\bar{z}}{\bar{u}_\alpha}\right), \tag{7.1.43}$$

where $\bar{u}_{0\alpha}(...) = u_{0\alpha}(...)$ are some arbitrary functions (see definitions (7.1.38), (7.1.5)) of the arguments $(t - \bar{z}/\bar{u}_\alpha)$.

As was mentioned above, the complete solutions of initial motion problem (7.1.4), (7.1.5) can be found after substituting of the solution (7.1.43) into (7.1.36). For example, let us choose the boundary conditions in the following simplest form

$$u_\alpha(0,t) = \bar{u}_\alpha(0,t) = u_{0\alpha} = \text{const}. \tag{7.1.44}$$

It is obvious that the solution for the averaged velocity in this case can be obtained as

$$\bar{u}_\alpha = u_{0\alpha}. \tag{7.1.45}$$

2.   FIELD PROBLEM. HIERARCHICAL ASYMPTOTIC INTEGRATION OF THE CONTINUITY EQUATION

The Maxwell equations (6.1.6) are used below as the basic equations for the field problem. Except for that, continuity equation (6.1.10) (which, as is well known, follows from the Maxwell equations) will be used also. Therein the general *field problem* in the framework of the discussed theory, in turn, is subdivided into two separate, more particular problems. They are the *problem of continuity of the electron beam* and the *field problem, in itself.* Let us begin the study of the field problem from the first one.

2.1 Continuity Equation of the Two-Velocity Electron Beam

Using relationships (6.1.11), let us reformulate continuity equation (6.1.10) in the more convenient form:

$$\left\{ \frac{\partial}{\partial t} + \left(u_\alpha \frac{\partial}{\partial z} \right) \right\} n_\alpha = -n_\alpha \frac{\partial u_\alpha}{\partial z}, \tag{7.2.1}$$

which we will solve with the following boundary conditions

$$z = 0, \quad n_\alpha(z,t) = n_{0\alpha}(t). \tag{7.2.2}$$

Here n_α is the concentration of electron plasma of α-th partial beam.

Then we use some result of the preceding Section. Namely, using solution (7.1.36), we construct the expression for partial derivative in the right part of (7.2.1):

$$\frac{\partial u_\alpha}{\partial z} = \frac{\partial \bar{u}_\alpha}{\partial z}$$

$$+ \frac{q_\alpha k_3}{2m_\alpha \bar{\Omega}^2 \bar{\gamma}_\alpha^3} \left(1 - \frac{3\bar{\gamma}_\alpha^2 \bar{u}_\alpha \bar{\Omega}}{c^2 k_3} \right) \left(\frac{\partial \bar{u}_\alpha}{\partial z} \right) \sum_{m=1}^{N} \left(\frac{E_{3m}}{im} e^{imp_3} + \text{c. c.} \right)$$

$$+ \frac{q_\alpha}{2m_\alpha \bar{\Omega} \bar{\gamma}_\alpha^3} \left(\sum_{m=1}^{N} \left(\frac{\partial E_{3m}}{\partial z} \frac{1}{im} e^{imp_3} + \text{c. c.} \right) - k_3 \sum_{m=1}^{N} \left(E_{3m} e^{imp_3} + \text{c. c.} \right) \right)$$

$$- \frac{q_\alpha k_3}{2m_\alpha \bar{\Omega}^2 \bar{\gamma}_\alpha^3} \left\{ \frac{q_\alpha}{2m_\alpha \bar{\gamma}_\alpha^3} \left(\frac{\bar{\Omega}}{k_3} - \frac{3\bar{\gamma}_\alpha^2 \bar{u}_\alpha}{c^2} \right) \sum_{m=1}^{N} \left(\frac{E_{3m}}{im} e^{imp_3} + \text{c. c.} \right) \right.$$

$$\times \sum_{m=1}^{N} \left(E_{3m} e^{imp_3} + \text{c. c.} \right) - \bar{u}_\alpha \sum_{m=1}^{N} \left(\frac{\partial E_{3m}}{\partial z} \frac{1}{im} e^{imp_3} + \text{c. c.} \right)$$

$$- \sum_{m=1}^{N} \left(\frac{\partial E_{3m}}{\partial t} \frac{1}{im} e^{imp_3} + \text{c. c.} \right) - \frac{q_\alpha k_3}{m_\alpha \bar{\Omega} \bar{\gamma}_\alpha^3} E_{30} \sum_{m=1}^{N} \left(\frac{E_{3m}}{im} e^{imp_3} + \text{c. c.} \right) \right\}$$

$$\tag{7.2.3}$$

where the functional dependency $\bar{u}_\alpha$ is determined by relationship (7.1.43). Substituting expressions (7.2.3) and (7.1.43) into (7.2.1), we get an equation with respect to the sought function $n_\alpha\,(z,t)$ only. Then we will use this equation as the basic one. The method of averaged characteristic is used for finding of its solution.

2.2   Averaged Characteristics and the Averaged Quasi-Linear Equation

Let us construct the equation system for characteristics — that is the first step of the used calculational procedure. Separating the oscillation phase p_3, as a parameter, we can rewrite this system in the standard form like (6.3.2)

$$
\begin{aligned}
\frac{dn_\alpha}{dt} &= -n_\alpha\frac{\partial u_\alpha}{\partial z}, \\
\frac{dz}{dt} &= u_\alpha, \\
\frac{dp_3}{dt} &= \omega_3 - k_3 u_\alpha.
\end{aligned}
\tag{7.2.4}
$$

Analogously with the preceding motion problem, we introduce further the concepts of slow time

$$
\tau' = \frac{1}{\xi}t,
\tag{7.2.5}
$$

and slow coordinate

$$
\chi' = \frac{1}{\xi}z,
\tag{7.2.6}
$$

respectively. Accomplishing set of required calculational procedures (which where described previously repeatedly) we have the set of equations that can be written in the standard form like

$$
\begin{aligned}
\frac{dn_\alpha}{dt} &= -n_\alpha\frac{\partial u_\alpha}{\partial z} = \frac{1}{\xi}X^n\,(n_\alpha,\chi,\tau,p_3) \\
\frac{d\chi}{dt} &= \frac{1}{\xi}u_\alpha = \frac{1}{\xi}X^\chi\,(n_\alpha,\chi,\tau,p_3) \\
\frac{d\tau}{dt} &= \frac{1}{\xi} = \frac{1}{\xi}X^\tau\,(n_\alpha,\chi,\tau,p_3) \\
\frac{dp_3}{dt} &= \omega_3 - k_3 u_\alpha = \Omega\,(n_\alpha,\chi,\tau) + \frac{1}{\xi}Y\,(n_\alpha,\chi,\tau,p_3)\,,
\end{aligned}
\tag{7.2.7}
$$

where we have omitted the 'prime' symbol for simplicity; the function $Y\,(n_\alpha,\chi,\tau,p_3)$ and other notations are self-evident. Therein we consider that really the function $\partial u_\alpha / \partial z$ is known because it can be determined from relationship (7.2.3).

Below we perform the analogue to the procedure of constructing the problem large parameters of the preceding problem. As a result it can be found that two of such large parameters could be constructed:

$$[\xi_i] = [\xi_4, \xi_5, \xi_6] \,, \tag{7.2.8}$$

where

$$\xi_4 \sim \chi_n \frac{\left|\frac{dp_3}{dt}\right|}{\left|\frac{dn_\alpha}{dt}\right|} \gg 1, \quad \xi_5 \sim \chi_\chi \frac{\left|\frac{dp_3}{dt}\right|}{\left|\frac{d\chi}{dt}\right|} \gg 1, \quad \xi_6 \sim \chi_\tau \frac{\left|\frac{dp_3}{dt}\right|}{\left|\frac{d\tau}{dt}\right|} \gg 1 \tag{7.2.9}$$

and $\chi_n, \chi_\chi, \chi_\tau$ are corresponding normalizing parameters (the large parameters $\xi_{4,5,6}$ must be dimensionless values). The less one of the parameters $\xi_{4,5,6}$ is chosen as the problem large parameter:

$$\xi = \min (\xi_4, \xi_5, \xi_6) \,. \tag{7.2.10}$$

The asymptotic solutions of the problem (7.2.4) are constructed during the next step of the used calculational scheme. We will seek these solutions in the form of Krylov–Bogolyubov substitutions

$$n_\alpha = \bar{n}_\alpha + \sum_{n=1}^{\infty} \frac{1}{\xi^n} u_n^{(n)} \left(\bar{\chi}, \bar{\tau}, \bar{n}_\alpha, \bar{p}_3\right)$$

$$\chi = \bar{\chi} + \sum_{n=1}^{\infty} \frac{1}{\xi^n} u_\chi^{(n)} \left(\bar{\chi}, \bar{\tau}, \bar{n}_\alpha, \bar{p}_3\right)$$

$$\tag{7.2.11}$$

$$\tau = \bar{\tau} + \sum_{n=1}^{\infty} \frac{1}{\xi^n} u_\tau^{(n)} \left(\bar{\chi}, \bar{\tau}, \bar{n}_\alpha, \bar{p}_3\right)$$

$$p_3 = \bar{p}_3 + \sum_{n=1}^{\infty} \frac{1}{\xi^n} v^{(n)} \left(\bar{\chi}, \bar{\tau}, \bar{n}_\alpha, \bar{p}_3\right),$$

where all averaged values are found from the equations of the first hierarchy:

$$\frac{d\bar{n}_\alpha}{dt} = \sum_{n=1}^{\infty} \frac{1}{\xi^n} A_n^{(n)}\left(\bar{\chi}, \bar{\tau}, \bar{n}_\alpha, \bar{p}_3\right)$$

$$\frac{d\bar{\chi}}{dt} = \sum_{n=1}^{\infty} \frac{1}{\xi^n} A_\chi^{(n)}\left(\bar{\chi}, \bar{\tau}, \bar{n}_\alpha, \bar{p}_3\right)$$

$$\frac{d\bar{\tau}}{dt} = \sum_{n=1}^{\infty} \frac{1}{\xi^n} A_\tau^{(n)}\left(\bar{\chi}, \bar{\tau}, \bar{n}_\alpha, \bar{p}_3\right) \qquad (7.2.12)$$

$$\frac{d\bar{p}_3}{dt} = \bar{\Omega} + \sum_{n=1}^{\infty} \frac{1}{\xi^n} B^{(n)}\left(\bar{\chi}, \bar{\tau}, \bar{n}_\alpha, \bar{p}_3\right),$$

where $\bar{\Omega} = \omega_3 - k_3 \bar{u}_\alpha\left(\bar{\chi}, \bar{\tau}\right)$.

Then we use the standard hierarchy scheme described in Chapter 4 and already applied to the preceding problem. Restricting ourselves to the first approximation (on $1/\xi$) only we obtain the following expressions for the unknown functions in (7.2.11) and (7.2.12):

$$A_n^{(1)} = \frac{1}{2\pi} \int_0^{2\pi} X^n\left(\bar{\chi}, \bar{\tau}, \bar{n}_\alpha, \bar{p}_3\right) d\bar{p}_3 = -\bar{n}_\alpha\left(\bar{\chi}, \bar{\tau}\right) \frac{\partial u\left(\bar{\chi}, \bar{\tau}\right)}{\partial \bar{\chi}}; \qquad (7.2.13)$$

$$u_n^{(1)} = \frac{1}{\bar{\Omega}} \int \left(X^n - A_n^{(1)}\right) d\bar{p}_3 = \frac{\bar{n}_\alpha q_\alpha k_3}{2\bar{\Omega}^2 m_\alpha \bar{\gamma}_\alpha^3} \left(\sum_{m=1}^{N} \frac{E_{3m}}{im} e^{im\bar{p}_3} + \text{c.\,c.}\right) \qquad (7.2.14)$$

$$A_\chi^{(1)} = \frac{1}{2\pi} \int_0^{2\pi} X^\chi\left(\bar{\chi}, \bar{\tau}, \bar{u}_\alpha, \bar{p}_3\right) d\bar{p}_3 = \frac{1}{\xi} \bar{u}_\alpha, \qquad (7.2.15)$$

$$u_\chi^{(1)} = \frac{1}{\bar{\Omega}} \int \left(X^\chi - A_\chi^{(1)}\right) d\bar{p}_3 = 0; \qquad (7.2.16)$$

$$A_\tau^{(1)} = \frac{1}{2\pi} \int_0^{2\pi} X^t\left(\bar{\chi}, \bar{\tau}, \bar{u}_\alpha, \bar{p}_3\right) d\bar{p}_3 = \frac{1}{\xi}, \qquad (7.2.17)$$

$$u_\tau^{(1)} = \frac{1}{\bar{\Omega}} \int \left(X^\tau - A_\tau^{(1)}\right) d\bar{p}_3 = 0; \qquad (7.2.18)$$

$$B^{(1)} = 0; \qquad (7.2.19)$$

$$V^{(1)} = \frac{q_\alpha k_3}{2\bar{\Omega}^2 m_\alpha \bar{\gamma}_\alpha^3} \left(\sum_{m=1}^{N} \frac{E_{3m}}{m^2} e^{im\bar{p}_3} + \text{c.\,c.}\right). \qquad (7.2.20)$$

Analogously we have relevant results for the second approximation $(1/\xi^2)$:

$$A_n^{(2)} = 0; \qquad (7.2.21)$$

$$
u_n^{(2)} = \frac{\bar{n}_\alpha q_\alpha^2 k_3^2}{4\bar{\Omega}^4 m_\alpha^2 \bar{\gamma}_\alpha^6} \left[\sum_{m=1}^{N} \left(E_{3m} e^{im\bar{p}_3} + \text{c. c.} \right) \sum_{m=1}^{N} \left(\frac{E_{3m}}{m^2} e^{im\bar{p}_3} + \text{c. c.} \right) \right.
$$

$$
\left. + \frac{1}{2} \left(\sum_{m=1}^{N} \left(\frac{E_{3m}}{im} e^{im\bar{p}_3} + \text{c. c.} \right) \right)^2 \right]
$$

$$
+ \frac{\bar{n}_\alpha q_\alpha k_3}{2\bar{\Omega}^3 m_\alpha \bar{\gamma}_\alpha^3} \left(\left(1 - \frac{3\bar{\gamma}_\alpha^2 \bar{u}_\alpha \bar{\Omega}}{c^2 k_3} \right) \left(\frac{1}{\xi} \frac{\partial \bar{u}_\alpha}{\partial \bar{\chi}} \bar{u}_\alpha + \frac{1}{\xi} \frac{\partial \bar{u}_\alpha}{\partial \bar{\tau}} \right) - \frac{q_\alpha k_3 E_{30}}{m_\alpha \bar{\gamma}^3 \bar{\Omega}} \right)
$$

$$
\times \sum_{m=1}^{N} \left(\frac{E_{3m}}{m^2} e^{im\bar{p}_3} + \text{c. c.} \right)
$$

$$
+ \frac{\bar{n}_\alpha q_\alpha^2 k_3^2}{8\bar{\Omega}^4 m_\alpha^2 \bar{\gamma}_\alpha^6} \left(1 - \frac{3\bar{\gamma}_\alpha^2 \bar{u}_\alpha \bar{\Omega}}{c^2 k_3} \right) \left(\sum_{m=1}^{N} \left(\frac{E_{3m}}{im} e^{im\bar{p}_3} + \text{c. c.} \right) \right)^2
$$

$$
+ \frac{\bar{n}_\alpha q_\alpha k_3}{2\bar{\Omega}^3 m_\alpha \bar{\gamma}_\alpha^3} \left(\frac{\omega_3}{k_3} + \bar{u}_\alpha \right) \sum_{m=1}^{N} \left(\frac{1}{\xi} \frac{\partial E_{3m}}{\partial \chi} \frac{1}{m^2} e^{im\bar{p}_3} + \text{c. c.} \right)
$$

$$
+ \frac{\bar{n}_\alpha q_\alpha k_3}{\bar{\Omega}^3 m_\alpha \bar{\gamma}_\alpha^3} \sum_{m=1}^{N} \left(\frac{1}{\xi} \frac{\partial E_{3m}}{\partial \tau} \frac{1}{m^2} e^{im\bar{p}_3} + \text{c. c.} \right)
$$

$$
+ \frac{\bar{n}_\alpha q_\alpha k_3}{2\bar{\Omega}^3 m_\alpha \bar{\gamma}_\alpha^3} \left(\frac{1}{\xi} \frac{\partial \bar{u}_\alpha}{\partial \chi} \right) \sum_{m=1}^{N} \left(\frac{E_{3m}}{m^2} e^{im\bar{p}_3} + \text{c. c.} \right)
$$

$$
+ \frac{\bar{n}_\alpha q_\alpha^2 k_3^2}{8\bar{\Omega}^4 m_\alpha^2 \bar{\gamma}_\alpha^6} \left(\sum_{m=1}^{N} \left(\frac{E_{3m}}{im^2} e^{im\bar{p}_3} + \text{c. c.} \right) \right)^2. \qquad (7.2.22)
$$

$$A_\chi^{(2)} = 0; \qquad (7.2.23)$$

$$
u_\chi^{(2)} = - \frac{q_\alpha}{2\bar{\Omega}^2 m_\alpha \bar{\gamma}_\alpha^3} \left(\sum_{m=1}^{N} \frac{E_{3m}}{m^2} e^{im\bar{p}_3} + \text{c. c.} \right); \qquad (7.2.24)
$$

$$A_\tau^{(2)} = 0; \qquad (7.2.25)$$

$$u_\tau^{(2)} = 0; \qquad (7.2.26)$$

$$B^{(2)} = 0. \qquad (7.2.27)$$

Substituting (7.2.13), (7.2.16), and (7.2.19) into system (7.2.12), we can rewrite the latter in the form:

$$\frac{d\bar{n}_\alpha}{dt} = -\bar{n}_\alpha \frac{\partial \bar{u}_\alpha}{\partial z},$$

$$\frac{d\bar{z}}{dt} = \bar{u}_\alpha, \qquad (7.2.28)$$

$$\frac{d\bar{p}_3}{dt} = \omega_3 - k_3 \bar{u}_\alpha = \bar{\Omega}.$$

Regarding equations (7.2.28) as characteristics equations of a corresponding averaged quasilinear equation with partial derivatives like (6.2.15) (quasilinear equation of the first hierarchical level), we can easily construct such an equation:

$$\left(\frac{\partial}{\partial t} + \bar{u}_\alpha \frac{\partial}{\partial \bar{z}} \right) \bar{n}_\alpha = -\bar{n}_\alpha \frac{\partial \bar{u}_\alpha}{\partial \bar{z}}. \qquad (7.2.29)$$

Equation (7.2.29) can be solved by means of one of the known analytical or numerical methods. An example of such scheme of solution will be given somewhat below. Here let us turn to the problem of back transformation.

2.3   Back Transformation

Substituting the expressions (7.2.14), (7.2.18), (7.2.20) into Krylov–Bogolyubov substitutions (7.2.11), we can construct the formal solutions for the sought complete solutions of the initial problem:

$$n_\alpha \left(\bar{\chi}, \bar{\tau}, \bar{p}_3 \right) = \bar{n}_\alpha \left(\bar{\chi}, \bar{\tau} \right) + \frac{1}{\xi} u_n^{(1)} \left(\bar{\chi}, \bar{\tau}, \bar{n}_\alpha, \bar{p}_3 \right) + \frac{1}{\xi^2} u_n^{(2)} \left(\bar{\chi}, \bar{\tau}, \bar{n}_\alpha, \bar{p}_3 \right)$$

$$(7.2.30)$$

Below we use the algorithm of back transformations described previously in Chapter 6, Subsection 4.2. First we turn our attention on the fact that the differences

$$n_\alpha - \bar{n}_\alpha = \sum_{n=1}^{\infty} \frac{1}{\xi^n} u_u^{(n)} \left(\bar{\chi}, \bar{\tau}, \bar{n}_\alpha, \bar{p}_3 \right)$$

$$\chi - \bar{\chi} = \sum_{n=1}^{\infty} \frac{1}{\xi^n} u_\chi^{(n)} \left(\bar{\chi}, \bar{\tau}, \bar{n}_\alpha, \bar{p}_3 \right)$$

$$\tau - \bar{\tau} = \sum_{n=1}^{\infty} \frac{1}{\xi^n} u_\tau^{(n)} \left(\bar{\chi}, \bar{\tau}, \bar{n}_\alpha, \bar{p}_3 \right) \qquad (7.2.31)$$

$$p_3 - \bar{p}_3 = \sum_{n=1}^{\infty} \frac{1}{\xi^n} v^{(n)} \left(\bar{\chi}, \bar{\tau}, \bar{u}_\alpha, \bar{p}_3 \right).$$

are proportional to $1/\xi$ (see (6.2.19) and corresponding comments). Using this, we perform corresponding expansion in the Taylor series (see (6.2.20):

$$n_\alpha \left(\chi, \tau, p_3 \right) = n_\alpha \left(\bar{\chi}, \bar{\tau} \right) \Big|_{\substack{\bar{\chi}=\chi \\ \bar{p}_3=p_3}}$$

$$+ \left\{ \frac{\partial n_\alpha \left(\bar{\chi}, \bar{\tau}, \bar{n}_\alpha, \bar{p}_3 \right)}{\partial \bar{\chi}} \left(\sum_{n=1}^{\infty} \frac{1}{\xi^n} u_\chi^{(n)} \right) + \frac{\partial u_\alpha \left(\bar{\chi}, \bar{\tau}, \bar{n}_\alpha, \bar{p}_3 \right)}{\partial \bar{\tau}} \left(\sum_{n=1}^{\infty} \frac{1}{\xi^n} u_\tau^{(n)} \right) \right.$$

$$\left. + \frac{\partial n_\alpha \left(\bar{\chi}, \bar{\tau}, \bar{n}_\alpha, \bar{p}_3 \right)}{\partial \bar{p}_3} \left(\sum_{n=1}^{\infty} \frac{1}{\xi^n} v^{(n)} \right) \right\} \Bigg|_{\substack{\bar{\chi}=\chi \\ \bar{p}_3=p_3}} + O \left(\frac{1}{\xi^3} \right), \quad (7.2.32)$$

Taking into account the above-obtained result, after not very difficult transformations we can rewrite expression (7.2.2) in the following form:

$$
n_\alpha\left(\chi, \tau, p_3\right) = \left.\left(\bar{n}_\alpha + \frac{1}{\xi} u_n^{(1)} + \frac{1}{\xi^2} u_n^{(2)} - \frac{1}{\xi^2} \frac{\partial u_n^{(1)}}{\partial \bar{p}_3} v^{(1)}\right)\right|_{\substack{\bar{\chi}=\chi, \bar{p}_3=p_3 \\ \bar{\tau}=\tau}}
$$

$$
= \bar{n}_\alpha + \frac{\bar{n}_\alpha q_\alpha k_3}{2\bar{\Omega}^2 m_\alpha \bar{\gamma}_\alpha^3} \sum_{m=1}^{N}\left(\frac{E_{3m}}{im} e^{imp_3} + \text{c.c.}\right)
$$

$$
+ \frac{\bar{n}_\alpha q_\alpha^2 k_3^2}{8\bar{\Omega}^4 m_\alpha^2 \bar{\gamma}_\alpha^6}\left(3 - \frac{3\bar{u}_\alpha \bar{\gamma}_\alpha^2 \bar{\Omega}}{c^2 k_3}\right)\left(\sum_{m=1}^{N}\left(\frac{E_{3m}}{im} e^{imp_3} + \text{c.c.}\right)\right)^2
$$

$$
+ \frac{\bar{n}_\alpha q_\alpha k_3}{2\bar{\Omega}^3 m_\alpha \bar{\gamma}_\alpha^3}\left(1 - \frac{3\bar{u}_\alpha \bar{\gamma}_\alpha^2 \bar{\Omega}}{c^2 k_3}\right) \frac{\partial \bar{u}_\alpha}{\partial z} \sum_{m=1}^{N}\left(\frac{E_{3m}}{m^2} e^{imp_3} + \text{c.c.}\right)
$$

$$
+ \frac{q_\alpha k_3 E_{30}}{\bar{\Omega} m_\alpha \bar{\gamma}_\alpha^3}\left(\frac{3\bar{u}_\alpha \bar{\gamma}_\alpha^2 \bar{\Omega}}{c^2 k_3} - 2\right)
$$

$$
+ \frac{\bar{n}_\alpha q_\alpha k_3}{2\bar{\Omega}^3 m_\alpha \bar{\gamma}_\alpha^3}\left[\sum_{m=1}^{N}\left(\frac{\partial E_{3m}}{\partial t} \frac{2}{m^2} e^{imp_3} + \text{c.c.}\right)\right.
$$

$$
\left. + \left(\bar{u}_\alpha + \frac{\omega_3}{k_3}\right) \sum_{m=1}^{N}\left(\frac{\partial E_{3m}}{\partial z} \frac{1}{m^2} e^{imp_3} + \text{c.c.}\right)\right], \quad (7.2.33)
$$

where $\bar{n}_\alpha(z, t)$ is the averaged concentration of the electron plasma of α-th beam, it can be calculated from equation (7.2.29). Obtained here expression (7.2.33) can be regarded as an approximate solution of initial continuity problem (7.1.1). However, corresponding solution for the averaged concentration $\bar{n}_\alpha(z, t)$ should be found after eventual solving of the considered field problem. We have in view the above mentioned electric field problem. The latter will be discussed somewhat below. Here, we turn our attention to the solving of the averaged quasilinear equation (7.2.29).

2.4   Characteristics of
the Averaged Continuity Equation

So, let us come back to averaged quasilinear equation (7.2.29)

$$
\left(\frac{\partial}{\partial t} + \bar{u}_\alpha \frac{\partial}{\partial \bar{z}}\right) \bar{n}_\alpha = -\bar{n}_\alpha \frac{\partial \bar{u}_\alpha}{\partial \bar{z}}, \qquad (7.2.34)
$$

which we will solve for the following boundary conditions:

$$
\bar{z} = 0, \ \bar{n}_\alpha = \bar{n}_{0\alpha}(t). \qquad (7.2.35)
$$

The characteristics of equation (7.2.34) are

$$\frac{d\bar{n}_\alpha(\bar{z}, t)}{dt} = -\bar{n}_\alpha(\bar{z}, t) \frac{\partial \bar{u}_\alpha(\bar{z}, t)}{\partial z},$$

$$\frac{d\bar{z}}{dt} = \bar{u}_\alpha(\bar{z}, t). \tag{7.2.36}$$

Solving equations (7.2.36) by standard means we can find the sought solutions of the averaged quasilinear equation (7.2.29), (7.2.34).

3. FIELD PROBLEM. APPLICATION OF THE METHOD OF AVERAGED CHARACTERISTICS FOR ASYMPTOTIC INTEGRATION OF THE MAXWELL's EQUATIONS

3.1 Averaged Maxwell's Equations

Taking into consideration one-dimension spatial nature of the considered model (see above Subsection 1.1) we can rewrite the Maxwell equations in the following simplified form:

$$\frac{\partial E_3}{\partial t} = -4\pi \sum_\alpha q_\alpha n_\alpha u_\alpha;$$

$$\frac{\partial E_3}{\partial z} = 4\pi \sum_\alpha q_\alpha n_\alpha. \tag{7.3.1}$$

As well known in the electrodynamics, any of two last equations can be chosen for calculational in the considered case. It is no difficult be convinced that the other one always can be constructed by using the continuity equation. Taking this into account, let us use further the only equation

$$\frac{\partial E_3}{\partial z} = 4\pi \sum_\alpha q_\alpha n_\alpha, \tag{7.3.2}$$

instead system (7.3.1). Here the functional expressions $n_\alpha(E_3, z, t)$ and $u_\alpha(E_3, z, t)$ are found in the two preceding Subsections. We will solve equation (7.3.2) for the boundary conditions

$$z = 0, \quad E_3 = E_{03}(t). \tag{7.3.3}$$

As before, the method of averaged characteristics for the asymptotic integration of (7.3.2) is chosen as the basic one.

We regard (7.3.1) as an equation of the zero hierarchical level. Constructing a corresponding system for the characteristics, we perform the passage to the parametrical form of the writing. For this we separate out the fast oscillatative phase p_3 and construct a corresponding equation for it. As a result we obtain:

$$\frac{dE_3}{dz} = 4\pi \sum_{\alpha} q_\alpha n_\alpha,$$

$$\frac{dt}{dz} = 0, \tag{7.3.4}$$

$$\frac{dp_3}{dz} = -k_3.$$

Similarly with the preceding cases, the concepts of slow time

$$\tau' = \frac{1}{\xi} t, \tag{7.3.5}$$

and slow coordinate

$$\chi' = \frac{1}{\xi} z \tag{7.3.6}$$

are introduced. Accomplishing required set of transformations, it is no difficult to obtain:

$$\frac{dE_3}{dz} = 4\pi \sum_{\alpha} q_\alpha n_\alpha = \frac{1}{\xi} X^E \left(\chi, E_3, p_3 \right),$$

$$\frac{d\tau}{dz} = 0 \Rightarrow t \equiv \tau = t_0,$$

$$\frac{d\chi}{dz} = \frac{1}{\xi} = \frac{1}{\xi} X^\chi \left(\chi, E_3, p_3 \right), \tag{7.3.7}$$

$$\frac{dp_3}{dz} = -k_3.$$

Here as before we have neglected the double prime symbol for simplicity.

As we did before, we separate out the large parameters of the problem:

$$[\xi_i] = [\xi_7, \xi_8] \tag{7.3.8}$$

where

$$\xi_7 \sim \chi_E \frac{\left|\frac{dp_3}{dz}\right|}{\left|\frac{dE_3}{dz}\right|} \gg 1, \quad \xi_8 \sim \chi_8 \frac{\left|\frac{dp_3}{dz}\right|}{\left|\frac{d\chi}{dz}\right|} \gg 1, \qquad (7.3.9)$$

where χ_7, χ_8 are corresponding normalizing factors. A smaller one of the large parameters χ_7, χ_8 is chosen as the problem's large parameter:

$$\xi = \min(\xi_7, \xi_8). \qquad (7.3.10)$$

According to the general scheme of the method of averaged characteristics, we seek the solutions of equations (7.3.4) in the form

$$E_3 = \bar{E}_3\left(\bar{\chi}, \bar{E}_3, \bar{p}_3, t\right) + \sum_{n=1}^{\infty} \frac{1}{\xi^n} u_E^{(n)}\left(\bar{\chi}, \bar{E}_3, \bar{p}_3, \dot{z}\right),$$

$$\chi = \bar{\chi} + \sum_{n=1}^{\infty} \frac{1}{\xi^n} u_\chi^{(n)}\left(\bar{\chi}, \bar{E}_3, \bar{p}_3, t\right), \qquad (7.3.11)$$

$$\bar{p}_3 = \bar{p}_3 + \sum_{n=1}^{\infty} \frac{1}{\xi^n} v^{(n)}\left(\bar{\chi}, \bar{E}_3, \bar{p}_3\right),$$

where for determination of the averaged values we have the system of equations for the first hierarchical level like

$$\frac{d\bar{E}_3}{dz} = \sum_{n=1}^{\infty} \frac{1}{\xi^n} A_E^{(n)}\left(\bar{\chi}, \bar{E}_3, t\right),$$

$$\frac{d\bar{\chi}}{dz} = \sum_{n=1}^{\infty} \frac{1}{\xi^n} A_\chi^{(n)}\left(\bar{\chi}, \bar{E}_3, t\right), \qquad (7.3.12)$$

$$\frac{d\bar{p}_3}{dz} = -k_3 + \sum_{n=1}^{\infty} \frac{1}{\xi^n} B^{(n)}\left(\bar{\chi}, \bar{E}_3, t\right).$$

Here, let us recall once more that averaged spaces in this and the previous problems are different. This means that the values $\bar{z}$ and $\bar{p}_3$ differ from the analogous values in the preceding problems.

As before, we restrict ourselves by keeping only terms $\sim 1/\xi$ in calculational of the auxiliary functions in expressions (7.3.11), (7.3.12):

$$A_E^{(1)} = 4\pi \sum_\alpha q_\alpha \bar{n}_\alpha, \qquad (7.3.13)$$

$$u_E^{(1)} = 4\pi \sum_\alpha \left(\frac{q_\alpha^2}{2\bar{\Omega}^2 m_\alpha \bar{\gamma}_\alpha^3} \sum_{m=1}^{N} \left(\frac{E_{3m}}{m^2} e^{imp_3} + \text{c. c.} \right) \right), \qquad (7.3.14)$$

$$A_\chi^{(1)} = 0 \qquad (7.3.15)$$

$$u_\chi^{(1)} = 0 \qquad (7.3.16)$$

$$B^{(1)} = 0 \qquad (7.3.17)$$

$$v^{(1)} = 0 \qquad (7.3.18)$$

$$A_E^{(2)} = 4\pi \sum_\alpha \left(\frac{3\bar{n}_\alpha q_\alpha^3 k_3^2}{4\bar{\Omega}^4 m_\alpha^2 \bar{\gamma}_\alpha^6} \left(1 - \frac{\bar{u}_\alpha \bar{\gamma}_\alpha^2 \bar{\Omega}}{c^2 k_3} \right) \sum_{m=1}^N \left(\frac{2E_{3m} E_{3m}^*}{m^2} \right) \right); \quad (7.3.19)$$

$$u_E^{(2)} = \sum_\alpha \left\{ \frac{3\pi \bar{n}_\alpha q_\alpha^3 k_3}{2\bar{\Omega}^4 m_\alpha^2 \bar{\gamma}_\alpha^6} \left(1 - \frac{\bar{u}_\alpha \bar{\gamma}_\alpha^2 \bar{\Omega}}{c^2 k_3} \right) \right.$$

$$\times \sum_{\substack{m=1,j=1 \\ m \neq j}}^{N,N} \left(-\frac{E_{3m} E_{3j}}{imj\,(m+j)} e^{i(m+j)\bar{p}_3} + \frac{E_{3m} E_{3j}^*}{imj\,(m-j)} e^{i(m-j)\bar{p}_3} + \text{c.c.} \right)$$

$$+ \frac{2\pi \, \bar{n}_\alpha q_\alpha^2}{\bar{\Omega}^3 m_\alpha \bar{\gamma}_\alpha^3} \left(\frac{1}{\xi} \frac{\partial \bar{u}_\alpha}{\partial \bar{\chi}} - \frac{k_3 q_\alpha E_{30}}{\bar{\Omega} m_\alpha \bar{\gamma}_\alpha^3} \left(2 - \frac{3\bar{u}_\alpha \bar{\gamma}_\alpha^2 \bar{\Omega}}{c^2 k_3} \right) \right)$$

$$- \frac{4\pi \, \bar{n}_\alpha q_\alpha^2}{\bar{\Omega}^3 m_\alpha \bar{\gamma}_\alpha^3}$$

$$\times \left(\sum_{m=1}^N \frac{1}{im^3} \frac{1}{\xi} \left(\frac{\partial E_{3m}}{\partial \tau} + \frac{1}{2} \left(\frac{\omega_3}{k_3} + \bar{u}_\alpha - \frac{\bar{\Omega}}{k_3} \right) \frac{\partial E_{3m}}{\partial \bar{\chi}} \right) e^{im\bar{p}_3} + \text{c.c.} \right)$$

$$+ \left(\frac{1}{\xi} \frac{\partial \bar{n}_\alpha}{\partial \bar{\chi}} \right) \frac{2\pi \, q_\alpha^2}{\bar{\Omega}^2 m_\alpha \bar{\gamma}_\alpha^3 k_3} \left(\sum_{m=1}^N \frac{E_{3m}}{im^3} e^{im\bar{p}_3} + \text{c.c.} \right) \right\}, \quad (7.3.20)$$

$$A_\chi^{(2)} = 0; \qquad (7.3.21)$$

$$u_\chi^{(2)} = 0; \qquad (7.3.22)$$

$$B^{(2)} = 0. \qquad (7.3.23)$$

Substituting (7.3.13), (7.3.15) and (7.3.17) into (7.3.12) we obtain the system of equations for the first hierarchical level in more specific form:

$$\frac{d\bar{E}_3}{dz} = 4\pi \sum_\alpha q_\alpha \bar{n}_\alpha,$$

$$\frac{dt}{dz} = 0; \qquad (7.3.24)$$

$$\frac{d\bar{p}_3}{dz} = -k_3.$$

Then we regard these equations as characteristic equation of some averaged quasilinear equation with partial derivatives:

$$\frac{\partial \bar{E}_3}{\partial z} = 4\pi \sum_\alpha q_\alpha \bar{n}_\alpha, \qquad (7.3.25)$$

that is equation with partial derivatives for the first hierarchical level. We can treat equation (7.3.25) as *the averaged Maxwell's equation*.

It should be mentioned that the values $\bar{u}_\alpha$ and $\bar{n}_\alpha$ in (7.3.25) are some known functions *in the averaged space of the considered field problem*. It should be emphasized that, in contrast to the situation which takes place in the previous cases, the values $\bar{u}_\alpha$ and $\bar{n}_\alpha$ in the considered case can be essentially different from the analogous values obtained in the preceding Subsections.

The calculational scheme of solution of equation (7.3.25) will be presented somewhat later. Now let us turn to the problem of back transformations.

3.2 Back Transformations

We will follow the calculational schemes of back transformations that were described more than once in this and the previous Chapters (see, for example, (6.2.17)–(6.2.20), and corresponding comments). Namely, we rewrite the Krylov–Bogolyubov substitutions in the form

$$E_3\left(\bar{\chi}, \bar{p}_3, t\right) = \bar{E}_3\left(\bar{\chi}, t\right) + \frac{1}{\xi} u_E^{(1)}\left(\bar{\chi}, \bar{n}_\alpha, \bar{p}_3, t\right) + \frac{1}{\xi^2} u_E^{(2)}\left(\bar{\chi}, \bar{n}_\alpha, \bar{p}_3, t\right),$$
$$\qquad (7.3.26)$$

and then the estimation of the differences could be represented as

$$E_3 - \bar{E}_3 = \sum_{n=1}^{\infty} \frac{1}{\xi^n} u_E^{(n)}\left(\bar{\chi}, \bar{n}_\alpha, \bar{p}_3, t\right)$$

$$\chi - \bar{\chi} = \sum_{n=1}^{\infty} \frac{1}{\xi^n} u_\chi^{(n)}\left(\bar{\chi}, \bar{n}_\alpha, \bar{p}_3, t\right) \qquad (7.3.27)$$

$$p_3 - \bar{p}_3 = \sum_{n=1}^{\infty} \frac{1}{\xi^n} v^{(n)}\left(\bar{\chi}, \bar{u}_\alpha, \bar{p}_3, t\right).$$

Using the circumstance that differences (7.3.27) are proportional to $1/\xi \ll 1$, we accomplish corresponding expansion in the Taylor series in (7.3.26). In what follows we take into consideration that $u_\chi^{(1)} = u_\chi^{(2)} = 0$, and $v^{(1)} = v^{(2)} = 0$. As a result we can write:

$$E_3\left(z,p_3,t\right)=E_3\left(\bar{\chi},\bar{p}_3,t\right)\Big|_{\substack{\bar{\chi}=\chi=\frac{1}{\xi}z\\ \bar{p}_3=p_3}},\qquad(7.3.28)$$

or, after some transformation, we obtain the sought solutions for the initial problem (7.3.2), which we can represent in the form like

$$E_3\left(z,p_3,t\right)=\bar{E}_3\left(z,p_3,t\right)+u_E^{(1)}\left(z,p_3,t\right)+u_E^{(2)}\left(z,p_3,t\right).\qquad(7.3.29)$$

3.3   Solving the Averaged Quasilinear Equation for the Electric Field

Solutions for the averaged electric field $\bar{E}_3\left(z,t\right)$ can be found from equation of the first hierarchical level (7.3.25):

$$\frac{\partial\bar{E}_3}{\partial\bar{z}}=4\pi\sum_{\alpha}q_{\alpha}\bar{n}_{\alpha},\qquad(7.3.30)$$

for the boundary conditions

$$\bar{z}=0,\bar{E}_3=\bar{E}_{030}\left(t\right).\qquad(7.3.31)$$

Let us use the method of characteristics (see previously Chapter 6, Section 3) for the solving of the equation (7.3.30).

According to the scheme of the characteristics method, the system for characteristics should formulated on its first step:

$$\frac{d\bar{E}_3}{d\bar{z}}=4\pi\sum_{\alpha}q_{\alpha}\bar{n}_{\alpha}=g_1\left(\bar{z},t\right),$$

$$\frac{dt}{d\bar{z}}=0.\qquad(7.3.32)$$

The first integrals of system (7.3.32) could be found in the form:

$$\begin{aligned}C_1&=t;\\ C_2&=\bar{E}_3\left(\bar{z},C_1\right)-g_2\left(\bar{z},C_1\right),\end{aligned}\qquad(7.3.33)$$

where

$$g_2\left(z,t\right)=\int g_1\left(z,t\right)dz.\qquad(7.3.34)$$

The constants $C'_{1,2}$ could be found after substituting boundary conditions (7.3.31) into (7.3.33):

$$C_1' = t_0;$$
$$C_2' = \bar{E}_{030}\left(C_1'\right) - g_2\left(0, C'\right). \tag{7.3.35}$$

Substitution of results (7.3.33) into (7.3.35) yields

$$\bar{E}_3\left(\bar{z}, t\right) = \bar{E}_{030}\left(C_1\right) + g_2\left(\bar{z}, C_1\right) - g_2\left(0, C_1\right), \tag{7.3.36}$$

$$\text{or } \bar{E}_3\left(\bar{z}, t\right) = \bar{E}_{030}\left(t_0\right) + \int\left(g_1\left(\bar{z}, t_0\right) - g_1\left(0, t_0\right)\right) d\bar{z}. \tag{7.3.37}$$

where $\bar{E}_{030}$ and $g_{1,2}$ are some arbitrary functions (of the stated arguments) that do not contradict to the definition (7.3.32) and boundary condition (7.3.31). Substitution of any of expressions (7.3.37) into (7.3.29) solves the stated initial problem (7.3.1). However, obtained this way solutions possess non-evident form. We should then reduce a non-evident solutions like (7.3.29) to evident form (7.1.3). Corresponding expressions for the harmonics $E_{3m}\left(z, t\right)$ in (7.1.3) for the beam space charge waves should be formulated for this.

3.4 Truncated Equations for the Harmonic Amplitudes of a Space Charge Wave

Let us turn to the problem of determining the complex amplitudes E_{3m}. We equalize for this definition for the electric wave field (7.1.3) and the obtained result like (7.3.29). Then, we equalize the same coefficients for equal exponents. As a result, the following equations could be obtained:

$$
E_{3m} = \sum_{\alpha}\left\{\frac{4\pi\,\bar{n}_\alpha(q_\alpha)^2}{m_\alpha\bar{\Omega}_\alpha^2\bar{\gamma}_\alpha^3}\left(\frac{E_{3m}}{m^2}\right)\right.
$$
$$
+ \frac{3\pi\,\bar{n}_\alpha q_\alpha^3 k_3}{\bar{\Omega}_\alpha^4 m_\alpha^2\bar{\gamma}_\alpha^6}\left(\frac{\bar{u}_\alpha\bar{\gamma}_\alpha^2\bar{\Omega}_\alpha}{c^2 k_3} - 1\right)
$$
$$
\times\left\langle\sum_{\substack{l=1,j=1\\l\neq j}}^{N,N}\left(-\frac{E_{3l}E_{3j}}{ilj\left(l+j\right)}e^{i(l+j)p_3} + \frac{E_{3l}E_{3j}^*}{ilj\left(l-j\right)}e^{i(l-j)p_3} + \text{c.c.}\right)\right\rangle_m
$$

$$+ \frac{E_{3m}}{im^3} \frac{4\pi \, \bar{n}_\alpha q_\alpha^2}{m_\alpha \bar{\Omega}_\alpha^3 \bar{\gamma}_\alpha^3} \left(\frac{\partial \bar{u}_\alpha}{\partial z} - \frac{k_3 q_\alpha E_{30}}{\bar{\Omega}_\alpha m_\alpha \bar{\gamma}_\alpha^3} \left(2 - \frac{3\bar{u}_\alpha \bar{\gamma}_\alpha^2 \bar{\Omega}_\alpha}{c^2 k_3} \right) \right)$$

$$- \frac{4\pi \, \bar{n}_\alpha q_\alpha^2}{\bar{\Omega}_\alpha^3 m_\alpha \bar{\gamma}_\alpha^3} \frac{1}{im^3} \left(2 \frac{\partial E_{3m}}{\partial t} + \left(\frac{\omega_3}{k_3} + \bar{u}_\alpha - \frac{\bar{\Omega}_\alpha}{k_3} \right) \frac{\partial E_{3m}}{\partial z} \right)$$

$$+ \frac{\partial \bar{n}_\alpha}{\partial z} \frac{4\pi \, q_\alpha^2}{m_\alpha \bar{\Omega}_\alpha^2 \bar{\gamma}_\alpha^3 k_3} \frac{E_{3m}}{im^3} \Bigg\}, \qquad (7.3.38)$$

where the expression $\langle \cdots \rangle_m = \frac{1}{2\pi} \int_0^{2\pi} \{ \cdots \} e^{imp_3} dp_3$ designates the procedure of separation out the m-th harmonic. We confine ourselves for simplicity only accounting the nonlinear terms no higher second order. The further generalization of obtained result (7.3.38) could be made by simple keeping all higher nonlinearities, which appear in the second approximation $1/\xi^2$ (i.e., on the lengths $L_{max} \sim \xi^2$ and time intervals $\tau_{max} \sim \xi^2$ — see the following Subsection for more details).

System (7.3.38) could be analyzed simpler if we use the linear dispersion relationship for the two-stream instability. Let use rewrite for this expressions (7.3.38) for the first harmonic in the linear approximation:

$$E_{3m} \left(1 - \frac{1}{m^2} \sum_\alpha \frac{\omega_{p\alpha}^2}{\bar{\Omega}_\alpha^2 \bar{\gamma}_\alpha^3} \right) = 0, \qquad (7.3.39)$$

where $\omega_{p\alpha}^2 = 4\pi \bar{n}_\alpha (q_\alpha)^2 / m_\alpha$ is the plasma frequency for α-th partial beam, $\bar{\gamma}_\alpha = \left(1 - \bar{u}_\alpha^2 / c^2 \right)^{-1/2}$, as before, is its relativistic factor. It is readily seen from (7.3.39) that nontrivial solution in this case could be got if the following condition is satisfied:

$$1 - \frac{1}{m^2} \sum_\alpha \frac{\omega_{p\alpha}^2}{(\omega_3 - \bar{u}_\alpha k_3)^2 \bar{\gamma}_\alpha^3} = 0. \qquad (7.3.40)$$

where $\bar{\Omega}_\alpha = \omega_3 - \bar{u}_\alpha k_3$. This is the sought linear dispersion equation [11–25] for the relativistic two-stream system. The correlation like $k_3 = k_3(\omega_3)$ could be found from (7.3.40) (see Chapter 13, Volume II for more details).

Let us shortly discuss dynamics of the harmonic amplitudes in the case, if we neglect the influence of the quasi-stationary longitudinal electric field E_{30}. This means that we can neglect in (7.3.38) by the terms proportional E_{30} as well as the terms $\sim \partial \bar{n}_\alpha / \partial z, \partial \bar{u}_\alpha / \partial z$ (because they are also proportional E_{30}):

$$E_{30} = 0,$$

$$\bar{u}_\alpha\left(z,t\right) = \bar{u}_\alpha\left(0,t\right) \equiv u_{0\alpha} = \text{const},$$
$$\bar{n}_\alpha\left(z,t\right) = \bar{n}_\alpha\left(0,t\right) \equiv n_{0\alpha} = \text{const}. \tag{7.3.41}$$

Then, accounting (7.3.39) and (7.3.40), we yield the following equation system from (7.3.38)

$$A\frac{1}{im^3}\frac{\partial E_{3m}}{\partial t} + B\frac{1}{im^3}\frac{\partial E_{3m}}{\partial z} = C$$

$$\times \left\langle \sum_{\substack{l=1,j=1 \\ l \neq j}}^{N,N} \left(-\frac{E_{3l}E_{3j}}{ilj\left(l+j\right)}e^{i(l+j)p_3} + \frac{E_{3l}E_{3j}^*}{ilj\left(l-j\right)}e^{i(l-j)p_3} + \text{c.c.}\right) \right\rangle_m,$$

$$\tag{7.3.42}$$

$$A = \sum_\alpha \left\{ \frac{2\omega_{p\alpha}^2}{\bar{\Omega}_\alpha^3 \bar{\gamma}_\alpha^3} \right\}, \quad B = \sum_\alpha \left\{ \frac{\omega_{p\alpha}^2}{\bar{\Omega}_\alpha^3 \bar{\gamma}_\alpha^3}\left(\frac{\omega_3}{k_3} + \bar{u}_\alpha - \frac{\bar{\Omega}_\alpha}{k_3}\right)\right\},$$

$$C = \sum_\alpha \left\{ \frac{3\omega_{p\alpha}^2 q_\alpha k_3}{4\bar{\Omega}_\alpha^4 m_\alpha \bar{\gamma}_\alpha^6}\left(\frac{\bar{u}_\alpha \bar{\gamma}_\alpha^2 \bar{\Omega}_\alpha}{c^2 k_3} - 1 \right)\right\}.$$

System (7.3.42) describes the nonlinear dynamics of the harmonic amplitudes. Analyzing its mathematical construction, we might be surprised by their resemblance with corresponding results like (8.2.16) which can be obtained in the framework of the method of slowly varying amplitudes (see further Chapter 8). This means that we, in principle, can regard the proposed calculational technique as a *new specific version of the slowly varying amplitude* method. As relevant comparative analysis shows, the proposed new version has a number of very important (from practical point of view) advantages. First of all, because it allows to obtain the equations like to (7.3.42) for arbitrary harmonic number m. The analogous procedure accomplished in framework of the traditional slowly varying amplitude method turns out to be too labor-intensive in such situation. Moreover, it cannot be really realized generally quite often.

3.5   Some Commentaries for the Obtained Results

Let us discuss the above mentioned possibility of accounting formally an unlimited number of the electron harmonics in the framework of the method considered. Saying about this, we, however, have in view a potential possibility only. The proposed algorithm indeed has no limitations, in principle, on the number of harmonics considered. But this does not means that we really should always deal with unrestricted number of harmonics. The matter is that the obtained solutions like (7.3.38) possess the approximate nature. I.e., they are obtained with some given precision, which, in our case, is determined by an accepted supposition about the number of approximation n (in the terms of the small parameter $1/\xi$). In this Chapter we have used the second approximation on $1/\xi$. This circumstance determines the real number of harmonics, which accounting makes sense for the accepted accuracy. It is necessary to increase the calculational precision in the case if we would consider more harmonics. For instance, to accomplish the analogous calculations, but in the next approximation on $1/\xi$. However, let us discuss this topic in more detail.

As was mentioned before, the peculiarity of the discussed type of methods (see, for example, Chapter 4) is that the precision of solutions which should be obtained, is determined beforehand. This means that if we want to accomplish calculations in some n-th approximation, the method used guarantee the given precision for some m' first harmonics only. This number of harmonics can be obtained from the following obvious condition:

$$\frac{E_{m'}}{E_1} \leqslant \frac{1}{\xi^n} << 1. \qquad (7.3.43)$$

The precision of calculational of the harmonics $m > m'$ cannot be guaranteed in this case. Or, in other words, the precision of calculational of all higher harmonics $m > m'$ can become lower than the precision that the given approximation can guarantee. Hence if we would like to increase the number of considered harmonics m we must increase the number of approximation n. So really the number of harmonics considered is not unlimited in the framework of the method of averaged characteristics. But let us stress once more that this limitation is connected with the precision of the problem only, and it has nothing to do with specific method limitations.

Besides the limitations discussed above on the number of considered harmonics, some others limitations exist in the considered theory. They are the limitations on the electron pulse duration τ_{max} and maximal length of the system L_{max}, respectively. These limitations arise from the characteristic feature of the used hierarchical asymptotic methods (see, for example, Chapter 4 for more details). Namely, they originate from the fundamental theorems like the Bogolyubov's theorem (see Chapter4, Subsection 2.3). According to the latter, we can guarantee the given precision of solutions for some limited time τ_{max}, and some limited length L_{max} only:

$$\tau_{\max} \leqslant a_\tau \xi^n, \quad L_{\max} \leqslant a_L \xi^n, \qquad (7.3.44)$$

where a_τ, a_L are some normalization constants. This means that results obtained, for instance, for the case $n = 2$ are valid only for the time $\tau_{\max}$ and the length $L_{\max}$:

$$\tau_{\max} \leqslant a_\tau \xi^2, \quad L_{\max} \leqslant a_L \xi^2, \qquad (7.3.45)$$

respectively. Taking into account limitations similar (7.3.43)–(7.3.45) is very important for a practice, especially in the cases when a numerical analysis is accomplished.

References

[1] A.A. Ruhadze, L.S. Bogdankevich, S.E. Rosinkii, V.G. Ruhlin. *Physics of high-current relativistic beams.* Atomizdat, Moscow, 1980.

[2] R.C. Davidson. Theory of nonlinear plasmas. Mass: Benjamin, Reading, 1974.

[3] A.G. Sitenko, V.M. Malnev. *Principles of the plasma theory.* Naukova Dumka, Kiev, 1994.

[4] J. Weiland, H. Wilhelmsson. *Coherent nonlinear interactions of waves in plasmas.* Pergamon Press, Oxford, 1977.

[5] L.A. Vainstein, V.A. Solnzev. *Lectures on Microwave electronics.* Sov. Radio, Moscow, 1973.

[6] V.I. Gaiduk, K.I. Palatov, D.M. Petrov. *Principles of microwave physical electronics.* Sov. Radio, Moscow, 1971.

[7] A.F. Alexandrov, L.S. Bogdankevich, A.A. Ruhadze. *Principles of plasma electrodynamics.* Vyschja Shkola, Moscow, 1978.

[8] A.N. Kondratenko, V.M. Kuklin. *Principles of plasma electronics.* Energoatomizdat, Moscow, 1988.

[9] B.E. Zshelezovskii. *Electron beam parametric microwave amplifiers.* Nauka, Moscow, 1971.

[10] V.V. Kulish, V.I. Savchenko. Method of averaged characteristics in the nonlinear theory of two-stream instability two-stream instability. *Gerald of Sumy State University, ser. Physics and Mathematics*, 4, 2002.

[11] V.V. Kulish. *Methods of averaging in nonlinear problems of relativistic electrodynamics.* World Federation Publishers, Atlanta, 1998.

[12] V.V. Kulish, S.A. Kuleshov, A.V. Lysenko. Nonlinear self-consistent theory of superheterodyne and free electron lasers. *The International journal of infrared and millimeter waves*, 14(3):451–568, 1993.

[13] N.Ja. Kotzarenko, V.V. Kulish. About the effect of superheterodyne amplification electromagnetic waves in the plasma-beam system. *Radiotechnika i Electronika (Sov.)*, 25(11):2470–2471, 1980.

[14] Perekupko V.A., Silivra A.A., Kotzarenko N.Ja., V.V. Kulish. Patent of USSR No. 8335259, 19. Priority of 28.01.80.

[15] G. Bekefi, K.D. Jacobs. Two-stream fels. *J. Appl. Phys.*, 53:4113–4121, 1982.

[16] O.N. Bolonin, V.V. Kulish, B.P. Pugachev. Superheterodyne amplification of electromagnetic waves in the system of two relativistic electron beams. *Ukrainian Physical Journal*, 33(10):1465–1468, 1988.

[17] M. Botton, A. Ron. Two-stream instability in fels. *IEEE Trans. of Plasma Sciences*, 18(3):416–423, 1990.

[18] V.V. Kulish, V.E. Storizshko. Free electron laser. Patent of USSR No. 1809934. Priority of 18.07.90.

[19] V.V. Kulish, B.P. Pugachev. To the theory of effect superheterodyne amplification waves in plasma of two-stream system. *Fizyka Plazmy (Sov.)*, 17(6):96–705, 1991.

[20] V.V. Kulish. To the theory of two-stream free electron lasers of klystron type. *Ukrainian Physical Journal*, 36(1):28–33, 1991.

[21] V.V. Kulish. To the theory of relativistic electron-wave free electron lasers. *Ukrainian Physical Journal*, 36(5):686–693, 1991.

[22] H. Wilhelmsson. Double beam free electron laser. *Physica Scripta*, 44:603–605, 1991.

[23] V.V. Kulish. Physics of two-stream free electron lasers. *Gerald of Moscow State University, ser. Physics and Astronomy*, 33(3):74–78, 1992.

[24] V.V. Kulish, V.E. Storizshko. Free electron laser. Patent of USSR No. 1837722. Priority of 13.10.92.

[25] V.V. Kulish. Superheterodyne electron-wave free electron laser. *The International Journal of Infrared and Millimeter Waves*, 14(3), 1993.

Chapter 8

HIERARCHICAL SYSTEMS WITH PARTIAL DERIVATIVES. SOME OTHER ASYMPTOTIC METHODS

As follows from the materials of the two previous Chapters, the hierarchical asymptotic methods, like the method of averaged characteristics, could be quite effective for solving nonlinear electrodynamic wave problems. At the same time, we have a possibility of seeing that these methods have some limitations with respect to their areas of application. In particular, it had been made clear that essential difficulties appear in the case of asymptotic integration of systems with number of variables more than three, etc.. As had been found, these limitations, predominantly, are determined by using some specific 'improving' of the method of characteristics, which is used here as one of the key elements (see Chapter 6, Section 3 for more details).

However, we note that the method of averaged characteristics is not the only method of the class discussed. There is a number of other methods which can also be effective, especially in the situations [1–12], when the method of averaged characteristics is not expedient. Some of them (such as the *method of slowly varying amplitudes* [1–5, 7–9]) are well known and widespread in practical research. Others (such as the *Mitropol'skii method* [10,11] or the *method of hierarchical transformation of coordinates* [12]) are not as popular, although they can also be quite useful in some specific calculational situations.

It should be mentioned that the traditional version of the method of slowly varying amplitudes is found to be suitable only for special calculational situations when the right-hand sides of some considered equations are described by some periodic and quasi-harmonic functions [11–15]. The systems of such a type are called *Rabinovich standard systems*. On the other hand, in studying the plasma-like systems researchers very often face specific physical phenomena which are found to be far

behind the mentioned standard situation. For instance, they are the
nonlinear generation of quasi-stationary fields and currents (which are
non-periodic), the degeneration mechanisms of wave resonant interactions, when the same resonance condition is satisfied for more than one
couple of waves, etc. [16–22]. This means that the method of slowly
varying amplitudes should be essentially improved (modernized) in the
case of its application for the electrodynamic nonlinear plasma-like wave
resonant problems. The main purpose of such modernization must be
relevant extending the application area for the mentioned standard version.

The goal of this Chapter is to give for the readers some general information about some of these 'other methods'. Below, the main attention
is paid to the method of slowly varying amplitudes (including its so called
modernized version). Besides that, we will discuss some general ideas of
the Mitropol'skii method and the method of hierarchical transformation
of coordinates.

1. MAIN IDEAS OF THE METHOD OF SLOWLY VARYING AMPLITUDES

The method of slowly varying amplitudes was been first proposed
and used by M.A. Leontovich [13]. The method was applied for solving
the known problem concerning nonlinear evolution of an electromagnetic wave propagating in the Earth atmosphere. The main idea of the
proposed method is, in principle, analogous to the Van der Pol method
(see in Chapter 3, Section 4 for more details). Namely, the wave amplitude owing to the weak nonlinearity of considered models should change
slowly (see the concept of weak nonlinearity in Subsection 2.1, Chapter 1). Or, in other words, the wave amplitude in such a situation clearly
demonstrates properties of a slowly varying value. This allows us to reduce the initial problem of the wave propagation to the problem of the
slow amplitude evolution. In fact, the latter 'slow' problem is essentially
simpler than the initial 'fast oscillatative' problem. Owing to this and
to the simplicity of the discussed calculational idea in itself the method
of slowly varying amplitudes became rather popular in electrodynamics,
radiophysics, nonlinear optics, etc. [4–9]. This original (simple) version
of the discussed method is called the *simplified version of slowly varying
amplitude method*.

Further generalization and improvement of the method was accomplished by a number of authors. However, it should be mentioned that
A.V. Gaponov-Grehov, M.I. Rabinovich, and L.A. Ostrovsky [1–3] have
achieved most remarkable successes in this field. As a result of their
efforts the method acquired the required level of validity and strictness.

This 'improved version' of the method was named the *rigorous version of the method of slowly varying amplitudes*.

Unfortunately, the essential complication of calculational procedures turned out to be an unpleasant price for achieving the rigor and versatility mentioned. On the other hand, there are a lot of practical situations in which these rigor and versatility are not required. That is why below we will discuss both these versions.

1.1 General Calculational Scheme of the Slowly Varying Amplitudes Method

The main calculational idea of the method of slowly varying amplitudes is illustrated in Fig. 8.1.1. Analogously with the method of averaged characteristics (see Fig. 6.2.1 and relevant commentaries) we can separate out the three following stages of the asymptotic integration of the equations like (7.1.1) (see Fig. 8.1.1). The first is the *straight transformation*. The goal of this stage is the transformation of an initial equation with partial derivatives (zeroth hierarchical level) into the so called *truncated* equation for slowly varying amplitudes (the first hierarchical level). The second stage is obtaining the solutions of the truncated equation. The third stage is constructing the solutions for the initial (zeroth) hierarchical level (on the basis of the solutions found earlier for the first hierarchical level). Let us discuss some calculational peculiarities of these stages in more detail. Such specific calculational technologies, however, will be described somewhat below. Here we restrict ourselves by discussions of the method general ideas only.

So we begin with the first stage of the calculational scheme considered. Let us choose some dynamical wave system described by the following differential matrix equation (see definition (3.3.1)):

$$A\frac{\partial U}{\partial t} + (BP)U + CU = R(U, \partial U/\partial t, (PU), z, t), \qquad (8.1.1)$$

which was previously called the *general standard form* (for the method of slowly varying amplitudes).

Here A, B, C are square matrices of size $n \times n$, $U = U(z,t)$ is some vector-function in Euclidean n-dimensional space R^n with coordinates $\{z_1, z_2, \ldots, z_n\}$, i.e. $\forall z \in R^n$ $z = (z_1, z_2 \ldots, z_n)^T$, $\forall z_i \in (-\infty, +\infty)$, $i \in (1, 2, \ldots, n)$, P is some linear differential operator in the space R^n, $R(\ldots)$ is a given weak nonlinear vector-function, t is some scalar variable (for instance, the laboratory time). As a result of the straight transformations we performed some equation for the first hierarchical level (i.e., truncated equation) like the following

$$A\frac{\partial U^{(1)}}{\partial t} + (BP^{(1)})U^{(1)} + CU^{(1)} = R^{(1)}(U^{(1)}, \frac{\partial U^{(1)}}{\partial t}, (P^{(1)}U^{(1)}), z^{(1)}, t),$$
$$(8.1.2)$$

should be obtained. Here superscript (1) denote that this value belongs
to the first hierarchical level.

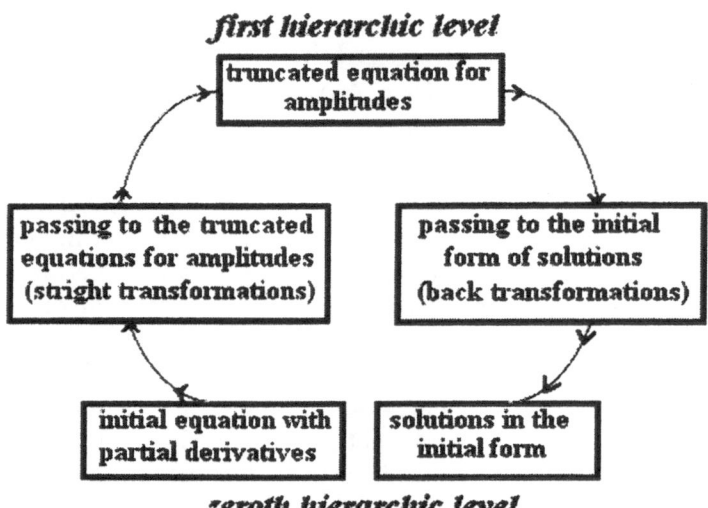

Figure 8.1.1. General calculational scheme of the slowly varying amplitude method

The main idea of the method discussed, as mentioned above, consists
in using the main property of the weak nonlinear wave systems (see
Chapter 1, Subsections 2.1 and 2.4). Namely, the dynamics of the weak
nonlinear wave system (8.1.1) is *almost* identical to the dynamics of
some *equivalent linear system*

$$A\frac{\partial U}{\partial t} + (BP)U + CU = 0 \qquad (8.1.3)$$

over a few wave oscillations. Some remarkable differences in behavior of
both system (linear and weak nonlinear ones) appear after *occurrence of
many wave oscillations* in the system. This means that in principle, we
could choose basis solution (*generated form of solution*)

$$U_k = \left\{ \sum_{s=1}^{q} \psi_k^s \exp\left[i\left(\omega_s t - \vec{k}_s\left(\omega_s\right)\vec{r}\right)\right] + \text{c. c.} \right\}, \qquad (8.1.4)$$

for description of the system nonlinear dynamics, which is obtained for
some equivalent linear system like (8.1.3). Here $\psi_k^s = \psi_{0k}^s \exp\left(i\varphi^s\right)$ is

the *complex amplitude* of the s-th wave (see the concept of complex amplitudes in Chapter 1, Subsection 2.4), ω_s is the cyclic frequency, and $\vec{k}_s(\omega_s)$ is the wave vector of the s-th wave; $\vec{r}$ is the vector of spatial coordinate. Differences between these solutions lies only in the following: the values, which are constants in the linear case (8.1.4), become slowly varying in the corresponding nonlinear solutions. As can be easily seen, we have two such constants that characterize the linear wave process (8.1.4). Namely, they are the *real amplitude* ψ_{0k}^s and *the initial phase* φ^s, respectively (see Chapter 1, Subsection 2.4). It is important to note that these values play the role of *slowly varying real amplitudes* and *slowly varying initial phases* in the discussed method. Thus the concept of complex amplitude $\psi_k^s = \psi_{0k}^s \exp(i\varphi^s)$ (see Chapter 1, Subsection 2.4) can also be used for description of the nonlinear system dynamics. In connection with this we, to be exact, talk about the *method of slowly varying complex amplitudes*. But, in fact, further we will use the shortened version of this term: the *method of slowly varying amplitudes*.

In what follows we should additionally apply some resonant condition in the case of the *wave resonant system*.

Substituting solution (8.1.4) and assuming the amplitudes to be the slowly varying functions

$$\psi_k^s = \psi_k^s(\vec{r}, t), \tag{8.1.5}$$

we can obtain the so called *truncated equation for amplitudes* (equation for the first hierarchical level) like (8.1.2). Thus we consider that

$$U^{(1)} = \begin{pmatrix} \psi_1^s \\ \psi_2^s \\ \dots \\ \psi_q^s \end{pmatrix}. \tag{8.1.6}$$

Thus the first stage of the considered calculational scheme, in turn, consists of the following five successive steps (substages):

a) constructing an equivalent linear equation (comparison equation);

b) solving of this linear equation;

c) formulation of the resonant conditions (in the case the system is resonant wave one);

d) constructing of a form of the sought nonlinear solutions of the initial problem on the basis of obtained linear solutions;

e) forming an equation for the slowly varying amplitudes.

The solving of the obtained truncated equation for slowly varying amplitudes like (8.1.2) is the second of the three above mentioned stages (see Fig. 8.1.1). Using of any suitable solution method, including the method of characteristics (see Chapter 6, Section 3), can be utilized for this.

The third stage is accomplishing of the back transformations. In contrast to the methods, similar to the method of averaged characteristic (see Chapter 6), this calculational procedure is found here to be rather simple and obvious. Indeed, the solutions of the initial problem (8.1.1) can easily be obtained using known solutions for the slowly varying amplitudes by simple substitution of these solutions in the linear (quasilinear) solutions like (8.1.4).

In the following let us illustrate the calculational scheme described above on the simplest example of the problem of *parametric amplification* of a wave [4, 5]. Therein later in this Section we will restrict ourselves by studying of the simplified version of the slowly varying amplitude method.

1.2 Simplified Version of the Slowly Varying Amplitude Method. Example: Effect of Parametric Amplification of a Wave

The model. We choose the simplest quadratic-nonlinear spatially one-dimensional model for illustration of the most characteristic features of the slowly varying amplitudes method. It is supposed that two waves of arbitrary physical nature are propagated along z-axis. Let us assume that the nonlinear dynamics of the first of them (one that is characterized by some dimensionless value u_1) can be described by the equation

$$\frac{\partial u_1}{\partial t'} + v_1 \frac{\partial u_1}{\partial z'} = u_1 u_2. \tag{8.1.7}$$

It can be easily seen that (8.1.7) is the simplest case of the standard form (8.1.1). Here t', z' and v_1 are dimensionless time, dimensionless spatial coordinate, and dimensionless phase velocity, respectively. Further we will omit the prime in the values t' and z' for the sake of simplicity.

The second wave that is characterized by the value u_2 plays a role of the *pumping wave*. The amplitude of this wave is considered to be much greater then the first wave $|u_1| << |u_2|$ (so called *approximation of strong pumping* [5, 7, 19]). We can suppose that, in contrast with the first wave, the amplitude of the second wave could be considered approximately constant. So we can consider the field of the pumping wave as a given one. The functional dependency $u_2(z,t)$ we choose in the form

$$u_2 = U_2 \exp\{i(\omega_2 t - k_2 z)\} + \text{c. c.} = U_2 \exp(ip_2) + \text{c. c.}, \tag{8.1.8}$$

where all definitions are obvious considering (8.1.4).

The first stage of calculational procedure. According was had been set forth in the preceding Subsection, we should pass to the first hierarchical level (see Fig. 8.1.1) at the first stage of the described calculational algorithm. In turn, the constructing of the equivalent linear equation like (8.1.3)

$$\frac{\partial u_1}{\partial t} + v_1 \frac{\partial u_1}{\partial z} = 0 \qquad (8.1.9)$$

is the first step of the first stage (first substage). We will seek for the solution of (8.1.9) in the form (*generating form of the solution*)

$$u_1 = U_1 \exp\left\{i\left(\omega_1 t - k_1 z\right)\right\} + \text{c. c.,} \qquad (8.1.10)$$

where the dispersion law (see Chapter 1, Subsection 3.6 and below in Chapter 8, Section 2) could be found after substitution of (8.1.10) into (8.1.9). After trivial calculational this law could be easily found:

$$k_1 = \omega_1/v_1.$$

Below we consider that the relevant solution of the initial nonlinear problem (8.1.7) should be sought in a form analogous to (8.1.10). The difference is only in the fact that the complex amplitude U_1 in this case turns out to be slowly varying value relative to the variables t and z. Let us simplify the model even more by supposition that it is spatially homogeneous with respect to wave amplitudes U_1 and U_2. Then the basic form of solution for the nonlinear problem can be written

$$u_1 = U_1\left(t\right) \exp\left\{i\left(\omega_1 t - k_1 z\right)\right\} + \text{c. c.} = U_1\left(t\right) \exp\left(ip_1\right) + \text{c. c.,} \qquad (8.1.11)$$

where $U_1\left(t\right)$ is the unknown (sought) slowly varying amplitude as a function on time t. Corresponding truncated equation like (8.1.2) should be constructed for determining of the slowly varying amplitude $U_1\left(t\right)$. Let us construct such equation.

Substitution of solution (8.1.11) into the left side of initial equation (8.1.7) yields:

$$\left(\frac{\partial}{\partial t} + v_1 \frac{\partial}{\partial z}\right) u_1 = i\left(\omega_1 - v_1 k_1\right) U_1 \exp\left(ip_1\right)$$

$$+ \exp\left(ip_1\right)\left(\frac{\partial}{\partial t} + v_1 \frac{\partial}{\partial z}\right) U_1 = \frac{dU_1}{dt} \exp\left(ip_1\right) + \text{c. c.} \qquad (8.1.12)$$

This means that expression (8.1.7) can be rewritten in the form

$$\frac{dU_1}{dt} \exp{(ip_1)} + \text{c. c.}$$
$$= (U_1 \exp{(ip_1)} + U_1^* \exp{(-ip_1)})(U_2 \exp{(ip_2)} + U_2^* \exp{(-ip_2)})$$
$$= U_1 U_2 \exp{\{i(p_1 + p_2)\}} + U_1 U_2^* \exp{\{i(p_1 - p_2)\}}$$
$$+ U_1^* U_2 \exp{\{-i(p_1 - p_2)\}} + U_1^* U_2^* \exp{\{-i(p_1 + p_2)\}}. \quad (8.1.13)$$

Two types of solutions for the equation (8.1.13) can be found there. The first solutions are non-resonant ones. They are not interesting for our study. Therefore, here we omit its analysis. The second are resonant solutions. They are essentially more interesting as well from the physical and calculational points of view. So let us discuss this case in more detail.

The resonant state of the system in the case of Lagrange description can be determined (see introduction for this Chapter and Chapter 1, Subsection 2.3) as a closeness of velocities of oscillatative phases change for proper and stimulated waves. One can easily see that this definition is equivalent to the *closeness of phases of proper and stimulated waves* in the case of Euler description (because variables z and t are independent values here). In our case the proper wave is connected with the phase p_1. In turn, the stimulated waves are described by the terms in the right side of expression (8.1.13). As can be easily seen, four combinational waves with the phases $\pm(p_1 \pm p_2)$, that correspond to four stimulated waves, can exist in the considered system. According to the given definition, the resonant state of the system could be reached in the case when the exponent $\exp{(ip_1)}$ in the left side of equation (8.1.13) is equal to one of the combinative exponents $\exp{\{\pm i(p_1 \pm p_2)\}}$ in the right side. Simplest analysis shows that interaction of two of the stimulated waves with the phases $-(p_1 \pm p_2)$ only with the proper wave p_1 can have the resonant nature. Thus the two remaining waves interact in the non-resonant manner. Generalizing, we can say that both mentioned resonant conditions could be written in the following form:

$$p_1 = -p_1 + \sigma p_2 \quad \text{or} \quad 2p_1 = \sigma p_2, \quad (8.1.14)$$

where $\sigma = \pm 1$ is the sign function. Some another form of writing of the resonant condition (8.1.14) is also used traditionally in literature (see, for instance, [4–13]):

$$2\omega_1 = \sigma \omega_2; \quad 2k_1 = \sigma k_2. \quad (8.1.15)$$

The expressions like (8.1.15) are called the *condition of parametric resonance* (see the classification of resonances in Chapter 4, Subsection 1.3).

Below we take into consideration that in the case of occurrence of the resonance we can neglect all other non-resonant terms in the right side of equation (8.1.13). The explanation is quite simple. However, let us discuss this topic in more detail. It is well known [4–9,19] that one of the conditions like (8.1.15) usually corresponds to increase of amplitude U_1 (on time t, in the considered case). This means that the non-resonant amplitudes appear inevitably in the situation when the amplitude of this resonant wave becomes essentially higher. So, in such case we can indeed neglect the influence of all non-resonant waves comparing with resonant waves.

Taking the latter into account and equating the expressions with the same (resonant) exponents in the left and right sides of the equation (8.1.13), we can formulate the sought equation for the slowly varying amplitude $U_1(t)$:

$$\frac{dU_1}{dt} = U_1^* U_2 \delta_{\sigma=+1} + U_1^* U_2^* \delta_{\sigma=-1}, \qquad (8.1.16)$$

where $\delta_{\sigma=\pm1}$ are the Kronecker symbols.

Thus the main task of the first stage of the discussed calculational procedure is fulfilled. So we can pass to the second stage (see Fig. 8.1.1) that is finding the solutions of equation (8.1.16).

The second stage of the calculational procedure. Let us pass to the real form of writing of equation (8.1.8). For this we use the definitions of the amplitudes $U_{1,2}$ as complex values:

$$U_j = U_{j0} \exp\left(i\varphi_j\right) = U_{j0}\left(\cos\varphi_j + i\sin\varphi_j\right), \qquad (8.1.17)$$

where $j = 1,2$. Substituting (8.1.17) into (8.1.16) and separating the real and imaginary parts in (8.1.16), we can obtain the following system of real equations:

$$\frac{dU_{10}}{dt} = U_{10}U_{20}\cos\Phi;$$
$$\frac{d\Phi}{dt} = -2U_{20}\sin\Phi, \qquad (8.1.18)$$

where $\Phi = 2\varphi_1 - \sigma\varphi_2$ is the *phase mismatch*. We divide the first of equations (8.1.18) by $U_{10}/2$

$$\frac{2}{U_{10}}\frac{dU_{10}}{dt} = 2U_{20}\cos\Phi. \tag{8.1.19}$$

Then by carrying out some obvious transformations we reduce it to the form

$$\frac{d}{dt}\left(\ln\left(U_{10}^2\right)\right) = 2U_{20}\cos\Phi. \tag{8.1.20}$$

In the following we turn our attention to the second of equations (8.1.18). We multiply both parts of this equation by the value $\cos\Phi/\sin\Phi$ and after simple transformations it can be easily obtained that

$$\frac{d}{dt}\left(\ln\left(\sin\Phi\right)\right) = -2U_{20}\cos\Phi. \tag{8.1.21}$$

Addition of equations (8.1.20) and (8.1.21) allows us to find the following system integral:

$$\ln\left(U_{10}^2\sin\Phi\right) = \text{const} = \ln C_1, \tag{8.1.22}$$

or, in the more acceptable form,

$$U_{10}^2\sin\Phi = C_1. \tag{8.1.23}$$

The integral (8.1.23) says that multiplication of the functions U_{10}^2 and $\sin\Phi$ is constant for any instant of time, including the initial instant $t = 0$. It can easily be seen that in the case

$$t = 0, \quad \Phi = \Phi_0 = n\pi, \tag{8.1.24}$$

where $n = 0,1,2,\ldots$, the constant $C_1 = 0$ for any time $t > 0$. It is interesting that, as analysis show, the effect of amplification of the first wave (8.1.11) can be maximal namely in the case (8.1.24). Therefore, further we restrict ourselves by the studying case (8.1.24) only. Accepting the supposition $C_1 = 0$, we can rewrite system (8.1.18) as

$$\begin{aligned}\frac{dU_{10}}{dt} &= U_{10}U_{20}\left(-1\right)^n; \\ \frac{d\Phi}{dt} &= 0,\end{aligned} \tag{8.1.25}$$

because $U_{10}^2 > 0$ always (see (8.1.23)). Equations (8.1.25) can be easily integrated:

$$U_{10} = U_{10}(0) \exp\{(-1)^n U_{20}t\},$$
$$\Phi = n\pi = \text{const}.$$
$$(8.1.26)$$

As follows from (8.1.26), *parametric resonant amplification* takes place for $n = 2k$. Correspondingly, we obtain the *parametric resonant damping* for the case $n = 2k + 1$.

According to the ideology accepted in this book, solutions (8.1.25) can be regarded as the solutions for the first hierarchical level. We must pass to the initial (zero hierarchical level — see Fig. 8.1.1) in order to obtain the solutions in the initial form. So, let us turn to the problem of back transformations.

The third stage of the calculational procedure. The third stage of the calculational procedure (back transformation) could be accomplished very simply. We found the sought nonlinear solution for the zeroth hierarchical level after substitution of solutions (8.1.26) into (8.1.17) and then into (8.1.11):

$$u_1 = U_{10}(0) \exp\{(-1)^n U_{20}t\} \exp\left\{\frac{i}{2}(\sigma\varphi_2 + n\pi)\right\} \exp\{i(\omega_1 t - k_1 z)\}$$
$$+ \text{c.c.} \quad (8.1.27)$$

Thus the problem stated above of the effect of parametric amplification of a signal wave in some weak nonlinear non-dispersion medium is solved.

2. TRADITIONAL VARIANT OF THE SLOWLY VARYING AMPLITUDES METHOD. RIGOROUS VERSION

As was mentioned above, the *traditional version of the slowly varying amplitude* method is intended for application in studying weak nonlinear wave resonant problems that occur in arbitrary distributed systems of different physical nature. Thus traditionally it is supposed only wave type processes realize. Standard system which describes such situation can be represented by the second of equations (3.3.21) for $U_0 = 0$. In the alternative case, we talk about the *modernized version of the method of slowly varying amplitudes* discussed below in Section 3.

Then let us turn the reader attention that the simplified version of the method of slowly varying amplitudes, discussed above in the preceding Section, sometimes turns out to be too rough for practical applications.

Firstly, because the proposed calculational procedures of the straight and back transformations (see Fig. 8.1.1) have not any satisfactory substantiation. This procedure can be effective, as a practice show. As a rule such situation is characteristic in the cases when we have to do with lowest nonlinear approximations (quadratic, as it takes place in the above described example) [4, 5, 7]. In other cases we risk to lose some terms in right parts of the truncated equations for amplitudes.

Secondly, because the simplified scheme very often does not give satisfactory answer about the methods of analysis of systems with different velocities of slow changing of amplitudes (and, sometimes, the medium parameters) along the different spatial coordinates. This is important inasmuch as such physical situations are not rare for real practice [4]. The *rigorous version of the method of slowly varying amplitudes* is intended for namely such situations. Besides that, it allows more confidently to perform the procedures of straight and back transformation.

2.1   Case of Spatially One-Dimensional Model

Let us start with the studying of the *spatially one-dimensional model*. The standard system (8.1.1) is chosen as initial one. We consider transversely unbounded homogeneous (in the plane XY) model in which partial plane waves propagate along the *z-axis*. In this case, all space-coordinate dependencies of the vector U components can be written as $U(t, z)$, i.e., the model is spatially one-dimensional. Operator P in (8.1.1) we choose in the simplest differential form

$$P \equiv \partial/\partial z, \tag{8.2.1}$$

We neglect effects such as 'nonlinear generation of the quasi-stationary fields' (see Chapters 12, 13, Volume II). Formally, this is equivalently to neglecting slowly quasi-stationary varying part of the vector U in (3.3.21), i.e., $U_0 = 0$. Assuming $U = \tilde{U}$ (which means that the wavy fields are presented in the model only), one-dimensional version of initial standard equation can be written in the so called Rabinovich's *standard form* [1–3]:

$$A\frac{\partial U}{\partial t} + B\frac{\partial U}{\partial z} + CU = \sum_{n=1}^{\infty} \varepsilon^n R^{(n)}\left(U, \frac{\partial U}{\partial t}, \frac{\partial U}{\partial z}, t, z, \tau, \chi\right). \tag{8.2.2}$$

Here, besides the laboratory 'ordinary' time t and the 'ordinary' coordinate z the *slow time* τ and *slow coordinate* χ are introduced additionally, i.e.,

$$\tau = t/\xi; \quad \chi = z/\xi, \tag{8.2.3}$$

Equation (8.2.2) can be represented in the scalar form

$$\sum_{l=1}^{p} \left(a_{kl}\frac{\partial}{\partial t} + b_{kl}\frac{\partial}{\partial z} + c_{kl} \right) U_l = \sum_{n=1}^{\infty} \varepsilon^n R_k^{(n)} \left(U, \frac{\partial U}{\partial t}, \frac{\partial U}{\partial z}, t, z, \tau, \chi \right).$$
$$\tag{8.2.4}$$

The next important step of the method of slowly varying amplitudes (see above Section 1) is the procedure of forming the sets of slow and fast variables. The so called *generating form of solution* (see (8.1.4)) can be obtained from the *equivalent linear system* (see (8.1.3)):

$$\sum_{l=1}^{n} \left(a_{kl}\frac{\partial}{\partial t} + b_{kl}\frac{\partial}{\partial z} + c_{kl} \right) U_l \,|_{\varepsilon=0} = 0. \tag{8.2.5}$$

We write the solution to equation (8.2.5) as $\sim \exp\{i(\omega t - kz)\}$. The system (8.2.5) has nontrivial solutions if the condition

$$D(\omega, k) = \det [a_{kl}\omega - b_{kl}k - ic_{kl}] = 0 \tag{8.2.6}$$

holds. This formula is called *the linear dispersion equation* determining linear dispersion properties of the system. The function $D(\omega, k)$ is the *dispersion function*. Solution of the equation (8.2.6) allows finding the *law of dispersion* $k = k(\omega)$ for the relevant waves of the system (see definitions (1.3.9)–(1.3.11) and relevant commentaries). The waves described by condition (8.2.6) relate to *proper waves* of the system. The main characteristic of these waves is that they propagate through the system after turning off the wave source. Electromagnetic waves in the vacuum or some non-conductive medium, waves on water surface, acoustic waves of different types, etc., can serve as examples. Apart from the proper waves in the system, as mentioned above in Section 2 there exist *stimulated (induced) waves*, too. These waves are excited in every point of the system by some distributed force. Properties of such waves are completely determined by characteristic features of the acting force. In contrast with the proper waves, the stimulated ones vanish after turning off the source, i.e., the stimulated force. The condition existing of improper waves can be formulated as an alternative to (8.2.6).

$$D(\omega, k) = \det [a_{kl}\omega - b_{kl}k - ic_{kl}] \neq 0, \tag{8.2.7}$$

i.e., cyclic frequency ω and wave number ω are independent quantities for such wavy objects.

Using (8.2.6) solution for linear system (8.2.5) can be written as (see above (8.1.4))

$$U|_{\varepsilon=0} = \sum_{m=1}^{r} \psi_k^m \exp\left[i\left(\omega t - k_m\left(\omega\right) z\right)\right] + \text{c.c.}, \qquad (8.2.8)$$

here ψ_k^m are amplitudes to be found. Substituting (8.2.8) in (8.2.5) yield a system of algebraic equations

$$\sum_{l=1}^{p-1} g_{kl}^m \psi_l^m = g_{lp}^m \psi_p^m, \qquad (8.2.9)$$

where $g_{kl}^m = a_{kl}\omega_m - b_{kl}k_m - ic_{kl}.$. We construct the solution to the system (8.2.9) in terms of Kramer formula with first $p-1$ equations being linearly independent, i.e.,

$$\psi_l^m = D_l^m / D_{pp}^m. \qquad (8.2.10)$$

The determinant

$$D_l^m = \sum_{k=1}^{p-1} A_{kl}^m g_{kp}^m \psi_p^m = \psi_p^m \sum_{k=1}^{p-1} A_{kl}^m g_{kp}^m = \psi_p^m \sum_{k=1}^{p-1} \left(\partial D_{pp}^m / \partial g_{kl}^m\right) g_{kp}^m,$$
$$(8.2.11)$$

is obtained substituting the column of free terms $\left\{g_{kp}^m \psi_p^m\right\}$ for *l-th* column of the matrix $[g_{kl}^m]$ of the order $(p-1) \times (p-1)$. Here D_{pp}^m is minor of the element g_{pp}^m in the determinant D^m, i.e.

$$D_{pp}^m = \partial D^m / \partial g_{pp}^m \qquad (8.2.12)$$

A_{kl} is the algebraic complement of the element g_{kl} in the determinant D_{pp}.

We construct solutions of initial standard system (8.2.2), (8.2.4) in terms of (8.2.8), regarding the complex amplitudes ψ_p^m as slowly varying functions, i.e., as the function of second hierarchical level. Therefore, the next stage of hierarchical procedure is constructing relevant dynamical operators connecting dynamical variables of the first and the second hierarchical levels. For this we use linear solutions (8.2.8) as the initial basis. Besides that, we must define the wave resonance and account it. Here we remind that nonlinear resonant processes are main causes of slowly varying of dynamical variables at second hierarchical level (see Chapter 1, Subsection 2.3).

Suppose q proper waves (8.2.8) (here q is finite) interact with the wave of stimulated (induced) oscillations. We impose the wave resonant condition for one of the stimulated wave only given by non-linearity of the medium and q proper waves (see illustration example (8.1.15) and corresponding commentaries), i.e.,

$$\omega_d = \sum_{s-1}^{q} n_{sd}\omega_s; \quad k_d = \sum_{s=1}^{q} n_{sd}k_s, \qquad (8.2.13)$$

where ω_d and k_d are the frequency and the wave number of stimulated wave, $n_{sd} = \pm 1, \pm 2, \cdots$. Within (8.2.7) resonant wave condition (i.e., closeness of the frequencies and wave numbers of proper and stimulated waves) can be given as

$$D\left(\omega, k\right)\Bigg|_{\substack{\omega \approx \omega_d \\ k \approx k_d}} \approx D\left(\omega_d, k_d\right) \cong 0. \qquad (8.2.14)$$

The condition (8.2.14) holds only for a finite number of waves since the system always has dispersion.

For the above reason we look for the solution of (8.2.4) for the slowly varying amplitudes generally as

$$U_k = \left\{ \sum_{s=1}^{q} \psi_k^s\left(\omega_s\right) A^s\left(\tau, \chi\right) \exp\left[i\left(\omega_s t - k_s\left(\omega_s\right)z\right)\right] - \text{c. c.} \right\}$$
$$+ \sum_{n=1}^{q} \varepsilon^n W_k^{(n)}\left(A, t, z, \tau, \chi\right), \quad (8.2.15)$$

where $A^s = \alpha^s \exp\left(i\varphi^s\right)$ is slowly varying complex amplitude of s-th wave normalized by ψ_k^s. Within above concept of functional hierarchical operator (2.2.7) we consider (8.2.15) as definition of the operator $\hat{U}^{(1)}$ for the wave resonant problem (8.2.2), (8.2.4). The next stage of hierarchical calculational is determining unknown variables of the first hierarchical level (i.e., they are slowly varying functions $A^s\left(\tau, \chi\right)$ and $W_k^{(n)}\left(A, t, z, \tau, \chi\right)$). For this purpose we use hierarchical resemblance principle, i.e., we construct the dynamical equation of the first hierarchical level in the similar (in mathematical structure) to equation (8.2.4) form (see also (8.1.2) and corresponding commentaries):

$$\sum_{l=1}^{p} \left(a_{kl}^{(1)} \frac{\partial}{\partial \tau} + b_{kl}^{(1)} \frac{\partial}{\partial \chi} + c_{kl}^{(1)} \right) A_l (\tau, \chi)$$

$$= \sum_{n=1}^{\infty} \varepsilon^{n+1} \hat{R}_k^{(n)} \left(A, \frac{\partial A}{\partial \tau}, \frac{\partial A}{\partial \chi}, \tau, \chi \right), \quad (8.2.16)$$

where $a_{kl}^{(1)}, b_{kl}^{(1)}, c_{kl}^{(1)}$ are relevant elements of matrices of the first hierarchical level, $\hat{R}_k^{(n)}$ is the corresponding nonlinear functions of the first hierarchical level. Then it is convenient to normalize equation (8.2.16) with respect to matrix element $a_{kl}^{(1)}$ in the way that this equation is reduced to the form

$$\frac{\partial A^s}{\partial t} = \sum_{n=1}^{\infty} \varepsilon^n F_n^s (A, \tau, \chi), \quad (8.2.17)$$

where F_n^s are nonlinear partial differential operators, yet unknown. Here we took into consideration the definitions (8.2.3). Thus asymptotic solutions to (8.2.16) reduce to expressions for unknown functions $W_k^{(n)}$ and operators F_n^s. In order to find $W_k^{(n)}$ we substitute (8.2.15) into initial system (8.2.4) and use (8.2.5), (8.2.17). Equating the coefficients near the same powers of ε in the left- and right-hand parts yields a system of interrelated equations

$$\sum_{l=1}^{p} \left(a_{kl} \frac{\partial}{\partial t} + b_{kl} \frac{\partial}{\partial z} + c_{kl} \right) W_l^{(n)} = h_k^{(n)} (\tau, \chi, z, t), \quad (8.2.18)$$

where

$$h_k^{(1)} = - \left\{ \sum_{l=1}^{p} \sum_{s=1}^{q} \exp [ip_s] \left(a_{kl} F_1^s + b_{kl} \frac{\partial A^s}{\partial \chi} \right) \psi_l^s + \text{c. c.} \right\}$$

$$+ \tilde{R}_k^{(1)} \left(U^{(0)}, t, z, \tau, \chi \right) ;$$

$$h_k^{(2)} = -\left\{ \sum_{l=1}^{p} \sum_{s=1}^{q} \exp\left[ip_s\right] \psi_l^s a_{kl} F_2^s + \text{c. c.} \right\}$$

$$- \sum_{l=1}^{p} \left(a_{kl} \frac{\partial}{\partial \tau} + b_{kl} \frac{\partial}{\partial \chi} \right) W_l^{(1)} + \tilde{R}_k^{(2)} \left(U^{(0)}, t, z, \tau, \chi \right)$$

$$+ \sum_{l=1}^{p} \frac{\partial}{\partial U_l} \tilde{R}_k^{(1)} \big|_{U=U^{(0)}} W_l^{(1)}$$

$$+ \sum_{l=1}^{p} \frac{\partial}{\partial (U_l)_t /} \tilde{R}_k^{(1)} \big|_{(U)_t/=(U^{(0)})_t/} \left(\frac{\partial U_l^{(0)}}{\partial \tau} + \frac{\partial W_l^{(1)}}{\partial t} \right)$$

$$+ \sum_{l=1}^{p} \frac{\partial}{\partial (U_l)_z /} \tilde{R}_k^{(1)} \big|_{(U)_z/=(U^{(0)})_z/} \left(\frac{\partial U_l^{(0)}}{\partial \chi} + \frac{\partial W_l^{(1)}}{\partial z} \right);$$

. . . .

$$h_k^{(n)} = -\left\{ \sum_{l=1}^{p} \sum_{s=1}^{q} \exp\left[ip_s\right] \psi_l^s a_{kl} F_n^s + \text{c. c.} \right\}$$

$$- \sum_{l=1}^{p} \left(a_{kl} \frac{\partial}{\partial \tau} + b_{kl} \frac{\partial}{\partial \chi} \right) W_l^{(n-1)}$$

$$+ \tilde{R}_k^{(n)} \left(U^{(0)}, t, z, \tau, \chi \right) + \sum_{l=1}^{p} \frac{\partial}{\partial U_l} \tilde{R}_k^{(n-1)} \big|_{U=U^{(0)}} + \dots \qquad (8.2.19)$$

and so on, and $U^{(0)}$ is determined by (8.2.15) for $W^{(n)} = 0$. Since $R_k^{(n)}$ is periodic, the expressions for $h_k^{(n)}$ can be expanded in the Fourier series, i.e.,

$$h_k^{(n)} = \sum_{s=1}^{q} H_k^{(n)s} (\tau, \chi) \exp\left[ip_s\right] + \sum_{d=q+1}^{d_0} H_k^{(n)d} (\tau, \chi) \exp\left[ip_d\right] + \text{c. c.},$$

$$(8.2.20)$$

where $p_{s,d} = \omega_{s,d} t - k_{s,d} z$ are the phases. The first group of addends relates to the proper waves of the system. The second sum originates from the non-linearity of the medium and external wave fields that are not synchronous with those of the first group (see (8.2.7)). For these we have

$$D\left(\omega_d, k_d\right) \neq 0. \qquad (8.2.21)$$

The functions $H_k^{(n)s,d}$ are determined by Fourier coefficients

$$H_k^{(n)s,d} = \left\langle h_k^{(n)} \exp\left[-ip_{s,d}\right] \right\rangle$$

$$= \frac{1}{(2\pi)^{d_0}} \int_0^{2\pi} \cdots \int_0^{2\pi} h_k^{(n)}(\tau, \chi, z, t) \exp\left[-ip_{s,d}\right] dp_1 \ldots dp_{d_0}. \quad (8.2.22)$$

The functions $W_k^{(n)}$ can be written as

$$W_k^{(n)} = \sum_{s=1}^{q} W_k^{(n)s}(\tau, \chi) \exp\left[ip_s\right] + \sum_{d=q+1}^{d_0} W_k^{(n)d}(\tau, \chi) \exp\left[ip_d\right]. \quad (8.2.23)$$

We substitute (8.2.20) and (8.2.23) into (8.2.18), equalize coefficients near the same exponential functions find the system of inhomogeneous equations for $W_l^{(n)s,d}$:

$$\sum_{l=1}^{p} q_{kl}^{s,d} W_l^{(n)s,d} = -i H_k^{(n)s,d}. \quad (8.2.24)$$

The dispersion function does not vanish for the improper (stimulated) waves (see formula (8.2.21)). Then, within Kramer formulas, we find the amplitudes $W_k^{(n)d}$:

$$W_l^{(n)d} = -i \sum_{j=1}^{p} A_{jk}^d H^{(n)} \Big/ D(\omega_d, k_d), \quad (8.2.25)$$

where A_{jk}^d is the algebraic complement of k-th element g_{jk}^d in the determinant $D(\omega_s, k_s)$. It is more difficult to find $W_k^{(n)s}$ because the determinant $D(\omega_s, k_s)$ vanishes. The solutions $W_k^{(n)s}$ are bounded only provided the vector $H_k^{(n)s}$ is orthogonal to its null vector ξ_k^{*s} corresponding to Hermitian conjugate matrix $[q_{kl}]$, i.e.,

$$\sum_{k=1}^{p} \xi_k^s H_k^{(n)s} = 0. \quad (8.2.26)$$

Functions $H_k^{(n)}$, containing unknown operators F_n^s (and derivative-dependent functions $W_k^{(n-1)}$ determined by preceding approximation), are unknown. That is why relations (8.2.26) are equations for F_n^s. We derive expressions for functionals in question substituting (8.2.19) and (8.2.20) into (8.2.26). In the first approximation, we find from (8.2.19), (8.2.20) that

$$H_k^{(1)s} = -\sum_{l=1}^{p} \psi_l^s \left(a_{kl} F_1^s + b_{kl} \frac{\partial A^s}{\partial \chi} \right) + \left\langle R_k^{(1)} \exp[-ip_s] \right\rangle. \quad (8.2.27)$$

Then using (8.2.26) we obtain

$$\frac{\partial A^s}{\partial t} + v_s \frac{\partial A^s}{\partial z} = \varepsilon f^{(1)s}, \quad (8.2.28)$$

where

$$f^{(1)s} = \sum_{k=1}^{p} A_{kl}^s \frac{\left\langle R_k^{(1)} \exp[-ip_s] \right\rangle}{(\partial D^s / \partial \omega^s) \psi_l^s} \quad (8.2.29)$$

is derived from the functional relation

$$\frac{\xi_k^s}{\sum_{k=1}^{p} (\xi_k^s a_{kl} \psi_l^s)} = \frac{A_{kl}^s}{(\partial D^s / \partial \omega_s) \psi_l^s}, \quad (8.2.30)$$

v_s is group velocity (see definition (1.3.3) and corresponding comments),

$$v_s = \frac{d\omega_s}{dk_s} = -\frac{(\partial D^s / \partial k_s)}{(\partial D^s / \partial \omega_s)} = \frac{\displaystyle\sum_{k=1}^{p} \sum_{l=1}^{p} (\xi_k^s b_{kl} \psi_l^s)}{\displaystyle\sum_{k=1}^{p} \sum_{l=1}^{p} (\xi_k^s a_{kl} \psi_l^s)}. \quad (8.2.31)$$

The equation (8.2.29) can be rewritten in equivalent form:

$$\frac{\partial A^s}{\partial \tau} + v_s \frac{\partial A^s}{\partial \chi} = f^{(1)s}. \quad (8.2.32)$$

Relations (8.2.30), (8.2.31) are obtained accounting for i) $\psi_l^s = c_1 A_{k'l}^s$ and $\xi_k^s c_2 A_{kl'}^s$, where c_1, c_2 are arbitrary constants and $A_{k'l}^s$, $A_{kl'}^s$ are algebraic complements of relevant matrix elements $[q_{ij}^s]$; ii) $A_{k'l}^s A_{kl'}^s = A_{kl}^s A_{k'l'}^s$ provided condition $D(\omega_s, k_s) = 0$ holds.

One can be confident that equation (8.2.31) is particular case (in the first approximation) of the general equation of second hierarchical level (8.2.16). This means that:

a) when the function $f^{(1)s}$ is periodic in τ or χ (or in τ and χ simultaneously) the above calculational procedure can be used once more with respect to the equation (8.2.31);

b) the basic hierarchical principles hold; hence, classic version of slowly varying amplitude method is one of hierarchical methods.

Analogous situation realizes for higher approximations, too. We show this for truncated equations of second approximation. In order to construct the latter with respect to A^s, we find functions $W_k^{(1)s}$ from the system (8.2.24). To do this we employ the expression (8.2.25) and write the functions $H_k^{(1)s}$ as

$$H_k^{(1)s} = \sum_{l=1}^{p} (a_{kl}v_s - b_{kl}) \, \psi_1^s \frac{\partial A^s}{\partial \chi} + \bar{H}_k^{(1)s}, \qquad (8.2.33)$$

where

$$\bar{H}_k^{(1)s} = \left\langle R_k^{(1)} \exp\left[-ip_s\right] \right\rangle = \sum_{l=1}^{p} \psi_1^s a_{kl} f^{(1)s}. \qquad (8.2.34)$$

Suppose the solution to equation (8.2.24) is

$$W_l^{(1)s} = W_{l1}^{(1)s} + W_{l2}^{(1)s}. \qquad (8.2.35)$$

Moreover, we assume the conditions

$$\sum_{l=1}^{p} g_{kl} W_{l1}^{(1)s} = -i \sum_{l=1}^{p} (a_{kl}v_s - b_{kl}) \, \psi_l^s \frac{\partial A^s}{\partial \chi}; \qquad \sum_{l=1}^{p} g_{kl} W_{l2}^{(1)s} = -i H_k^{(1)s}. \qquad (8.2.36)$$

Using (8.2.9), we find

$$i \frac{\partial A^s}{\partial \chi} \frac{d}{dk_s} \left(\sum_{l=1}^{p} g_{kl}^s \psi_l^s \right) = 0, \qquad (8.2.37)$$

and

$$W_{l1}^{(1s)} = i \frac{\partial A^s}{\partial \chi} \frac{d\psi_l}{dk_s} \qquad (8.2.38)$$

is the solution of the first equation of (8.2.36). We employ the Kramer rule and de l'Hopital rule for the indeterminacy of the type 0/0 to find from the second equation (8.2.36) that

$$W_{12}^{(1)s} = -i \left(\sum_{j=1}^{p} \frac{\partial}{\partial \omega_s} \left(A_{jl}^s \bar{H}^{(1)s} \right) \right) \Big/ (\partial D^s / \partial \omega_s). \qquad (8.2.39)$$

Thus after all the transformations we can write

$$W_l^{(1)s} = i \frac{d\psi_l^s}{dk_s} \frac{\partial A^s}{\partial \chi} - i \left(\sum_{j=1}^{p} \frac{\partial}{\partial \omega_s} \left(A_{jl}^s \bar{H}^{(1)s} \right) \right) \Big/ (\partial D^s / \partial \omega_s). \quad (8.2.40)$$

Similarly, to deriving the expression for $H_k^{(1)s}$, we employ (8.2.27) and (8.2.28) to obtain

$$
\begin{aligned}
H_k^{(2)s} = & -\sum_{l=1}^{p} \psi_l^s a_{kl} F_2^s - \sum_{l=1}^{p} \left(a_{kl} \frac{\partial}{\partial \tau} + b_{kl} \frac{\partial}{\partial \chi} \right) W_l^{(1)} \\
& + \left\langle \tilde{R}_k^{(2)} \exp\left[-ip_s\right] \right\rangle + \left\langle \sum_{l=1}^{p} \frac{\partial}{\partial u_l} \tilde{R}_k^{(1)} W_k^{(1)} \exp\left[-ip_s\right] \right\rangle \\
& + \left\langle \sum_{l=1}^{p} \frac{\partial}{\partial (U_l)_t} \tilde{R}_k^{(1)} \left(\frac{\partial U_l^{(0)}}{\partial \tau} + \frac{\partial W_k^{(1)}}{\partial t} \right) \exp\left[-ip_s\right] \right\rangle \\
& + \left\langle \sum_{l=1}^{p} \frac{\partial}{\partial (U_l)_z'} \tilde{R}_k^{(1)} \left(\frac{\partial U_l^{(0)}}{\partial \chi} + \frac{\partial W_l^{(1)}}{\partial z} \right) \exp\left[-ip_s\right] \right\rangle. \qquad (8.2.41)
\end{aligned}
$$

We substitute (8.2.41) into (8.2.26) with regard for (8.2.40) and use (8.2.32). Thus we obtain equations for slowly varying wave amplitudes accurate to the second-order terms in ε:

$$\frac{\partial A^s}{\partial t} + v_s \frac{\partial A^s}{\partial z} + \frac{1}{2} + \frac{d^2 \omega_s}{d^2 k_s} \frac{\partial^2 A^s}{\partial z^2} = \varepsilon f^{(1)s} + \varepsilon^2 f^{(2)s}$$

$$+ \varepsilon^2 \left\{ \frac{(-i)}{2} \frac{\sum\limits_{k=1}^{p} \sum\limits_{l=1}^{p} (\xi_k^s a_{kl} d\psi_l^s / dk_s)}{\sum\limits_{k=1}^{p} \sum\limits_{l=1}^{p} (\xi_k^s a_{kl} \psi_l^s)} \frac{\partial f^{(1)s}}{\partial \chi} \right.$$

$$+ i \sum\limits_{k=1}^{p} \sum\limits_{l=1}^{p} \xi_k^s \left[a_{kl} \sum\limits_{j=1}^{p} \frac{\partial}{\partial \omega_s} \left(A_{jl}^s \partial \bar{H}_l^{(1)s} / \partial \tau \right) \right.$$

$$\left. + b_{kl} \sum\limits_{j=1}^{p} \frac{\partial}{\partial \omega_s} \left(A_{jl}^s \partial \bar{H}_l^{(1)s} / \partial \chi \right) \right] \Bigg/ \left[(\partial D^2 / \partial \omega_s) \sum\limits_{k=1}^{p} \sum\limits_{l=1}^{p} (\xi_k^s a_{kl} \psi_l^s) \right]$$

$$+ \sum\limits_{k=1}^{p} \xi_k^s \left[\left\langle \sum\limits_{l=1}^{p} \left(\frac{\partial}{\partial U_l} \tilde{R}_k^{(1)} W_1^{(1)} + \frac{\partial}{\partial (U_l)_t'} \tilde{R}_k^{(1)} \left(\frac{\partial U_l^{(0)}}{\partial \tau} + \frac{\partial W_l^{(1)}}{\partial t} \right) \right. \right. \right.$$

$$\left. \left. \left. + \frac{\partial}{\partial (U_l)_z'} \tilde{R}_k^{(1)} \left(\frac{\partial U_l^{(0)}}{\partial \chi} + \frac{\partial W_l^{(1)}}{\partial z} \right) \right) e^{-ip_s} \right\rangle \right] \Bigg/ \left[\sum\limits_{k=1}^{p} \sum\limits_{l=1}^{p} (\xi_k^s a_{kl} \psi_l^s) \right] \Bigg\}.$$

$$(8.2.42)$$

Here $f^{(2)s}$ is obtained similarly to $f^{(1)s}$ (see formula (8.2.29). Deriving (8.2.42), we use the relation

$$\frac{d^2 \omega_s}{dk_s^2} = \left[\sum\limits_{k=1}^{p} \sum\limits_{l=1}^{p} (-a_{kl} v_s + b_{kl}) (\xi_k^s d\psi_l^s / dk_s + \psi_k^s d\xi_l^s / dk_s) \right]$$

$$\times \left[\sum\limits_{k=1}^{p} \sum\limits_{l=1}^{p} (\xi_k^s a_{kl} \psi_l^s) \right]^{-1}$$

$$= 2 \left[\sum\limits_{k=1}^{p} \sum\limits_{l=1}^{p} (-a_{kl} + b_{kl}) (\xi_k^s d\psi_l^s / dk_s) \right] \left[\sum\limits_{k=1}^{p} \sum\limits_{l=1}^{p} (\xi_k^s a_{kl} \psi_l^s) \right]^{-1}.$$

$$(8.2.43)$$

Within (8.2.31), the latter can be obtained from

$$\frac{d}{dk_s} \left[\sum\limits_{k=1}^{p} \sum\limits_{l=1}^{p} (-a_{kl} + b_{kl}) \xi_k^s \psi_l^s \right] = 0. \qquad (8.2.44)$$

Equations of higher approximations can be derived in a similar manner.

The case of slowly varying parameters of the medium. Recall that above calculational scheme is constructed for the simplest one-dimensional model. Equation (8.2.17) was derived with matrices A, B and C independent of time and coordinates. However, this restriction is not principal and is used only for simplicity. One can be confident that above derivation of equations for slowly varying wave amplitudes can be generalized onto the case with matrices A, B and C weakly dependent of coordinates and time. Such situation can be treated as model with slowly varying parameters of the medium (within which the considered wavy processes occur). We carry out the procedure analogous to the above derivation and obtain an equation for slowly varying amplitudes in the first approximation with respect to ε, i.e.,

$$
\frac{\partial A^s}{\partial t} + v_s\left(\tau, \chi\right) \frac{\partial A^s}{\partial z} = \varepsilon f^{(1)s}
$$

$$
- A \left[\sum_{k=1}^{p} \sum_{l=1}^{p} A_{kl}^s \left(a_{kl} \left(\partial \psi_l^s / \partial \tau \right) + b_{kl} \left(\partial \psi_l^s / \partial \chi \right) \right) \right] \Big/ \left[\left(\partial D^s / \partial \omega \right) \psi_l^s \right].
$$

$$
(8.2.45)
$$

In a similar way one obtains truncated equations of higher-order approximations. This methodological aspect is illustrated in Chapters 12, 13, Volume II, for isochronous free electron laser (the amplifier of Dopplertron type).

The other similar generalization of discussed calculational scheme can be realized, too. So, in the above calculational algorithm we assume that

a) the resonant waves have constant amplitudes (i.e., are neither damped nor growing) in the *linear approximation* ($\varepsilon \to 0$);

b) the resonance band cannot contain simultaneously more than one spectral component of each wave, i.e. wave resonant interactions in the system are *non-degenerated.*

In electrodynamics of plasma-like systems there are more models with violated assumptions than that with satisfied ones. For example, signal wave and space charge wave in parametric free electron laser (see Chapters 12, 13, Volume II) are damped in linear approximation due to dissipation processes. The beam waves in the superheterodyne free electron lasers are unstable (i.e., growing) even in the linear approximation. Several space charge waves can be in the state of combination synchronism simultaneously under the Compton interaction modes in the free electron lasers (both parametric and superheterodyne — see Chapters 12, 13, Volume II) and so on. It is easy to verify that above procedures can be extended to these cases as well. However, the calculations be-

come too involved to be considered here without loss of consistency. To
avoid this, we illustrate such generalizations for several particular cases
in Chapter 12, Volume II.

2.2   Classification of Transversely Inhomogeneous
Models

Suppose the parameters of the model slowly vary in the transverse
plane. This may imply, for the physical characteristic cases in calculational practice, transverse non-uniformity of the pump and signal fields
in free electron laser, inhomogeneity of the density of relativistic electron
beam, etc.. Two distinctive possibilities occur:

a) transverse inhomogeneity (initial or developed) determined by the
amplitude derivatives (with respect to coordinates x and y) of the same
order of magnitude as the derivative with respect to z-coordinate — the
model with moderate inhomogeneity,

$$\frac{\partial A^s}{\partial x}, \frac{\partial A^s}{\partial y} \sim \frac{\partial A^s}{\partial z}; \qquad (8.2.46)$$

b) the inhomogeneity much more pronounced in transverse directions
than in the longitudinal one — the model with strong inhomogeneity,

$$\frac{\partial A^s}{\partial x}, \frac{\partial A^s}{\partial y} >> \frac{\partial A^s}{\partial z}. \qquad (8.2.47)$$

Respectively, the case

$$\frac{\partial A^s}{\partial x}, \frac{\partial A^s}{\partial y} << \frac{\partial A^s}{\partial y}, \qquad (8.2.48)$$

as almost transversely homogeneous model is equivalent to the situation
considered above in this subsection.

We note that ideology of constructing asymptotic solutions for models
(8.2.46) and (8.2.47) is similar to the above reasoning. That is why
we consider brief comments of calculational procedure reproducing the
previous cases.

2.3   Model with Moderate Inhomogeneity

Let us consider a non-one-dimensional model with quasi-plane waves.
In this case, operator P (compare with (8.2.1)) takes the form

$$P = \vec{\nabla}^*, \qquad (8.2.49)$$

where $\vec{\nabla}^*$ is *n-dimensional* nabla operator with the components $\rightrightarrows \nabla$.
Then the standard form (i.e., (3.3.21) for $U_0 = 0$) reduces to

$$A\frac{\partial U}{\partial t} + B\left(\vec{\nabla}^{*}U\right) + CU = \sum_{n=1}^{\infty} \varepsilon^{n} R^{(n)}\left(U, \frac{\partial U}{\partial t}, \left(\vec{\nabla}U\right), \tau, \rho, t, \vec{r}\right),$$

(8.2.50)

where $\vec{\rho} = \varepsilon\vec{r}$ is the slow radius-vector. The standard systems (8.2.50) are also called *Rabinovich standard form* [1–3]. We consider the case of constant matrices A, B, and C. In the scalar form equation (8.2.50) can be written as

$$\sum_{l=1}^{p}\left(a_{kl}\frac{\partial}{\partial t} + \left(\vec{b}_{kl}\vec{\nabla}\right) + c_{kl}\right)U_{l} = \sum_{n=1}^{\infty} \varepsilon^{n} R_{k}^{(n)}\left(U, \frac{\partial U}{\partial t}, \left(\vec{\nabla}U\right), \tau, \vec{\rho}, t, \vec{r}\right),$$

(8.2.51)

with $a_{kl}, \vec{b}_{kl}, c_{kl}$ be elements of matrices A, B, and C. We assume solutions of (8.2.51) to be of the same form as (8.2.15), i.e.,

$$U_{k} = \left\{\sum_{s=1}^{q}\psi_{k}^{s}\left(\omega_{s}\right) A^{s}\left(\tau, \vec{\rho}\right) \exp\left[ip_{s}\right] + \text{c. c.}\right\} + \sum_{n=1}^{\infty} \varepsilon^{n} W_{k}^{(n)}\left(A, \tau, \vec{\rho}, t, \vec{r}\right).$$

(8.2.52)

Here $A^{s}\left(\tau, \vec{\rho}\right) = A^{s}\left(\tau, \vec{\rho}\right) \exp\left[i\varphi^{s}\left(\tau, \vec{\rho}\right)\right]$, $p_{s} = \left(\omega_{s}t - \vec{k}_{s}\left(\omega_{s}\right)\vec{r}\right)$ is the wave phase. We carry out the same procedure as in previous subsection to obtain the equation of the first approximation (see (8.2.28))

$$\frac{\partial A^{s}}{\partial t} + \vec{v}_{s}\vec{\nabla}A^{s} + \varepsilon f^{(1)s},$$

(8.2.53)

where

$$\vec{v}_{s} = \frac{d\omega_{s}}{d\vec{k}_{s}} = \frac{\displaystyle\sum_{k=1}^{p}\sum_{l=1}^{p}\left(\xi_{k}^{s}\vec{b}_{kl}\psi_{l}^{s}\right)}{\displaystyle\sum_{k=1}^{p}\sum_{l=1}^{p}\left(\xi_{k}^{s}a_{kl}\psi_{l}^{s}\right)},$$

(8.2.54)

is the group velocity.

Equations of the second approximation can be constructed analogously to (8.2.53); those for complex amplitudes are

$$\frac{\partial A^s}{\partial t} + \vec{v}_s \vec{\nabla} A^s + \frac{1}{2} \left(\frac{\partial^2 \omega_s}{\partial^2 k_x} \frac{\partial^2 A^s}{\partial x^2} + \frac{\partial^2 \omega_s}{\partial^2 k_y} \frac{\partial^2 A^s}{\partial y^2} + \frac{\partial^2 \omega_s}{\partial^2 k_z} \frac{\partial^2 A^s}{\partial z^2} \right.$$

$$\left. + 2 \frac{\partial^2 \omega_s}{\partial k_x \partial k_y} \frac{\partial^2 A^s}{\partial x \partial y} + 2 \frac{\partial^2 \omega_s}{\partial k_x \partial k_z} \frac{\partial^2 A^s}{\partial x \partial z} \right)$$

$$+ \frac{\partial^2 \omega_s}{\partial k_y \partial k_z} \frac{\partial^2 A^s}{\partial y \partial z} = \varepsilon f^{(1)s} + \varepsilon^2 f^{(2)s} + \varepsilon^2 K \left(\left\langle R_k^{(1)} \exp\left[-i p_s\right] \right\rangle \right),$$

$$(8.2.55)$$

where $K \left(\left\langle R_k^{(1)} \exp\left[-i p_s\right] \right\rangle \right)$ contains the vector $\left\langle R_k^{(1)} \exp\left[-i p_s\right] \right\rangle$ of the first approximation (see (8.2.51)).

Equations (8.2.56) are derived assuming derivatives $\partial A^s/\partial t$, $\partial A^s/\partial x$, $\partial A^s/\partial y$, $\partial A^s/\partial z$ to be of the order of magnitude of ε while coefficients of the equations are of the order of ~ 1. In some models, however, this assumption does not hold. Nevertheless, the procedure can be easily extended to such cases, too.

2.4   Method of Parabolic Equation

Let us discuss the method of parabolic equation [2, 4]. This method is especially efficient in the study of effects produced by various mechanisms of wave beam diffraction in nonlinear media [4].

Suppose the s-th wave amplitude varies in the direction of wave propagation more slowly than in the perpendicular plane (condition (8.2.47)). Assuming waves propagating along z axis, i.e., $\vec{k} \cong \vec{k}_z$, we change the variables in equations (8.2.55):

$$\eta = t - z/v_z; \quad z' = z. \tag{8.2.56}$$

At the same time,

$$\partial A^s/\partial t = \partial A^s/\partial \eta; \quad \partial A^s/\partial z' - (\partial A^s/\partial \eta) \, v_z. \tag{8.2.57}$$

The relevant amplitudes rates of change can be estimated as

$$\frac{\partial A^s}{\partial \eta} \sim \frac{\partial A^s}{\partial x} \sim \frac{\partial A^s}{\partial y} \sim \varepsilon^{1/2}; \quad \frac{\partial A^s}{\partial z} \sim \varepsilon. \tag{8.2.58}$$

Then within (8.2.56) and (8.2.57) equations (8.2.55) reduce to

$$v_z \frac{\partial A^s}{\partial z'} + \frac{1}{2} \left(\frac{\partial^2 \omega_s}{\partial^2 k_x} \frac{\partial^2 A^s}{\partial x^2} + \frac{\partial^2 \omega_s}{\partial^2 k_y} \frac{\partial^2 A^s}{\partial y^2} + \frac{1}{v_z^2} \frac{\partial^2 \omega_s}{\partial^2 k_z} \frac{\partial^2 A^s}{\partial \eta^2} \right)$$

$$+ \frac{1}{2} \left(2 \frac{\partial^2 \omega_s}{\partial k_x \partial k_y} \frac{\partial^2 A^s}{\partial x \partial y} - \frac{1}{2} \frac{\partial^2 \omega_s}{\partial x \partial \eta} - \frac{1}{2} \frac{\partial^2 \omega_s}{\partial k_y \partial k_z} \frac{\partial^2 A^s}{\partial y \partial \eta} \right)$$

$$= \varepsilon f^{(1)s}, \quad (8.2.59)$$

where the terms of the order of ε are allowed for, and so on, and so forth.

3. MODERNIZED VERSION OF THE SLOWLY VARYING AMPLITUDE METHOD

3.1 Field Problem

As experience shows, attempts at using the conventional version of the slowly varying amplitude method in non-linear wave resonant problems of plasma-like systems show it is much more difficult task than it seems.

The first difficulty arises at the stage of problem formulating. In particular, when one tries to describe adequately treated model in terms of the traditional version of the slowly varying amplitude method. Thus we should to reduce the Maxwell equations (1.4.23), (6.1.6) to standard form like (8.2.2), (8.2.50). But two main problems arise here. The first concerns the manner of selecting of the *problem small parameter* ε. The second is connected with the circumstance that the vector U, strictly saying, is not periodic one (with respect to all fast phases) because it contains null slowly varying harmonics of fields and currents. The latter could be generated, for instance, as a result of nonlinear wave resonant interaction in the system (see Chapters 12 and 13 for more details). All this means that the considered method should be modernized in the case of electrodynamics applications. The result of such improving is referred as to the *modernized version of the slowly varying amplitude* method.

The *first characteristic feature* of the mentioned improving procedure is the taking into account of the mentioned quasi-stationary part of acting fields and currents (i.e., their null harmonics). The modernization in this case consist in that we should introduce the concept of so called *modernized standard system* like (3.3.21) instead standard systems (8.2.2), (8.2.50):

$$A\frac{\partial U_0}{\partial t} + \left(B\hat{\vec{\nabla}}\right)U_0 + CU_0 = \sum_{n=1}^{\infty} \varepsilon^n R_{U0}^{(n)}\left(U, \frac{\partial U}{\partial t}, \frac{\partial U}{\partial \vec{r}}, t, \vec{r}\right);$$

$$A\frac{\partial \tilde{U}_0}{\partial t} + \left(B\hat{\vec{\nabla}}\right)\tilde{U} + C\tilde{U} = \sum_{n=1}^{\infty} \varepsilon^n \tilde{R}_U^{(n)}\left(U, \frac{\partial U}{\partial t}, \frac{\partial U}{\partial \vec{r}}, t, \vec{r}\right);$$

(8.3.1)

where $\hat{\vec{\nabla}}$ is operator consisting of differential three-dimensional nabla-operator $\vec{\nabla}$ as components, $R_{0U}^{(n)}$ is a null harmonic of function $R_U^{(n)}$, $\tilde{R}_U$ is its oscillatory part, and the vector U is defined in the following manner

$$U = U_0 + \tilde{U}.$$

(8.3.2)

The second characteristic feature of the modernized version in the case of field problem is the use of the methods described above in Chapters 6, 7 for constructing the right parts of equations (8.3.1). We have in view the methods of averaged characteristics, kinetic equations, averaged quasi-hydrodynamic or current density equations.

The point is that a number of methodological problems are connected with the method of separating the small parameter ε in the right hand part of the Rabinovich standard forms (see (8.2.2), (8.2.50)). Usually, some largest amplitude of interacted waves (after their corresponding normalization) is taken to be ε at calculational practice [3–9]. However, this is important that such separation method, strictly speaking, has no any acceptable mathematical substantiation. It is based on some purely semi-qualitative considerations only. It should be mentioned that it turns out to be tolerable, as a rule, in the low-order nonlinear theories only. But the practical calculational situation becomes unacceptable quite often in the high-order nonlinear cases. Thus a number of calculational problems, which appear in practice, are results of incorrect use of the standard slowly varying amplitude method. Including, the *convergence problems*, the problem of *determining the confidence interval* (over which obtained solutions are correct), the problem of *losing* some terms in right parts of corresponding truncated equations, etc..

These problems could be solved in framework of the modernized version by use the methods, described earlier in Chapter 6, which are used for constructing the right parts of equation like (8.3.1). The methods of averaged characteristics, or averaged quasi-hydrodynamic or kinetic equations, or current density equations, can be used for this purposes. They are applied for solving the motion problem (i.e., current density and space charge problems, that is the same), which are characterized by the rigorous method of separation of the small parameter $1/\xi_1 \equiv \varepsilon$.

In turn, these solutions determine the mathematical arrangement of the corresponding right hand parts of equations. Let us turn the reader attention that the procedure of the small parameter separation is strictly correct in this case (see Chapters 4 and 6 for more details). This means automatically that the above noted difficulties are not typical for the modernized version of the considered method.

Let us mention once more that concrete realization of the calculational algorithm for construction of the modernized standard form like (8.3.1) is determined essentially by the method of solving the motion problem. As already mentioned, the latter, from the physical point of view, coincides with the current density or space charge problems.

Let us assume that the method of averaging kinetic equation is chosen as basic for calculational of the right parts of equations (8.3.1). This means that the vector U in (8.3.1), (8.3.2) should be constructed in this case in the form:

$$U = \begin{pmatrix} \vec{E} \\ \vec{D} \\ \vec{H} \\ \vec{B} \end{pmatrix}. \tag{8.3.3}$$

In what follows we turn to the averaged quasi-hydrodynamic or averaged current density or the averaged characteristic methods. It is remarkably that the space charge density ρ also serves as one of coordinates of vector U in this case:

$$U = \begin{pmatrix} \vec{E} \\ \vec{D} \\ \vec{H} \\ \vec{B} \\ \rho \end{pmatrix}. \tag{8.3.4}$$

But let us discuss this problem in more detail.

3.2   Current Density Problem

Let us choose the method of averaged current density equation as the basic one. Taking into consideration the characteristic structure of the Maxwell equations (1.4.23), (6.1.6), the vector-function

$$R_U = \sum_{n=1}^{\infty} \varepsilon^n R_U^{(n)} \left(U, \frac{\partial U}{\partial t}, \frac{\partial U}{\partial \vec{r}}, t, \vec{r} \right), \tag{8.3.5}$$

is some linear function of the vector of current density $\vec{j}$:

$$R_U = \hat{F}\hat{j} = R_{0U} + \tilde{R}_U; \qquad (8.3.6)$$

where $\hat{j}$ is the vector consisting of three-dimensional current density vectors $\vec{j}$ as components, $\hat{F}$ is relevant square constant matrix; $R_{0U} = \hat{F}\hat{j}_0$ and $\tilde{R}_U = \tilde{\hat{j}}$ are the same vectors consisting of null-harmonics $\hat{j}_0$ and non-null-harmonics $\tilde{\hat{j}}$ of the current density. Furthermore, we perform a Fourier series and power series expansions of the function $\vec{j}(\vec{r}, t)$ obtained by means the averaged current–density method (see Chapter 6, Section 6 for more details):

$$\vec{j}(\vec{r}, t) = \vec{j}_0(\vec{r}, t) + \tilde{\vec{j}}(r, t) = \sum_{n=1}^{\infty} \frac{1}{\xi_1^n} J_0^{(n)}(\vec{r}, t) + \sum_{n=1}^{\infty} \frac{1}{\xi_1^n} \tilde{J}^{(n)}(\vec{r}, t); \quad (8.3.7)$$

where all notations are obvious. It is readily seen that by using definition (8.3.5) and current density equation (6.6.1) we can complete system (8.3.1) by the equation

$$\frac{\partial R_U}{\partial t} + f\left(\vec{j}\right) \frac{\partial R_U}{\partial \vec{r}} = \Re\left(R_U, U, \frac{\partial U}{\partial t}, \frac{\partial U}{\partial \vec{r}}, \theta_1, \psi_1, \vec{r}, t\right). \qquad (8.3.8)$$

where $\vec{f}\left(\vec{j}\right)$ is some known function of current density vector $\vec{j}$.

We construct asymptotic solutions for R_U-function by means of the previously described calculational procedures. Then, solutions for the R_U-vector can be obtained in the form of relevant asymptotic convergent series:

$$R_U = \sum_{n=1}^{\infty} \frac{1}{\xi_1^n} R_U^{(n)} = \sum_{n=1}^{\infty} \frac{1}{\xi_1^n} R_{0U}^{(n)} + \sum_{n=1}^{\infty} \frac{1}{\xi_1^n} \tilde{R}_U^{(n)}; \qquad (8.3.9)$$

where

$$\sum_{n=1}^{\infty} \frac{1}{\xi_1^n} R_{0U}^{(n)} = R_{0U}; \quad \sum_{n=1}^{\infty} \frac{1}{\xi_1^n} \tilde{R}_U^{(n)} = \tilde{R}_U; \qquad (8.3.10)$$

(see definition (8.3.6)).

All this means automatically that the second of the two above mentioned difficulties of the traditional version of the slowly varying amplitude method could be overcame, in principle, by the proposed way. Indeed, taking into consideration the relationship

$$\varepsilon = 1/\xi_1, \qquad (8.3.11)$$

and substituting the obtained expansions like (6.6.17), (8.3.9) in modernized standard form (8.3.1), we have the standard form with completely determined right side. (Here, as earlier, $\xi_1 \gg 1$ is the largest scale parameter of relevant hierarchical series like (4.1.6)).

3.3 Current Density Problem in Framework of the Kinetic Approach

The calculational scheme for kinetic equation (Boltzmann equation (6.1.1)) has a number of specific peculiarities. Let us discuss some of them. According to (6.1.7), (6.1.8), the densities of current $\vec{j}$ and the space charge ρ could be represented in kinetic case as

$$\vec{j} = qn_0 \int_{-\infty}^{\infty} \frac{d\vec{r}}{dt} f(t, \vec{r}, \vec{P}) d^3 \mathcal{P}; \qquad (8.3.12)$$

$$\rho = qn_0 \int_{-\infty}^{\infty} f(t, \vec{r}, \vec{P}) d^3 \mathcal{P}; \qquad (8.3.13)$$

where we omit subscript α for simplicity. By virtue of asymptotic representations (6.7.5), (6.7.7) we can write (8.3.12), (8.3.13) in the form of the power series

$$\vec{j} = qn_0 \int_{\infty}^{\infty} \left\{ \sum_{n=1}^{\infty} \frac{1}{\xi_1^n} \left[\vec{A}_r^{(n)} \left[\bar{x}(x) \right] + \frac{d}{dt} \vec{U}_r^{(n)} \left[\bar{x}(x), \bar{\psi}_1(x) \right] \right] \right\}$$

$$\times \left\{ \bar{f} \left[\bar{x}(x) \right] + \sum_{n=1}^{\infty} \frac{1}{\xi_1^n} \vec{U}_f^{(n)} \left[\bar{x}(x), \bar{\psi}_1(x) \right] \right\} d^3 \mathcal{P}$$

$$= \sum_{n=1}^{\infty} \frac{1}{\xi_1^n} \vec{j}^{(n)}(\vec{r}, t) ; \quad (8.3.14)$$

$$\rho = qn_0 \int_{-\infty}^{\infty} \left\{ \bar{f} \left[\bar{x}(x) \right] + \sum_{n=1}^{\infty} \frac{1}{\xi_1^n} \vec{U}_f^{(n)} \left[\bar{x}(x), \bar{\psi}_1(x) \right] \right\} d^3 \mathcal{P}$$

$$= \bar{\rho}(\vec{r}, t) + \sum_{n=1}^{\infty} \frac{1}{\xi_1^n} \rho^{(n)}(\vec{r}, t) ; \quad (8.3.15)$$

where notations for $\vec{j}^{(n)}$ and $\rho^{(n)}$ are evident, x and $\bar{x}$ are the vectors of non-averaged and averaged slow variables, respectively. So the earlier discussed problem of separation out the problem small parameter $\varepsilon = 1/\xi_1$ (see the preceding Subsection) here also finds a satisfactory solution.

Thus the motion beam problem has some another calculational procedure in the kinetic version of self-consistent theory. The distinction, from the 'point of view of the slowly varying amplitude method' concerns only structure of the vector U (8.2.5), (8.3.4). It is obvious that the mentioned distinction is not important for general calculational scheme of the considered method.

4. METHOD OF HIERARCHICAL TRANSFORMATION OF COORDINATES

4.1 Main Idea of the Hierarchical Transformations

Previously we discussed some version of application of the averaged current–density equation method in framework of the modernized method of slowly varying amplitudes. Analysis shows (see, in particular, Chapters 12, 13, Volume II) the calculational procedures of the discussed class could be very promising for solving problems of various nonlinear theory of wave dynamical systems. For instance, electrodynamic systems with high intense electron or plasma beams very often are characterized by complicated dynamical surface configurations. Including intensive electron beams in relativistic two-stream systems, relativistic high current electron devices (free electron lasers (FELs), cancerotrons, gyrotrons, etc.), various high-current acceleration systems, such as linear induction accelerators, radio-frequency accelerators, EH-accelerators, etc.. The considered hierarchical methods could be used for overcoming the difficulties connected with such type of peculiarities. We take in view, in particular, the partial variety of the hierarchical ideology realization that is known as the *method of hierarchical transformation of coordinates* [12].

The proposed approach can be especially effective for solving of the earlier discussed nonlinear dynamical problems characterized by complex non-stationary boundary conditions on the beam surface. Including, the situations if the beam surface being a self-consistent function of the solved problem (i.e., it is also the subject of determining). It should be mentioned that this problem is rather old for the physics of intensive high-current electron and ion beams. However, its has no acceptable solution even today [14, 15]. This circumstance additionally stimulates researcher for developing of new, of principle, conceptual ideas of such kind.

Another topical problem of electrodynamics of plasma-like objects is the problem of determination of equilibrium states of systems, which are characterized by presence of some 'controlled small-scale instabilities' of some another types. But let us to discuss this problem in more detail.

It is well known that the theory of stable equilibrium of charged particle beams was elaborated mainly for the unperturbed 'electrodynamically cold' systems [14, 15], i.e. the systems within which any active wave processes do not occur. Usually, talking about the 'stable states of beams', namely, the 'electrodynamycally cold' systems are taken in view namely. However, some wider treatment of the beam stability problem is known, too. Here we talk about the conditional stability of 'hot' charged particle beams, as a whole. It is obvious that we have unstable, in principle, electron beams in the interaction region of power devices like FELs (see Section 1, Chapter 1 and Chapters 9–13, Volume II) or other similar. But this instability carries small-scale character because all processes here develop in the scale of a wave length λ. (This leads, in particular, to that the same small-scale oscillations characterize the beam surface oscillations, too.) On the other hand, the beam, as a whole, must be stable in the scale of transverse characteristic (design) size a also. Usually, it is typically for many traditional plasma-like systems: $\lambda << a$. It is clearly that such 'electrodynamically hot' electron beam possesses an averaged surface configuration different greatly from the 'cold' one. Correspondingly, the determination of the *averaged equilibrium states* in this case becomes very important because contrary to 'cold' system the latter in the 'hot' regime could be unstable with the characteristic 'large scale' $\sim a$. Really this means that the electron can not free pass through the system working bulk, i.e. a normal work of the device becomes impossible.

Let us show that rational solution for both discussed problems can be got within framework of the method of hierarchical transformation of co-ordinates. The essence of this calculational scheme consists in transforming of the Maxwell equations from the three-dimensional *non-averaged* space $\vec{r} = \{x, y, z\}$ into a three-dimensional also, but m-fold *averaged*, space $\vec{r}^{(m)} = \{x^{(m)}, y^{(m)}, z^{(m)}\}$. The use of the transformation formulas like (4.1.13), (4.1.15)

$$\vec{r}\left(\bar{\vec{r}}\right) = \bar{\vec{r}} + \sum_{n=1}^{\infty} \frac{1}{\xi_1^n} \vec{u}_{r1}^{(n)}\left(\bar{\vec{r}}, \bar{x}, \bar{\psi}_1\right) ;$$

$$\bar{\vec{r}}\left(\bar{\bar{\vec{r}}}\right) = \bar{\bar{\vec{r}}} + \sum_{n=1}^{\infty} \frac{1}{\xi_2^{(n)}} \vec{u}_{r2}^{(n)}\left(\bar{\bar{\vec{r}}}, \bar{\bar{x}}, \bar{\bar{\psi}}_2\right) ;$$

. . . .

$$\vec{r}^{(m-1)}\left(r'^{(m)}\right) = \vec{r}^{(m)} + \sum \frac{1}{\xi_m^n} \vec{u}_{rm}^{(n)}\left(\vec{r}^{(m)}, x^{(m)}, \psi_m^{(m)}\right) \qquad (8.4.1)$$

is characteristic feature of this method. Here $\bar{x}$ is the vector consisting of averaged slowly varying values of the first hierarchy (excluding components of the vector $\vec{r}$), $\bar{\bar{x}}$ is similar slowly varying vector of second hierarchy, and so on, and m is the largest hierarchy number.

The physical background of this transformation idea is the following. The Lagrange or Euler description forms usually are used to describe charged particle beams (see Chapter 1, Subsection 4.9 for more details). Let us remind that the first approach includes the problems of motion of *individual particles* as its inseparable part. Thus an observer is connected with the *observed particle* and registers its coordinate and velocity in time t. In contrast, one needs to fix a *spatial point* of radius-vector $\vec{r}$ and to register velocities of passed particles in the second approach. It is important that the coordinate of observed space point $\vec{r}$ always coincides with electron coordinates $\vec{r}_e$ in the first case. Hence the coordinate $\vec{r}_e$ depends on time t as $\vec{r}_e(t) = \vec{r}(t)$ that is characteristic feature of the Lagrange description. In contrast to this the variables $\vec{r}$ and t are reciprocally independent ones in the second case. This can be explained by that we have different particles in the same spatial point $\vec{r}$ in different instants of time t in framework of the Euler description.

Let us suppose that all observed space is filled by charged particles. Then, we may put in correspondence each spatial point for any instant of time with some charged particle, i.e., we have different particles in the same spatial point in different time t. But *all these particles move accordingly to the same motion dynamical law*. Or, in other words, particle motion is described by the same differential equations. The difference concerns only the initial conditions, which are 'personal' for each particle. Thus the wave nature of the studied object determines the coherentness of oscillations of different particles in different spatial points. This means that any observer in the coordinate system oscillated with all frequencies of particle motion should see the initial beam as a non-oscillative ensemble of the particles. Hence, *we can use the relationships like (8.4.1) for accomplishing the transformation of the spatial Euler's coordinates from the initial (non-averaged) to averaged space*. These observations are put in the ideological basis of the method of hierarchical transformation of coordinates. But let us discuss their in more detail.

Reverting to the above discussed beam problems we can reveal that both: the boundary condition problem as well as the equilibrium problem can be solved essentially simpler in m-fold averaged coordinate space. In this case, the equation system consisted of current–density and Maxwell's equations in the m-fold-averaged space does not contains any oscillations. All averaged fields here are represented by some effective quasi-stationary fields' only (see also about the effective fields in Chapter 5, Subsections 2.5 and 2.6). Hence, the initial non-linear resonant wave fast oscillation problem can be reducing to a spatially 'smooth' quasi-stationary problem. Because all small-scale oscillations here are not presented already that the boundary conditions in this averaged space are formulated in the terms of averaged smooth (i.e., non-oscillating) beam surface. It is obviously that the such m-fold averaged problems are much simpler than the initial (zeroth hierarchical level) fast-oscillative non-linear wave resonant problems.

It should be mentioned, however, that the method of hierarchical transformation is not too developed today. The general idea and calculational scheme have been described only in references [12]. Therefore, below in this Section we also confine ourselves by short discussion of the general idea and calculational scheme of this method.

4.2   Hierarchical Equations

Let us successively transform standard system (8.1.1) into the m-fold averaged coordinate space. Inasmuch as vector-functions $R_U^{(n)}$ contain (as components) current density vector $\vec{j}$ and with transformation formulas (6.6.17), (8.3.1), (8.4.1) we can write for the R_U-function the following transformation equations of the first hierarchy:

$$R_U = \sum_{n=1}^{\infty} \frac{1}{\xi_1^n} R_U^{(n)}\left(j_x, j_y, j_z\right) = R_U\left(\bar{\vec{j}}, \tilde{\vec{j}}\right) = \bar{R}_U + \tilde{R}_U; \qquad (8.4.2)$$

$$\frac{\partial \bar{R}_U}{\partial t} + \sum_{n=1}^{\infty} \frac{1}{\xi_1^n} \vec{A}_r^{(n)}\left(\vec{r}, \bar{\theta}_1, R_U\right) \frac{\partial \bar{R}_U}{\partial \vec{r}} = \sum_{n=1}^{\infty} \frac{1}{\xi_1^n} \vec{A}_R^{(n)}\left(\vec{r}, \bar{\theta}_1, \bar{R}_U\right); \quad (8.4.3)$$

$$\tilde{R}_U = \sum_{n=1}^{\infty} \frac{1}{\xi_1^n} \tilde{U}_R\left(\vec{r}, \bar{R}_U \bar{\theta}_1, \bar{\psi}_1,\right), \qquad (8.4.4)$$

where all the quantities have been defined earlier. Hence, we represent the vector U as

$$U = \bar{U}\left(\bar{x}\right) + \tilde{U}\left(\bar{x}, \bar{\psi}_1\right) \qquad (8.4.5)$$

where $\bar{x}$ is the vector of averaged slow quantities. Consequently, we can got the following presentation for modernized standard system (8.3.1) on the first hierarchical level ($\varepsilon = 1/\xi_1$, as before)

$$A\frac{\partial \bar{U}}{\partial t} + \left(B\bar{\tilde{\nabla}}\right)\bar{U} + C\bar{U} = \sum_{n=1}^{\infty}\frac{1}{\xi_1^n}\bar{R}_U^{(n)}\left(\bar{U}, \frac{\partial \bar{U}}{\partial t}, \frac{\partial \bar{U}}{\partial \bar{\bar{r}}}, t, \bar{r}\right); \qquad (8.4.6)$$

$$A\frac{\partial \tilde{U}}{\partial t} + \left(B\bar{\tilde{\nabla}}\right)\tilde{U} + C\tilde{U} = \sum_{n=1}^{\infty}\frac{1}{\xi_1^n}\tilde{R}_U^{(n)}\left(U, \frac{\partial U}{\partial t}, \frac{\partial U}{\partial \bar{r}}, t, \bar{r}\right); \qquad (8.4.7)$$

where $\bar{\tilde{\nabla}}$ is the differential operator constructed of the averaged three-dimentional nabla operators $\bar{\nabla}$ as elements. We obtain for the second hierarchical level

$$\bar{R}_U = \sum_{n=1}^{\infty}\frac{1}{\xi_1^n}\bar{R}_U^{(n)}; \qquad \bar{R}_U^{(n)} = \bar{\bar{R}} + \bar{\tilde{R}}; \qquad (8.4.8)$$

$$\frac{\partial \bar{\bar{R}}_U}{\partial t} + \sum_{n=1}^{\infty}\frac{1}{\xi_2^n}\bar{A}_r^{(n)}\left(\bar{\bar{r}}, \bar{R}_U\right)\frac{\partial \bar{\bar{R}}_U}{\partial \bar{\bar{r}}} = \sum_{n=1}^{\infty}\frac{1}{\xi_2^n}\bar{A}_R\left(\bar{\bar{r}}, \bar{R}_U\right); \qquad (8.4.9)$$

$$\bar{\tilde{R}}_U = \sum_{n=1}^{\infty}\frac{1}{\xi_2^n}\bar{\tilde{U}}_R^{(n)}\left(\bar{\bar{r}}, \bar{R}_U, \bar{\psi}_2\right), \qquad (8.4.10)$$

and so forth. Here notations of all the quantities are quite clear within the context of above accomplished discussions. We proceed the procedure of successive hierarchical transformations until the terminal hierarchical level m will be attained. Here we solve the constructed m-fold averaged truncated equation

$$\frac{\partial R_U^{(m)}}{\partial t} + \sum_{n=1}^{\infty}\frac{1}{\xi_m^n}A_r^{(n,m)}\left(\bar{r}^{(m)}, R_U^{(m)}\right)\frac{\partial R_U^{(m)}}{\partial \bar{r}^{(m)}}$$

$$= \sum_{n=1}^{\infty}\frac{1}{\xi_2^n}A_R^{(n,m)}\left(\bar{r}^{(m)}, R_U^{(m)}\right), \qquad (8.4.11)$$

and obtain corresponding solutions for functions $R_U^{(m)}$. The important peculiarity of such truncated equations is absence any periodic dependencies in both: the slow and fast oscillatative phases. In this regard equation (8.4.11) is quasi-stationary one, i.e., it describes only non-oscillatory (quasi-stationary) dynamics of the system considered.

Then we discuss standard system (8.4.6), (8.4.7) in this m-fold averaged space. We note that all resonances and oscillations are 'hidden'

in transformation formula (8.4.10). Equation (8.4.11) in its mathematical structure is much simpler than non-averaged initial equation (8.3.8).
Hence, all excited electromagnetic fields in this m-fold averaged system are quasi-stationary (i.e., non-oscillatative), too. We solve equation
(8.4.11) (by means some analytical numerical method) and construct
m-fold averaged the standard equation:

$$
\frac{\partial U^{(m)}}{\partial t} + \left(B\hat{\vec{\nabla}}^{(m)} \right) U^{(m)}
$$

$$
= \sum_{n=1}^{\infty} \frac{1}{\xi_m^n} R_U^{(n,m)} \left(U^{(m)}, \frac{\partial U^{(m)}}{\partial t}, \frac{\partial U^{(m)}}{\partial \vec{r}^{(n)}}, \vec{r}^{(m)}, t \right); \quad (8.4.12)
$$

where subscript and superscript m means multiplicity of hierarchi-
cal level, and $\hat{\vec{\nabla}}^{(m)}$ is differential operator consisting of m-fold averaged three-dimension nabla-operator $\vec{\nabla}^{(m)}$. Then we solve equation (8.4.12) and obtain relevant solutions for quasi-stationary vector-
function $U^{(m)}(\vec{r}^{(m)}, t)$.

We construct equations for functions of $(m-1)$ order of hierarchical
multiplicity at the following stage of calculational:

$$
R_U^{(m-1)} \left(\vec{r}^{(m)} \right) = R_U^{(m)} \left(\vec{r}^{(m)} \right) + \tilde{R}_U^{(m)} \left(\vec{r}^{(m)} \right); \quad (8.4.13)
$$

$$
\tilde{R}_U^{(m)} = \sum_{n=1}^{\infty} \frac{1}{\xi_m^n} U_R^{(n,m)} \left(\vec{r}^{(m)}, R_U^{(m)}, \tilde{\psi}_{m1} \right); \quad (8.4.14)
$$

$$
U^{(m-1)} = U^{(m)} + \tilde{U}^{(m)}; \quad (8.4.15)
$$

$$
\frac{\partial \tilde{U}^{(m)}}{\partial t} + \left(B\hat{\vec{\nabla}}^{(m)} \right) \tilde{U}^{(m)} = \sum_{n=1}^{\infty} \frac{1}{\xi_m^n} \tilde{R}_U^{(m,n)}; \quad (8.4.16)
$$

For the $(m-2)$ hierarchical level we obtain

$$
U^{(m-2)} = U^{(m-1)} + \tilde{U}^{(m-1)}; \quad (8.4.17)
$$

$$
\frac{\partial \tilde{U}^{(m-1)}}{\partial t} + \left(B\hat{\vec{\nabla}}^{(m-1)} \right) \tilde{U}^{(m-1)} = \sum_{n=1}^{\infty} \frac{1}{\xi_{m-1}^n} \tilde{R}_U^{(n,m-1)}; \quad (8.4.18)
$$

. . . .

and so on.

Following this way we eventually obtain the solution of initial problem (8.4.6), (8.4.7).

Thus hierarchical procedure of asymptotic integration of the modernized standard system (8.3.1) in the discussed case consists of the following steps:

1) the extending of averaged current density method on calculational of $R_U^{(m)}$-functions;

2) the transforming of coordinate $\vec{r}$ and relevant differential operators into m-fold averaged coordinate space (see below);

3) the solving of m-fold averaged equation for $R_U^{(m)}$-functions (see (8.4.11)) and constructing m- fold averaged standard equation like (8.4.12);

4) the solving equation (8.4.12) and forming the basis for obtaining relevant solutions of hierarchical $(m - 1)$-level, and so on;

5) the using successive inverse transformations into the initial (zeroth) hierarchical level.

Two important circumstances should be mentioned especially. The first is that solving standard system (8.4.16), (8.4.17) we use the method of averaged characteristics or the slowly varying amplitude method. The second concerns the level of the intensity of labour of these procedures.

The function R_U, in the case of 'ordinary' modernized slowly varying amplitude method (or the averaged characteristics method), contains all fast Euler's oscillation phases ψ_1, ψ_2, ... and θ_1, θ_2, ...(the combination phases θ_1, θ_2, ... are fast in framework of the Euler's description). Therefore, this function is rather complicated and corresponding calculational procedure for U immediately becomes too complicated also.

Thus the main advantage of the method of hierarchical transformation of coordinates is the following. Firstly, it allows to divide the initial too complex problem into a number of much simpler partial ones. Thus total labor-intensity of the calculations performed accordingly with the discussed method is less than the labor-intensity of the slowly varying amplitude method. Secondly, we got a promising way for solving the above discussed problem of boundary conditions and equilibrium states.

4.3   Averaged Operator $\vec{\bar{\nabla}}$

At last, we give the transformation procedure of the $\vec{\nabla}$-operator ($\vec{\nabla} = \partial/\partial\vec{r}$) into the averaged $\vec{\bar{\nabla}}$-operator ($\vec{\bar{\nabla}} = \partial/\partial\vec{\bar{r}}$). This problem might be formulated in the following manner:

$$\frac{\partial}{\partial\vec{r}} = \alpha(\vec{\bar{r}})\frac{\partial}{\partial\vec{\bar{r}}}, \ \alpha\left(\vec{\bar{r}}\right) =?. \tag{8.4.19}$$

It is obvious that

$$\frac{\partial}{\partial \vec{r}} = \frac{\partial \bar{\vec{r}}(\vec{r})}{\partial \vec{r}}\bigg|_{\vec{r}=\bar{\vec{r}}} \frac{\partial}{\partial \bar{\vec{r}}} \quad \text{i.e.,} \quad \alpha(\vec{r}) = \frac{\partial \bar{\vec{r}}(\vec{r})}{\partial \vec{r}}\bigg|_{\vec{r}=\bar{\vec{r}}} \tag{8.4.20}$$

In what follows we use the Krylov–Bogolyubov substitutions (8.4.1):

$$\vec{r}(\bar{\vec{r}}) = \bar{\vec{r}} + \sum_{n=1}^{\infty} \frac{1}{\xi_1^n} \vec{u}_r^{(n)}(\bar{\vec{r}}), \tag{8.4.21}$$

where the method calculational of functions $\vec{u}_r^{(n)}$ is described in Chapter 4. Furthermore, we take into account that

$$\vec{r}(\bar{\vec{r}}) - \bar{\vec{r}} = \sum_{n=1}^{\infty} \frac{1}{\xi_1^n} \vec{u}_r^{(n)}(\bar{\vec{r}}) \sim \frac{1}{\xi_1^n} << 1, \tag{8.4.22}$$

i.e. we get the definition of functions $\vec{u}_r^{(n)}$ through the non-averaged coordinates $\vec{r}$ by the method of back transformations (see Chapter 6, Subsection 2.2 and others):

$$\vec{u}_r^{(n)}(\vec{r}) = \vec{u}_r^{(n)}(\bar{\vec{r}}(\vec{r})) + \frac{\partial \vec{u}_r^{(n)}(\bar{\vec{r}})}{\partial \bar{\vec{r}}}\bigg|_{\bar{\vec{r}}=\vec{r}} \left(\sum_{n=1}^{\infty} \frac{1}{\xi_1^n} \vec{u}_r^{(n)}(\bar{\vec{r}})\right)\bigg|_{\bar{\vec{r}}=\vec{r}} +$$

$$+ \frac{1}{2} \frac{\partial^2 \vec{u}_r^{(n)}(\bar{\vec{r}})}{\partial \bar{\vec{r}}^2}\bigg|_{\bar{\vec{r}}=\vec{r}} \left(\sum_{n=1}^{\infty} \frac{1}{\xi_1^2} \vec{u}_r^{(n)}(\bar{\vec{r}})\right)^2\bigg|_{\bar{\vec{r}}=\vec{r}} + \dots \tag{8.4.23}$$

This means that using (8.4.21)–(8.2.24) we can write the following

$$\bar{\vec{r}}(\vec{r}) = \vec{r} - \sum_{n=1}^{\infty} \frac{1}{\xi_1^n} \vec{u}_r^{(n)}(\bar{\vec{r}}(\vec{r})) = \vec{r} - \sum_{n=1}^{\infty} \frac{1}{\xi_1^n} \bigg[\vec{u}_r^{(n)}(\vec{r}) +$$

$$+ \frac{\partial \vec{u}_r^{(n)}(\vec{r})}{\partial \bar{\vec{r}}} \left(\sum_{n=1}^{\infty} \frac{1}{\xi_1^n} \vec{u}_r^{(n)}(\vec{r})\right)\bigg] \tag{8.4.24}$$

i.e. required expression for the transformation function (8.4.24) $\alpha(\vec{r})$ eventually can be represented as

$$\alpha(\vec{r}) = 1 - \sum_{n=1}^{\infty} \frac{1}{\xi_1^n} \left[\frac{\partial \vec{u}_r^{(n)}(\vec{r})}{\partial \bar{\vec{r}}} + \frac{\partial^2 \vec{u}_r^{(n)}}{\partial^2 \bar{\vec{r}}} \left(\sum_{n=1}^{\infty} \frac{1}{\xi_1^n} \vec{u}_r^{(n)}(\vec{r})\right) + \dots \right].$$

$$\tag{8.4.25}$$

5.   MITROPOL'SKII METHOD

Previously, in this and preceding Chapters, we considered the asymptotic integration algorithm for a certain class of systems of weakly nonlinear partial differential equations. The common property of the discussed before methods is specific combination of the straight and back transformations. However, some other technical realizations of calculational schemes of such type are known also. In spite of that, some universal basic ideas of the averaging method are employed here too, formally, these algorithms essentially differs from the above discussed 'classical' calculational procedures. The discussed method is described in references [10, 11]. Taking into consideration the surname of their principal author, this method had been call as the *Mitropol'skii method.*

At the same time, the problems of electrodynamics of distributed-parameter systems sometimes can be reduced to the standard form that is characteristic for this method. So, let us further to discuss shortly some main ideas and calculational peculiarities of the Mitropol'skii method.

5.1   Reduction of a Partial Differential Equation to the Standard Form with Fast Rotating Phases

Initial equations.   Let us consider a special case of general equation (6.1.12) when this system of first-order differential equations may be reduced to an equation of the second order that is given by

$$A' \frac{\partial^2 u}{\partial t^2} + 2B' \frac{\partial^2 u}{\partial t \partial z} + C' \frac{\partial^2 u}{\partial z^2} + D \frac{\partial u}{\partial t} + E \frac{\partial u}{\partial t} + Gu$$
$$= \mu F \left(t, z, u, \frac{\partial u}{\partial t}, \frac{\partial u}{\partial z}, \mu \right) \quad (8.5.1)$$

where A', B', C', D, E, G are constant coefficients, $\mu = 1/\xi$ is the small parameter of the problem ($\mu << 1$), F is a given nonlinear function of its arguments. We assume that the constant coefficients A', B', C' satisfy the condition:

$$2B' \frac{\partial^2 u}{\partial t \partial z} - A'C' > 0 \quad (8.5.2)$$

Let us show that in the case (8.5.2), the partial differential equation (8.5.1) reduces to a system of ordinary differential equations. The asymptotic integration of these equations can be carried out by means of Bogolyubov–Zubarev method (see Chapter 4).

We follow the procedure described in [10, 11], carry out standard changes of variables, and thus reduce equation (8.5.1) to the canonical form:

$$\frac{\partial^2 u'}{\partial t^2} - a^2 \frac{\partial^2 u'}{\partial z^2} = \lambda u' + \mu F' \left(t, z, u', \frac{\partial u'}{\partial t}, \frac{\partial u'}{\partial z}, \mu \right) \qquad (8.5.3)$$

where u' is the new variable, a and λ are constant coefficients, F' is a known nonlinear function. For the sake of simplicity, in what follows we omit the primes in the notation of the quantities which enter (8.5.3).

We impose the boundary and initial conditions by means of the equations

$$u(0, t) = u(l, t) = 0; \qquad (8.5.4)$$

$$u|_{t=0} = f(z); \qquad \left.\frac{\partial u}{\partial t}\right|_{t=0} = \Phi(z); \qquad (8.5.5)$$

where $f(z)$ and $\Phi(z)$ are sufficiently smooth functions which satisfy all relevant conditions.

5.2   Basic Solutions

Let us find the form of the basic solutions of equation (8.5.3). To do this, we linearize the equations in a manner similar to the procedure of Sections 1 and 2. Namely, we put $\mu = 0$, so that equation (8.5.3) takes the form

$$\frac{\partial^2 u'}{\partial t^2} - a^2 \frac{\partial^2 u'}{\partial z^2} = \lambda u' \qquad (8.5.6)$$

We write the solutions of equation (8.5.6) (with boundary conditions (8.5.4), (8.5.5)) in terms of Fourier series, i.e.:

$$u(z, t) = \sum_{n=1}^{\infty} \{A_n \cos \omega_n t + B_n \sin \omega_n t\} \sin n \frac{\pi}{l} x, \qquad (8.5.7)$$

where

$$\omega_n = \sqrt{\left(n \frac{\pi a^2}{l^2} - \lambda \right)}, \quad \lambda < n \frac{\pi a^2}{l^2} \qquad (8.5.8)$$

is the normal oscillation frequency of the linear system, $n = 1,2,3,...$ are the numbers of Fourier harmonics, A_n and B_n are constant coefficients (amplitudes of the Fourier harmonics) determined by the conditions (8.5.4), (8.5.5).

5.3   Truncated Equations

Similarly with the method of slowly varying amplitudes (see Sections 1 and 2), we assume that the solutions preserve their form (i.e., reproduce (8.5.7)) for $\mu \neq 0$. The distinction is that now the amplitudes A_n and B_n are slowly varying functions. However, the calculational procedure is modified as compared to the previous cases. Namely, in the general case we write the solution of the equation (8.5.6) in the form:

$$u(t, z, \mu) = \sum_{n=1}^{\infty} z_n(t, \mu) \sin n \frac{\pi}{l} z, \qquad (8.5.9)$$

where z_n are unknown functions to be found. We substitute (8.5.9) in the equation (8.5.6) with regard for the initial conditions (8.5.5). Then we multiply the result obtained by $\sin(m\pi z/l)$ $(m=1, 2, 3, ...)$ and integrate from 0 to t. Thus we come to an infinite system of ordinary differential equations [10, 11]

$$\frac{d^2 z_n}{dt^2} + \omega_n^2 z_n = \mu F(t, z_1, z_2, ..., \dot{z}_1, \dot{z}_2, ..., \mu) \quad (n = 1, 2, 3, ...). \quad (8.5.10)$$

The system (8.5.10) satisfies the initial conditions

$$z|_{t=0} = f_n; \quad \left.\frac{dz}{dt}\right|_{t=0} = \Phi_n; \qquad (8.5.11)$$

where f_n and Φ_n are the Fourier coefficients of the functions $f(z)$ and $\Phi(z)$, respectively.

We introduce (instead of A_n and B_n) new slowly varying complex amplitudes x_n and fast phases ψ_n, thus we have

$$z_n = x_n \exp(i\psi_n) + x_{-n} \exp(-i\psi_n);$$
$$\frac{dz_n}{dt} = i\omega_n x_n \exp(i\psi_n) - i\omega_n x_{-n} \exp(-i\psi_n), \quad (\omega_{-n} = -\omega_n); \quad (8.5.12)$$
$$\psi_n = \omega_n t.$$

Then we differentiate the last expression of (8.5.12) with respect to t and compare the expressions obtained to (8.5.10). As a result we have the system of equations

$$\frac{dx_n}{dt}\exp\left(i\psi_n\right)+\frac{dx_{-n}}{dt}\exp\left(-i\psi_n\right)=0;$$

$$i\omega_n\frac{dx_n}{dt}\exp\left(i\psi_n\right)-i\omega_n\frac{dx_{-n}}{dt}\exp\left(-i\psi_n\right)=\mu F_n\left(\ldots\right); \qquad (8.5.13)$$

$$\frac{d\psi_n}{dt}=\omega_n.$$

Besides that, we observe that

$$x_n=x_n^*, \quad F_{-n}=F_n, \qquad (8.5.14)$$

where the asterisk symbolizes the complex conjugation operation.

We solve the system with respect to dx_n/dt and dx_{-n}/dt and thus obtain the required infinite (innumerable) system of ordinary differential equations for the complex amplitudes, i.e.,

$$\frac{dx_n}{dt}=\mu X_n\left(x_1,x_2,\ldots;\psi_1,\psi_2,\ldots;x_{-1},x_{-2},\ldots;\psi_{-1},\psi_{-2},\ldots;\mu\right);$$
$$\frac{d\psi_n}{dt}=\omega_n, \quad (n=\pm1,\ \pm2,\ \cdots), \qquad (8.5.15)$$

where

$$X_k\left(x_1,x_2,\ldots;\psi_1,\psi_2,\ldots;x_{-1},x_{-2},\ldots;\psi_{-1},\psi_{-2},\ldots;\mu\right)$$
$$=\frac{\exp[-i\psi_k]}{2i\omega_k}F_k\left(x_1\exp[i\psi_1]+x_{-1}\exp[i\psi_{-1}],\ldots;\right.$$
$$x_n\exp[i\psi_n]+x_{-n}\exp[i\psi_{-n}],\ldots;$$
$$i\omega_1 x_1\exp\left(i\psi_1\right)+i\omega_{-1}x_{-1}\exp\left(i\psi_{-1}\right),\ldots;$$
$$\left.i\omega_n x_n\exp\left(i\psi_n\right)+i\omega_{-n}x_{-n}\exp\left(i\psi_{-n}\right),\ldots;\mu\right).$$

It is obvious that the system of equations of the form (8.5.15) is a standard system with rotating phases. Therefore its asymptotic integration can be carried out in terms of one of algorithms described in Chapter 4. The truncation procedure for the infinite system (8.5.15) should be specialized for each problem under consideration.

6.   EXAMPLES OF REDUCING OF THE MAXWELL EQUATIONS TO THE STANDARD FORM FOR THE METHOD OF SLOWLY VARYING AMPLITUDES

Maxwell equations can be reduced to hierarchical standard form like to (8.1.1), (8.2.2), and (8.2.50). Let us demonstrate this at the examples of kinetic and quasi-hydrodynamic versions of slowly varying amplitude method. Thus we will take into account that both they are differed by the definitions for the vector U(see (8.3.3) and (8.3.4)) only.

6.1   Kinetic Version

We start with Maxwell equations (1.4.23), (6.1.6) supplemented by kinetic equation (6.1.1) and relevant definitions (6.1.7), (6.1.8). Comparing (1.4.23), (6.1.6) and (8.1.1), (8.2.2), and (8.2.50) we see that Maxwell equations (1.4.23), (6.1.6) in their original form do not satisfy standard equation requirement (8.1.1), (8.2.2), and (8.2.50). Therefore, the first obvious step is to standardize the set of equations (1.4.23), (6.1.6).

Let us represent the current density vector $\vec{J}$ and the charge density ρ in (1.4.23), (6.1.6) as sums of linear and nonlinear parts, respectively:

$$\vec{J} = \vec{J}^{(1)} + \vec{J}^{(2)}, \qquad \rho = \rho^{(1)} + \rho^{(2)}. \qquad (8.6.1)$$

In the most general case the linear parts $\vec{J}^{(1)}$ and $\rho^{(1)}$ can be written as

$$\begin{pmatrix} J_x^{(1)} \\ J_y^{(1)} \\ J_z^{(1)} \end{pmatrix} = \begin{pmatrix} \sigma_{11} & \sigma_{12} & \sigma_{13} \\ \sigma_{21} & \sigma_{22} & \sigma_{23} \\ \sigma_{31} & \sigma_{32} & \sigma_{33} \end{pmatrix} \begin{pmatrix} E_x \\ E_y \\ E_z \end{pmatrix}$$
$$+ \begin{pmatrix} d_{11} & d_{12} & d_{13} \\ d_{21} & d_{22} & d_{23} \\ d_{31} & d_{32} & d_{33} \end{pmatrix} \begin{pmatrix} B_x \\ B_y \\ B_z \end{pmatrix}; \qquad (8.6.2)$$

$$\rho^{(1)} = (g_1 \, g_2 \, g_3) \begin{pmatrix} E_x \\ E_y \\ E_z \end{pmatrix} + (h_1 \, h_2 \, h_3) \begin{pmatrix} B_x \\ B_y \\ B_z \end{pmatrix}. \qquad (8.6.3)$$

Explicit expressions for the matrices $[\sigma_{ij}]$, $[d_{ij}]$, $[g_{ij}]$, $[h_{ij}]$ can be easily derived from Maxwell's equations.

Fields. We arrange the components of the vectors $\vec{E}$, $\vec{H}$, $\vec{D}$ and $\vec{B}$ as a column vector

$$U = \begin{pmatrix} \vec{E} \\ \vec{H} \\ \vec{B} \\ \vec{D} \end{pmatrix}. \qquad (8.6.4)$$

We have for the operator $\hat{P}$ (see (8.1.1))

$$\hat{P} = \hat{\vec{\nabla}}. \qquad (8.6.5)$$

Maxwell's equations (1.4.23), (6.1.6) in the terms of above notations could be reduced to form like (8.1.1)

$$A \frac{\partial U}{\partial t} + (B\hat{P}) U + CU = R, \qquad (8.6.6)$$

where

$$A = \begin{pmatrix} [0] & [0] & [\frac{F}{c}] & [0] \\ [0] & [0] & [0] & [-\frac{F}{c}] \\ (0) & (0) & (0) & (0) \\ (0) & (0) & (0) & (0) \end{pmatrix};$$

$$B = \begin{pmatrix} \vec{k} & [0] & [0] & [0] \\ [0] & \vec{k} & [0] & [0] \\ (0) & (0) & \vec{b} & (0) \\ (0) & (0) & (0) & \vec{b} \end{pmatrix};$$

$$C = \begin{pmatrix} [0] & [0] & [0] & [0] \\ [-\frac{4\pi}{c}\sigma_{ij}] & [0] & [-\frac{4\pi}{c}d_{ij}] & [0] \\ (0) & (0) & (0) & (0) \\ (-4\pi g_i) & (0) & (-4\pi h_i) & (0) \end{pmatrix};$$

$$R = \begin{pmatrix} \{0\} \\ \{\frac{4\pi}{c} j_i^{(2)}\} \\ 0 \\ 4\pi\rho^{(2)} \end{pmatrix};$$

$$b = (\vec{e}_x\, \vec{e}_y\, \vec{e}_z); \quad \vec{k} = \begin{pmatrix} 0 & -\vec{e}_z & \vec{e}_y \\ \vec{e}_z & 0 & -\vec{e}_x \\ -\vec{e}_y & \vec{e}_x & 0 \end{pmatrix}, \quad F = \begin{pmatrix} 1 & 0 & 0 \\ 0 & 1 & 0 \\ 0 & 0 & 1 \end{pmatrix},$$

$$(8.6.7)$$

[0] are 3×3 zero matrices; (0) and {0} are three-dimensional zero-row vectors and zero-column vector, respectively. The function R can be represented as

$$R = \sum_{n=1}^{\infty} \frac{1}{\xi^n} R^{(n)} = \sum_{n=1}^{\infty} \varepsilon^n R^{(n)}.$$

6.2   Quasi-Hydrodynamic Case

Here the difference with previous kinetic case is only that the vector U includes the space charge density ρ, as a component (see (8.3.4)):

$$U = \begin{pmatrix} \vec{E} \\ \vec{H} \\ \vec{B} \\ \vec{D} \\ \rho \end{pmatrix}. \qquad (8.6.8)$$

Accomplishing the sequence of the above described calculational procedures we can get the following expressions for the matrices A, B, C, and the vector-function R:

$$A = \begin{pmatrix} [0] & [0] & (F/c) & [0] & [0] \\ [0] & [0] & [0] & (-F/c) & [0] \\ (0) & (0) & (0) & (0) & (0) \\ (0) & (0) & (0) & (0) & (0) \\ (0) & (0) & (0) & (0) & F \end{pmatrix};$$

$$B = \begin{pmatrix} \vec{k} & [0] & (F/c) & [0] & [0] \\ [0] & \vec{k} & [0] & [0] & [0] \\ (0) & (0) & \vec{b} & (0) & (0) \\ (0) & (0) & (0) & \vec{b} & (0) \\ \vec{b}(\sigma_{ij}) & (0) & \vec{b}(d_{ij}) & (0) & (0) \end{pmatrix};$$

$$C = \begin{pmatrix} [0] & [0] & [0] & [0] & [0] \\ (-4\pi/c)[\sigma_{ij}] & [0] & (-4\pi/c)[d_{ij}] & [0] & [0] \\ (0) & (0) & -4\pi(h_j) & (0) & (0) \\ (0) & (0) & (0) & (0) & -4\pi(g_i) \\ (0) & (0) & (0) & (0) & (0) \end{pmatrix};$$

$$
R = \begin{pmatrix} \{0\} \\ (4\pi/c)\vec{j}_i^{(2)} \\ 0 \\ \{0\} \\ \hat{N}_i \end{pmatrix} ;
$$

$$
b = (\vec{e}_x \ \vec{e}_y \ \vec{e}_z); \quad \vec{k} = \begin{pmatrix} 0 & -\vec{e}_z & \vec{e}_y \\ \vec{e}_z & 0 & -\vec{e}_x \\ -\vec{e}_y & \vec{e}_x & 0 \end{pmatrix} ; \quad F = \begin{pmatrix} 1 & 0 & 0 \\ 0 & 1 & 0 \\ 0 & 0 & 1 \end{pmatrix} ,
$$

$$(8.6.9)$$

where $\hat{N}_i = \hat{\vec{\nabla}}\vec{j}_i^{(2)}$ are known predetermined functions.

7. EXAMPLE: THE TWO-STREAM INSTABILITY IN A TWO-VELOCITY ELECTRON BEAM. THE METHODS OF AVERAGED QUASI-HYDRODYNAMIC EQUATION AND SLOWLY VARYING AMPLITUDES

The nonlinear theory of the two-stream instability is discussed previously in Chapter 7. The calculational technology of the method of averaged characteristic had been illustrated owing to this example there. Let us turn once more to the problem about nonlinear evolution of the two-stream instability in relativistic two-velocity system. However, we will use, at that time, a specific combination of the methods of quasi-hydrodynamic equation and slowly varying amplitudes for the same purpose. The chosen manner of illustration allows us to compare calculational features of both these approaches. As will happen below, such comparison could be very useful for deeper understanding calculational technologies of the method of averaged characteristics as well as the methods of quasi-hydrodynamic equation and slowly varying amplitudes, respectively.

7.1 Statement of the Problem

Let us choose the statement of the problem analogous to the described previously in Chapter 7, Subsection 1.1. The difference is only that we take into account here a possible presence of longitudinal quasi-stationary electric field. This field might be put to the model working bulk (such type of external fields is referred to as the *electrostatic support* [19–22] — see also Chapters 11–13, Volume II). Apart from that, the intrinsic electric field which can be generated as a result of non-

linear interaction of electrostatic waves is also taken into consideration
(see the corresponding discussion about the generated quasi-stationary
electric field in Chapter 7, Subsection 3.2). We will use instead definition (7.1.3) the following definition for the electric field in the considered
system

$$E_3 = E_0 + \frac{1}{2} \sum_{m=1}^{N} (E_m \exp{(imp_3)} + \text{c. c.}), \qquad (8.7.1)$$

where $E_0 = E_0(z,t)$ describes the resulted slowly varying longitudinal
quasi-stationary electric field that acts at electrons (the generated as
well as the external ones). All other notations are the same that were
given in Chapter 7 (see definition (7.1.3)).

We also take into consideration the following. As is well known (see
further Chapter 13, Volume II) four electron waves can exist in the two-
velocity electron system. They are the *increasing, decreasing (damped),
fast,* and *slow* electron waves. Thus the increasing wave only exerts main
influence at the amplification process. We neglect other three waves at
the system input that allows to simplify the solving the problem. At
the same time, as the analysis shows, such supposition does not exert
essential influence at the eventual results.

The general initial self-consistent problem is divided into two more
particular ones. Namely, they are the problem about motion of electron beam in the given electromagnetic field (*motion problem*) and the
problem about excitation of these fields for the given motion of electrons (*field problem*). The method of quasi-hydrodynamic equation (see
Chapter 6, Section 5) will be use for solution of the first problem. The
method of slowly varying amplitudes (see Sections 1 and 2) will be used
in the case of the field problem.

7.2   Motion Problem. The Averaged Quasi-hydrodynamic Equation

The quasi-hydrodynamic equation in the form (6.1.5), (7.1.1)

$$\left\{\frac{\partial}{\partial t} + \left(\vec{u}_\alpha \frac{\partial}{\partial \vec{r}}\right) + \nu\right\} \vec{u}_\alpha = \frac{q_\alpha}{m_\alpha \gamma_\alpha} \left\{\vec{E} + \frac{1}{c}\left[\vec{u}_\alpha \vec{B}\right] - \frac{\vec{u}_\alpha \left(\vec{u}_\alpha \vec{E}\right)}{c^2}\right\} -$$

$$- \frac{v_T^2}{3} \left\{\frac{1}{n_\alpha} \frac{\partial n_\alpha}{\partial \vec{r}} + \frac{\gamma_\alpha^2}{2c^2} \frac{\partial (\vec{u}_\alpha \vec{u}_\alpha)}{\partial \vec{r}}\right\} \qquad (8.7.2)$$

is chosen as the basic for the motion problem. Here all definitions are
given for (6.1.5), (7.1.1). Similarly with the case discussed in Chapter 7

we assume for simplicity: $\nu = 0$, $v_T = 0$, $\vec{B} = 0$. Then equation (8.7.2) could be essentially simplify:

$$\left(\frac{\partial}{\partial t} + u_\alpha \frac{\partial}{\partial z}\right) u_\alpha = \frac{q_\alpha E_3}{m_\alpha} \left(1 - \frac{u_\alpha^2}{c^2}\right)^{3/2}. \qquad (8.7.3)$$

Let us remind that the function $E_3 = E_3\,(z,t)$ is considered as a given one in framework of the motion problem.

We follow, as mentioned above, to the calculational scheme of the method of quasi-hydrodynamic equation (see Chapter 6, Section 5). Accordingly with the method we should to rewrite further equation (8.7.3) in the form of system of ordinary differential equations:

$$\frac{du_\alpha}{dt} = \frac{q_\alpha E_3}{m_\alpha}\left(1 - \frac{u_\alpha^2}{c^2}\right)^{3/2};$$

$$\frac{dz}{dt} = u_\alpha; \qquad (8.7.4)$$

$$\frac{dp_3}{dt} = \omega_3 - k_3\frac{dz}{dt} = \omega_3 - k_3 u_\alpha.$$

In what follows we perform the classification of variables into fast and slow ones, respectively. Taking into consideration the previously given definition for the latter (see, for instance, Chapter 1, Subsection 2.4), we may regard the velocity u_α and the coordinate z as slow variables for system (8.7.4). The variable p_3 we regard as the fast rotating phase ($p_3 \equiv \psi$, see, in particular, Chapter 4). This gives a possibility of separating out the problem large parameters:

$$\xi_1 = \left|\frac{dp_3}{dt}\right| \Big/ \left|\frac{du_\alpha}{v_0 dt}\right| \gg 1; \quad \xi_2 = \left|\frac{dp_3}{dt}\right| \Big/ \left|\frac{dz}{L_n dt}\right| \gg 1, \qquad (8.7.5)$$

where L_n and $v_0 = (u_{10} + u_{20})/2$ $(u_{\alpha 0} = u_\alpha\,(z = 0, t = 0))$ are the normalization constants. As a result of corresponding transformations, we can write equations (8.7.1) in the standard form like to (4.2.1):

$$\frac{dR}{dt} = \frac{1}{\xi}\mathcal{R}\,(R, \psi, \xi);$$

$$\frac{d\psi}{dt} = \omega(R) + \frac{1}{\xi}Y\,(R, \psi, \xi), \qquad (8.7.6)$$

where all notions in view (4.2.1) are self-evident. Characteristic feature of the considered problem is that the vector of slow variables R is here the two-dimensional one:

$$\mathcal{R} = \left\{ \begin{matrix} u_\alpha \\ z \end{matrix} \right\}. \tag{8.7.7}$$

Correspondingly, the vector-function in right part of the first of equations (8.7.6) is two-dimensional, too:

$$\frac{1}{\xi} \mathcal{R} (R, \psi, \xi)$$

$$= \left\{ \begin{matrix} \frac{q_\alpha}{m_\alpha} \left(E_0 + \frac{1}{2} \sum_{m=1}^{N} \left(E_m \exp\left(im\psi\right) + \text{c. c.} \right) \right) \left(1 - \frac{u_\alpha^2}{c^2} \right)^{3/2} \\ u_\alpha \end{matrix} \right. \tag{8.7.8}$$

Thus

$$\psi = p_3; \tag{8.7.9}$$

$$\omega(R) = \omega_3 - k_3 u_\alpha; \tag{8.7.10}$$

$$\frac{1}{\xi} Y (R, \psi, \xi) = 0. \tag{8.7.11}$$

The solutions for the system (8.7.6) could be written in the form of the Krylov–Bogolyubov substitutions like (7.1.10):

$$u_\alpha = \bar{u}_\alpha + \sum_{n=1}^{\infty} \frac{1}{\xi^n} U_{u_\alpha}^{(n)} \left(\bar{R}, \bar{p}_3 \right);$$

$$z = \bar{z} + \sum_{n=1}^{\infty} \frac{1}{\xi^n} U_z^{(n)} \left(\bar{R}, \bar{p}_3 \right); \tag{8.7.12}$$

$$p_3 = \bar{p}_3 + \sum_{n=1}^{\infty} \frac{1}{\xi^n} V^{(n)} \left(\bar{R}, \bar{p}_3 \right),$$

where the following equations of the first hierarchical level can be constructed for the averaged values:

$$\frac{d\bar{u}_\alpha}{dt} = \sum_{n=1}^{\infty} \frac{1}{\xi^n} A_{u_\alpha}^{(n)} \left(\bar{R} \right);$$

$$\frac{d\bar{z}}{dt} = \sum_{n=1}^{\infty} \frac{1}{\xi^n} A_z^{(n)} \left(\bar{R} \right); \tag{8.7.13}$$

$$\frac{d\bar{p}_3}{dt} = \omega(\bar{R}) + \sum_{n=1}^{\infty} \frac{1}{\xi^n} B^{(n)} \left(\bar{R} \right).$$

The functions $A^{(n)}$, $B^{(n)}$, $U^{(n)}$, $V^{(n)}$ we find accordingly with the algorithm described in Chapter 4, Section 2.

First approximation. First approximation on $1/\xi$. Then, accomplishing relevant calculations for the two-dimensional vector-function

$$A^{(1)} = \left\{ \begin{array}{c} A^{(1)}_{u_\alpha} \\ A^{(1)}_z \end{array} \right\} \tag{8.7.14}$$

we obtain:

$$\frac{1}{\xi}A^{(1)} = \frac{1}{2\pi}\int_0^{2\pi}\frac{1}{\xi}\mathcal{R}\left(\bar{R},\bar{\psi},\infty\right)d\bar{\psi}; \tag{8.7.15}$$

$$\frac{1}{\xi}A^{(1)}_{u_\alpha} = \frac{1}{2\pi}\int_0^{2\pi}\left\{ \frac{q_\alpha}{m_\alpha}\left(E_0 + \frac{1}{2}\sum_{m=1}^{N}\left(E_m\exp\left(im\bar{p}_3\right) + \text{c. c.}\right)\right) \right.$$
$$\left. \times\left(1 - \frac{\bar{u}_\alpha^2}{c^2}\right)^{3/2}\right\}d\bar{p}_3 = \frac{q_\alpha E_0}{m_\alpha\bar{\gamma}_\alpha^3} \tag{8.7.16}$$

$$\frac{1}{\xi}A^{(1)}_z = \frac{1}{2\pi}\int_0^{2\pi}\{\bar{u}_\alpha\}d\bar{p}_3 = \bar{u}_\alpha, \tag{8.7.17}$$

where $\bar{R} = \left\{ \begin{array}{c} \bar{u}_\alpha \\ \bar{z} \end{array} \right\}$, $\bar{\psi} = \bar{p}_3$, $\xi = \infty$, $\bar{\gamma}_\alpha = \left(1 - \bar{u}_\alpha^2/c^2\right)^{-1/2}$.

Analogous calculations for the two-dimensional function

$$U^{(1)} = \left\{ \begin{array}{c} U^{(1)}_{u_\alpha} \\ U^{(1)}_z \end{array} \right\} \tag{8.7.18}$$

yield:

$$\frac{1}{\xi}U^{(1)} = \frac{1}{\omega(\bar{R})}\int_{\bar{\psi}_0}^{\bar{\psi}}\left\{\frac{1}{\xi}\mathcal{R}\left(\bar{R},\bar{\psi},\infty\right) - \frac{1}{\xi}A^{(1)}\right\}d\bar{\psi} + \varphi_1(\bar{R}); \tag{8.7.19}$$

$$\frac{1}{\xi}U_{u_\alpha}^{(1)} = \frac{1}{\Omega}\int_{\bar{p}_{30}}^{\bar{p}_3}\left\{\frac{q_\alpha}{m_\alpha}\left(E_0 + \frac{1}{2}\sum_{m=1}^{N}(E_m\exp{(im\bar{p}_3)} + \text{c.\,c.})\right)\right.$$

$$\times\left.\left(1 - \frac{\bar{u}_\alpha^2}{c^2}\right)^{3/2} - \frac{q_\alpha E_0}{m_\alpha\bar{\gamma}_\alpha^3}\right\}d\bar{p}_3 + \varphi_{1u_\alpha}(\bar{R})$$

$$= \frac{1}{\Omega}\frac{q_\alpha}{m_\alpha\bar{\gamma}_\alpha^3}\left(\frac{1}{2}\sum_{m=1}^{N}\left(\frac{E_m}{im}\exp{(im\bar{p}_3)} + \text{c.\,c.}\right)\right);$$

$$\frac{1}{\xi}U_z^{(1)} = \frac{1}{\Omega}\int_{\bar{p}_{30}}^{\bar{p}_3}\{\bar{u}_\alpha - \bar{u}_\alpha\}\,d\bar{p}_3 + \varphi_{1z}(\bar{R}) = 0, \qquad (8.7.20)$$

where $\Omega = \xi\omega(\bar{R}) = \omega_3 - k_3\bar{u}_\alpha$. Here let us recall that the vector-function $\varphi_1(\bar{R})$, accordingly with the used calculational procedure, is the 'integration constant'. In our case we used the conditions (4.2.9), (4.2.12) for its rendering concrete. Analogous calculations for the functions $B^{(1)}, V^{(1)}$ allow to got:

$$\frac{1}{\xi}B^{(1)} = \frac{1}{2\pi}\int_0^{2\pi}\left\{\sum_j\left(\frac{\partial(\omega(\bar{R}))}{\partial\bar{R}_j}\frac{1}{\xi}U_j^{(1)}\right) + \frac{1}{\xi}Y(\bar{R},\bar{\psi},\infty)\right\}d\bar{\psi},$$

$$(8.7.21)$$

or

$$\frac{1}{\xi}B^{(1)} = \frac{1}{2\pi}\int_0^{2\pi}\left\{\frac{\partial\Omega}{\partial\bar{u}_\alpha}\frac{1}{\xi}U_{u_\alpha}^{(1)} + \frac{\partial\Omega}{\partial z}\frac{1}{\xi}U_z^{(1)} + 0\right\}d\bar{p}_3$$

$$= \frac{1}{2\pi}\int_0^{2\pi}\left\{\frac{\partial(\omega_3 - k_3\bar{u}_\alpha)}{\partial\bar{u}_\alpha}\frac{1}{\Omega}\frac{q_\alpha}{m_\alpha\bar{\gamma}_\alpha^3}\left(\frac{1}{2}\sum_{m=1}^{N}\left(\frac{E_m}{im}\exp{(im\bar{p}_3)} + \text{c.\,c.}\right)\right)\right.$$

$$\left. + \frac{\partial(\omega_3 - k_3\bar{u}_\alpha)}{\partial z}\cdot 0\right\}d\bar{p}_3 = 0; \qquad (8.7.22)$$

$$\frac{1}{\xi}V^{(1)} = \frac{1}{\omega(\bar{R})}\int_{\bar{\psi}_0}^{\bar{\psi}}\left\{\sum_j\left(\frac{\partial(\omega(\bar{R}))}{\partial\bar{R}_j}\frac{1}{\xi}U_j^{(1)}\right) + \frac{1}{\xi}Y(\bar{R},\bar{\psi},\infty)\right.$$

$$\left. - \frac{1}{\xi}B^{(0)}\right\}d\bar{\psi} + \varphi_2(\bar{R}). \quad (8.7.23)$$

$$\frac{1}{\xi}V^{(1)}$$

$$= \frac{1}{\Omega} \int_{\bar{p}_{30}}^{\bar{p}_3} \left\{ \frac{\partial \Omega}{\partial \bar{u}_\alpha} \frac{1}{\xi} U_{u_\alpha}^{(1)} + \frac{\partial \Omega}{\partial \bar{z}} \frac{1}{\xi} U_z^{(1)} + \frac{1}{\xi} Y(\bar{R}, \bar{p}_3, \infty) - B^{(1)} \right\} dp_3$$
$$+ \varphi_2(\bar{R})$$

$$= \frac{1}{\Omega} \int_{\bar{p}_{30}}^{\bar{p}_3} \left\{ \frac{\partial(\omega_3 - k_3 \bar{u}_\alpha)}{\partial \bar{u}_\alpha} \frac{1}{\Omega} \frac{q_\alpha}{m_\alpha \bar{\gamma}_\alpha^3} \left(\frac{1}{2} \sum_{m=1}^{N} \left(\frac{E_m}{im} e^{im\bar{p}_3} + \text{c. c.} \right) \right) \right.$$
$$\left. + \frac{\partial(\omega_3 - k_3 \bar{u}_\alpha)}{\partial \bar{z}} \cdot 0 + 0 - 0 \right\} dp_3 + \varphi_2(\bar{R})$$

$$= \frac{k_3}{\Omega^2} \frac{q_\alpha}{m_\alpha \bar{\gamma}_\alpha^3} \left(\frac{1}{2} \sum_{m=1}^{N} \left(\frac{E_m}{m^2} \exp(im\bar{p}_3) + \text{c. c.} \right) \right), \quad (8.7.24)$$

where the sum on j in (8.2.4) is performed on $j = \{u_\alpha, z\}$. The function $\varphi_2(\bar{R})$, as well as the function $\varphi_1(\bar{R})$, is also the 'integration constant'. The method their determining is the same.

Second approximation. In what follows let us turn to calculate the addends of the second approximation. This topic is very interesting in the illustrative example considered because we could come to rather unexpected results. In particular, we have that part of the second approximation addends in averaged equations (8.7.13) turn out to be equal zero. These means that, *in spite of the above obtained results (8.7.14)– (8.7.24) found in framework of the first approximation, they are valid for system length $L \sim 1/\xi^2$ and for the observation time $\tau \sim 1/\xi^2$* (see the discussion about the accuracy problem in Chapter 7, Subsection 3.5). Later we will illustrate this peculiarity by concrete calculations.

Following accordingly with the calculational scheme used we obtain corresponding expressions for the required functions in the second approximation $1/\xi^2$. Including, it could be found for the vector-function $A^{(2)}$:

$$A^{(2)} = \left\{ \begin{array}{c} A_{u_\alpha}^{(2)} \\ A_z^{(2)} \end{array} \right\}; \quad\quad (8.7.25)$$

$$\frac{1}{\xi^2}A^{(2)}_{u_\alpha,z} = \frac{1}{2\pi}\int_0^{2\pi}\left\{\frac{\partial\frac{1}{\xi}\mathcal{R}_{u_\alpha,z}}{\partial\bar{R}_j}\frac{1}{\xi}U_j^{(1)} + \frac{\partial\frac{1}{\xi}\mathcal{R}_{u_\alpha,z}}{\partial\bar{p}_3}\frac{1}{\xi}V^{(1)} + \frac{\partial\left(\frac{1}{\xi}\mathcal{R}_{u_\alpha,z}\right)}{\partial(1/\xi)}\right.$$

$$\left. -\frac{\partial\frac{1}{\xi}U^{(1)}_{u_\alpha,z}}{\partial\bar{R}_j}\frac{1}{\xi}A_j^{(1)} - \frac{\partial\frac{1}{\xi}U^{(1)}_{u_\alpha,z}}{\partial\bar{p}_3}\frac{1}{\xi}B^{(1)}\right\}d\bar{\psi}.\quad(8.7.26)$$

As readily seen, the five terms (integrals) in right side of expressions (8.7.26) determine the function $A^{(2)}$. Let us calculate successively each of these terms.

We start with the first term in the function $A^{(2)}_{u_\alpha}$:

$$\frac{1}{2\pi}\int_0^{2\pi}\left\{\frac{\partial\frac{1}{\xi}\mathcal{R}_{u_\alpha}}{\partial\bar{R}_j}\frac{1}{\xi}U_j^{(1)}\right\}d\bar{\psi}$$

$$=\frac{1}{2\pi}\int_0^{2\pi}\left\{\frac{\partial\frac{1}{\xi}\mathcal{R}_{u_\alpha}}{\partial\bar{u}_\alpha}\frac{1}{\xi}U^{(1)}_{\bar{u}_\alpha} + \frac{\partial\frac{1}{\xi}\mathcal{R}_{u_\alpha}}{\partial\bar{z}}\frac{1}{\xi}U_{\bar{z}}^{(1)}\right\}d\bar{p}_3$$

$$=\frac{1}{2\pi}\int_0^{2\pi}\left\{\frac{\partial\left(\frac{q_\alpha}{m_\alpha}\left(E_0 + \frac{1}{2}\sum\limits_{m=1}^{N}\left(E_m e^{im\bar{p}_3} + \text{c.c.}\right)\right)\left(1-\frac{\bar{u}_\alpha^2}{c^2}\right)^{\frac{3}{2}}\right)}{\partial\bar{u}_\alpha}\right.$$

$$\left.\times\left(\frac{1}{\Omega}\frac{q_\alpha}{m_\alpha\bar{\gamma}_\alpha^3}\left(\frac{1}{2}\sum\limits_{m=1}^{N}\left(\frac{E_m}{im}e^{im\bar{p}_3} + \text{c.c.}\right)\right)\right) + \frac{\partial\mathcal{R}_{u_\alpha}}{\partial\bar{z}}\cdot 0\right\}d\bar{p}_3$$

$$=\frac{1}{2\pi}\int_0^{2\pi}\left\{\left(E_0 + \frac{1}{2}\sum\limits_{m=1}^{N}\left(E_m e^{im\bar{p}_3} + \text{c.c.}\right)\right)\right.$$

$$\left.\times\left(\left(\frac{1}{2}\sum\limits_{m=1}^{N}\left(\frac{E_m}{im}e^{im\bar{p}_3} + \text{c.c.}\right)\right)\right)\right\}d\bar{p}_3$$

$$+\frac{1}{\Omega}\frac{q_\alpha}{m_\alpha\bar{\gamma}_\alpha^3}\frac{\partial}{\partial\bar{u}_\alpha}\left(\frac{q_\alpha}{m_\alpha}\left(1-\frac{\bar{u}_\alpha^2}{c^2}\right)^{\frac{3}{2}}\right) = 0,$$

because the integral on $\bar{p}_3$ is equal zero. The second term is also equal zero:

$$\frac{1}{2\pi} \int_0^{2\pi} \left\{ \frac{\partial \frac{1}{\xi} \mathcal{R}_{u\alpha}}{\partial \bar{p}_3} \frac{1}{\xi} V^{(1)} \right\} d\bar{\psi}$$

$$= \frac{1}{2\pi} \int_0^{2\pi} \left\{ \frac{\partial \left(\frac{q_\alpha}{m_\alpha} \left(E_0 + \frac{1}{2} \sum_{m=1}^{N} (E_m \exp\left(im\bar{p}_3\right) + \text{c. c.}) \right) \left(1 - \frac{\bar{u}^2}{c^2} \right)^{\frac{3}{2}} \right)}{\partial \bar{p}_3} \right.$$

$$\left. \times \frac{k_3}{\Omega^2} \frac{q_\alpha}{m_\alpha \bar{\gamma}_\alpha^3} \left(\frac{1}{2} \sum_{m=1}^{N} \left(\frac{E_m}{m^2} \exp\left(im\bar{p}_3\right) + \text{c. c.} \right) \right) \right\} d\bar{p}_3 = 0.$$

The third term is equal zero

$$\frac{1}{2\pi} \int_0^{2\pi} \left\{ \frac{\partial \left(\frac{1}{\xi} \mathcal{R}_{u\alpha} \right)}{\partial (1/\xi)} \right\} d\bar{\psi} = 0,$$

because (see formula (8.7.8)) the function

$$\frac{1}{\xi} \mathcal{R}_{u\alpha} = \frac{q_\alpha}{m_\alpha} \left(E_0 + \frac{1}{2} \sum_{m=1}^{N} (E_m \exp\left(im\psi\right) + \text{c. c.}) \right) \left(1 - \frac{u_\alpha^2}{c^2} \right)^{\frac{3}{2}}$$

does not depend on the large parameter ξ.

The fourth term in (8.7.26) is equal zero, too:

$$\frac{1}{2\pi} \int_0^{2\pi} \left\{ -\frac{\partial \frac{1}{\xi} U_{u\alpha}^{(1)}}{\partial \bar{R}_j} \frac{1}{\xi} A_j^{(1)} \right\} d\bar{\psi} = 0,$$

because the integral of function $U_{u\alpha}^{(1)}$ on $\bar{p}_3 \equiv \bar{\psi}$ equal zero (see correspondent commentary to (8.7.19) concerning the choosing of the function $\varphi_1(\bar{R})$), and $A_j^{(1)}$ do not depend on $\bar{p}_3$.

The fifth term in (8.7.26) is also equal zero:

$$\frac{1}{2\pi} \int_0^{2\pi} \left\{ -\frac{\partial \frac{1}{\xi} U_{u\alpha}^{(1)}}{\partial \bar{p}_3} \frac{1}{\xi} B^{(1)} \right\} d\bar{\psi} = 0,$$

because, as mentioned above, the integral of function $U_{u\alpha}^{(1)}$ on $\bar{p}_3 \equiv \bar{\psi}$ equal zero, and the function $B^{(1)}$ does not depend on $\bar{p}_3$.

Thus we find finally that

$$A_{u_\alpha}^{(2)} = 0. \tag{8.7.27}$$

In what follows let us turn to the calculating function $A_z^{(2)}$. Taking into consideration the structure of expression (8.7.25), we will calculate each of five integrals (terms) in the right side of (8.7.26). Then one is readily convinced that the first term is equal zero:

$$\frac{1}{2\pi} \int_0^{2\pi} \left\{ \frac{\partial \frac{1}{\xi} \mathcal{R}_z}{\partial \bar{R}_j} \frac{1}{\xi} U_j^{(1)} \right\} d\bar{\psi} =$$

$$= \frac{1}{2\pi} \int_0^{2\pi} \left\{ \frac{\partial \frac{1}{\xi} \mathcal{R}_z}{\partial \bar{u}_\alpha} \frac{1}{\xi} U_{\bar{u}_\alpha}^{(1)} + \frac{\partial \frac{1}{\xi} \mathcal{R}_z}{\partial \bar{z}} \frac{1}{\xi} U_{\bar{z}}^{(1)} \right\} d\bar{p}_3 =$$

$$= \frac{1}{2\pi} \int_0^{2\pi} \left\{ \frac{\partial \bar{u}_\alpha}{\partial \bar{u}_\alpha} \cdot \frac{1}{\xi} U_{\bar{u}_\alpha}^{(1)} + \frac{\partial \bar{u}_\alpha}{\partial \bar{z}} \frac{1}{\xi} U_{\bar{z}}^{(1)} \right\} d\bar{p}_3 = 0,$$

because integrals of functions $U_{\bar{u}_\alpha}^{(1)}$, $U_{\bar{z}}^{(1)}$ on $\bar{p}_3$ equal zero, and other factors do not depend on $\bar{p}_3$.

The third term in (8.7.26)

$$\frac{1}{2\pi} \int_0^{2\pi} \left\{ \frac{\partial \frac{1}{\xi} \mathcal{R}_z}{\partial \bar{p}_3} \frac{1}{\xi} V^{(1)} + \frac{\partial \left(\frac{1}{\xi} \mathcal{R}_z \right)}{\partial (1/\xi)} \right\} d\bar{\psi}$$

is equal zero because the function $\frac{1}{\xi} \mathcal{R}_z = u_\alpha$ does not depend neither on the average phase $\bar{p}_3$, nor of the parameter ξ.

The fourth and fifth terms in (8.7.26)

$$\frac{1}{2\pi} \int_0^{2\pi} \left\{ -\frac{\partial \frac{1}{\xi} U_z^{(1)}}{\partial \bar{R}_j} \frac{1}{\xi} A_j^{(1)} - \frac{\partial \frac{1}{\xi} U_z^{(1)}}{\partial \bar{p}_3} \frac{1}{\xi} B^{(1)} \right\} d\bar{\psi} = 0$$

because $U_z^{(1)} = 0$. Summarizing, we can write:

$$A_z^{(2)} = 0, \quad B^{(2)} = 0. \tag{8.7.28}$$

However, carrying out analogous calculations for the functions $U_{u,z\alpha}^{(2)}$ and $V^{(2)}$ (see Krylov–Bogolyubov substitutions (8.7.12)) we can observe that these addends are not equal to zero. Results of such calculations are given in Appendix. This means that, In contrast to the averaged components of motion, the oscillatative components calculated in the first approximation are valid on the length $L \sim 1/\xi$. Hence we should use in this case the functions $U_{u,z\alpha}^{(2)}$ and $V^{(2)}$ for extending the '*confidence interval'* for the length $L \sim 1/\xi^2$ (i.e., the length on which is guaranteed the accuracy of obtained asymptotic solutions).

Thus the nonlinear averaged dynamical of the system in the second approximation really is determined by correspondent functions calculated in the first approximation:

$$\frac{1}{\xi}A_{u_\alpha}^{(1)} = \frac{q_\alpha E_0}{m_\alpha \bar{\gamma}_\alpha^3}; \quad \frac{1}{\xi}A_z^{(1)} = \bar{u}_\alpha; \quad \frac{1}{\xi}U_z^{(1)} = 0; \quad \frac{1}{\xi}B^{(1)} = 0;$$

$$\frac{1}{\xi}U_{u_\alpha}^{(1)} = \frac{1}{\Omega}\frac{q_\alpha}{m_\alpha \bar{\gamma}_\alpha^3}\left(\frac{1}{2}\sum_{m=1}^{N}\left[\frac{E_m}{im}\exp(im\bar{p}_3) + \text{c. c.}\right]\right); \qquad (8.7.29)$$

$$\frac{1}{\xi}V^{(1)} = \frac{k_3}{\Omega^2}\frac{q_\alpha}{m_\alpha \bar{\gamma}_\alpha^3}\left(\frac{1}{2}\sum_{m=1}^{N}\left[\frac{E_m}{m^2}\exp(im\bar{p}_3) + \text{c. c.}\right]\right),$$

where $\Omega = \omega_3 - k_3\bar{u}_\alpha$, $\bar{\gamma}_\alpha = \left(\sqrt{1 - \bar{u}_a^2/c^2}\right)^{-1}$. But as it noted above (and that follows from the results given in Appendix) this conclusion is not valid for the oscillatative part of the problem.

Substituting expressions (8.7.29) into (8.7.13) we have the set of averaged ordinary equations (equations for the first hierarchical level):

$$\frac{d\bar{u}_\alpha}{dt} = \frac{q_\alpha E_0}{m_\alpha}\left(1 - \frac{\bar{u}_\alpha^2}{c^2}\right)^{3/2};$$

$$\frac{d\bar{z}}{dt} = \bar{u}_\alpha; \qquad (8.7.30)$$

$$\frac{d\bar{p}_3}{dt} = \omega_3 - k_3\bar{u}_\alpha.$$

Carrying out the inverse passage from total derivatives to partial ones in first of equations (8.7.30), and using the second equation, we obtain the averaged quasi-hydrodynamic equation

$$\left(\frac{\partial}{\partial t} + \bar{u}_\alpha\frac{\partial}{\partial \bar{z}}\right)\bar{u}_\alpha = \frac{q_\alpha E_0}{m_\alpha}\left(1 - \frac{\bar{u}_\alpha^2}{c^2}\right)^{3/2}. \qquad (8.7.31)$$

It is readily seen that in the simplest case $E_0 = 0$ equation (8.7.2) (which is obtained by the method of averaged quasi-hydrodynamic equation) can be reduced to equation (7.1.19) that is got by the method of averaged characteristics. This once more illustrates the previously formulated conclusion that the methods of averaged quasi-hydrodynamic equation and averaged characteristics are not independent ones. The difference is only that the partial derivatives in left side of initial equation in the case of the method of averaged characteristics can be 'rolled up' into the total derivative immediately, i.e., without use of the concept of characteristics.

Comparing equation (8.7.31) and initial equation (8.7.3) one can be convinced that the first is essentially simpler because the right side of the latter is non-oscillative. Various known solution methods (including the method of characteristics — see Chapter 6, Section 3) could be used for its integration. In the particular case of *stationary model* (when the averaged velocity $\bar{u} = \bar{u}(\bar{z})$ only, i.e., it does not depend on time t) equation (8.7.31) can be integrated, for instance, by the method of separation of variables. In what follows, we assume that the considered model is stationary and relevant solution $\bar{u} = \bar{u}(\bar{z})$ is known.

7.3   Motion Problem. The Back Transformations

In what follows, let us to discuss the calculational peculiarity of the back transformations. We suppose that relevant solution for averaged quasi-hydrodynamic equation (8.7.31) is known. Then we simplify the studied model assuming that the stationary state is established. Or, in other words, it is considered that the averaged velocity $\bar{u}$ does not depend on time t.

The formal solutions of equation (8.7.31) could be written in the form of the Krylov–Bogolyubov substitutions:

$$u_\alpha = \bar{u}_\alpha(\bar{z}) + U_{u_\alpha}^{(1)}\left(\bar{u}_\alpha(\bar{z}), \bar{z}, \bar{p}_3\right);$$
$$z = \bar{z} + U_z^{(1)}\left(\bar{u}_\alpha(\bar{z}), \bar{z}, \bar{p}_3\right), \tag{8.7.32}$$

where required expressions for $U_{u_\alpha}^{(1)}$ $U_z^{(1)}$ are given by formulas (8.7.29). Substituting the latter into (8.7.32), we have:

$$u_\alpha\left(\bar{u}_\alpha(\bar{z}), \bar{p}_3\left(\bar{z}, t\right)\right) = \bar{u}_\alpha(\bar{z}) + \frac{1}{\Omega}\frac{q_\alpha}{m_\alpha \bar{\gamma}_\alpha^3}\left(\frac{1}{2}\sum_{m=1}^{N}\left[\frac{E_m}{im}e^{im\bar{p}_3} + \text{c.c.}\right]\right);$$
$$z = \bar{z}. \tag{8.7.33}$$

(let us turn the reader attention that in the considered case, accordingly with the (8.7.29), $U_z^{(1)} = 0$).

Thus we got the expression for the averaged beam velocity u_α as a function of averaged values $\bar{u}(\bar{z})_\alpha, \bar{p}_3\left(\bar{z}, t\right)$, which, in turn, depend on $\bar{z}$ and t implicitly. Accordingly with the calculational procedure of the considered method, we should accomplish the back transformation for obtaining solutions into the initial form (see Chapter 6, Subsection 5.2). We perform for this the expansions of expressions (8.7.32) in the Taylor series for two variables $\bar{u}(\bar{z})_\alpha, \bar{p}_3 (\bar{z}, t)$, confining ourselves by only accounting terms no higher $1/\xi$:

$$
u_\alpha (z, t) = u_\alpha (\bar{u}(\bar{z}), \bar{p}_3)|_{\substack{\bar{z}=z \\ \bar{p}_3=p_3}}
$$

$$
+ \left\{ \frac{\partial}{\partial \bar{z}} (u_\alpha (\bar{u}(z), \bar{p}_3)) (z - \bar{z}) + \frac{\partial}{\partial \bar{p}_3} (u_\alpha (\bar{u}(z), \bar{p}_3)) (p_3 - \bar{p}_3) \right\} \Bigg|_{\substack{\bar{z}=z \\ \bar{p}_3=p_3}}
$$

$$
= u_\alpha (\bar{u}(\bar{z}), \bar{p}_3) \Bigg|_{\substack{\bar{z}=z \\ \bar{p}_3=p_3}}
$$

$$
+ \left\{ \frac{\partial}{\partial \bar{z}} (u_\alpha (\bar{u}(z), \bar{p}_3)) U_Z^{(1)} + \frac{\partial}{\partial \bar{p}_3} (u_\alpha (\bar{u}(z), \bar{p}_3)) V^{(1)} \right\} \Bigg|_{\substack{\bar{z}=z \\ \bar{p}_3=p_3}} .
$$

$$
\tag{8.7.34}
$$

Substituting then expressions (8.7.29) and (8.7.33) into (8.7.34), we find the looked for solution for the beam velocity $u_\alpha (z, t)$ in initial form (i.e., for the zeroth hierarchical level)

$$
u_\alpha (z, t) = \bar{u}_\alpha (z) + \frac{1}{\Omega(z)} \frac{q_\alpha}{m_\alpha \bar{\gamma}_\alpha^3(z)} \left(\frac{1}{2} \sum_{m=1}^{N} \frac{E_m}{im} \exp(imp_3) + \text{c. c.} \right)
$$

$$
+ \frac{k_3}{\Omega^3(z)} \left(\frac{q_\alpha}{m_\alpha \bar{\gamma}_\alpha^3(z)} \right)^2 \left(\frac{1}{2} \sum_{m=1}^{N} E_m \exp(imp_3) + \text{c. c.} \right)
$$

$$
\times \left(\frac{1}{2} \sum_{m=1}^{N} \frac{E_m}{m^2} \exp(imp_3) + \text{c. c.} \right). \tag{8.7.35}
$$

7.4   Field Problem. The Method of Slowly Varying Amplitudes

Analyzing the solution (8.7.35) obtained we see that in the accepted approximation (the second, in our case — see above expressions (8.7.27), (8.7.28)) is quadratic on the wave harmonic amplitudes E_m. The analogous calculations for higher approximation orders (see, for instance, Appendix) would be do for obtaining nonlinearities higher than quadratic.

However, as it known from the theory of two-stream (and its particular case — the plasma-beam) instability [14–18] the main source of nonlinearities here serves the field equations (Maxwell and continuity equations). Namely field nonlinearities mainly determine the multi-harmonic properties of the two-stream instability (see relevant discussion in the introduction to Chapter 7). In what follows, we will turn our attention to the field problem, taking into account what has been said.

The system consisted of continuity equation (6.1.10), (7.2.1) and Maxwell equations (6.1.6) is chosen as the initial one:

$$
\left(\frac{\partial}{\partial t} + u_\alpha \frac{\partial}{\partial z}\right) n_\alpha + n_\alpha \frac{\partial u_\alpha}{\partial z} = 0;
$$

$$
\frac{\partial E_3}{\partial z} = 4\pi \sum_{\alpha=1}^{3} q_\alpha n_\alpha,
$$

(8.7.36)

where, as before, $\alpha = 1,2$ corresponds to the partial electron beams, and $\alpha = 3$ is responsible for the immobile ion background, the intensity of longitudinal electric field is determined by definition (8.7.1). All other notations are given earlier for expressions (6.1.10), (7.2.1).

Let us represent solutions for the concentration n_α and beam velocity u_α (see solution (8.7.35) in the form of expansions in the Fourier series:

$$
n_\alpha = n_{\alpha 0} + \frac{1}{2} \sum_{m=1}^{N} (n_{\alpha m} \exp (imp_3) + \text{c. c.}),
$$

(8.7.37)

$$
u_\alpha = u_{\alpha 0} + \frac{1}{2} \sum_{m=1}^{N} (u_{\alpha m} \exp (imp_3) + \text{c. c.}).
$$

(8.7.38)

Thus the slowly varying amplitudes $n_{\alpha m} = n_{\alpha m}(z, t)$ further are regarded as sought values, and they can be easily found by use of solutions (8.7.35)

$$
u_{\alpha m} = \frac{1}{\pi} \int_{-\pi}^{\pi} \left\{u_\alpha \cdot e^{-imp_3}\right\} dp_3;
$$

$$
u_{\alpha 0} = \frac{1}{2\pi} \int_{-\pi}^{\pi} \left\{u_\alpha\right\} dp_3.
$$

(8.7.39)

Here, as before, $m = 1, 2, \ldots, N$.

Then we rewrite continuity equation in (8.7.36) in the form:

$$\frac{\partial n_\alpha}{\partial z} = -\frac{1}{u_\alpha}\left(n_\alpha\frac{\partial u_\alpha}{\partial z} + \frac{\partial n_\alpha}{\partial t}\right) = -\frac{1}{u_\alpha}\left(\frac{n_\alpha}{u_\alpha}\left(\frac{q_\alpha E_3}{m_\alpha\gamma_\alpha^3} - \frac{\partial u_\alpha}{\partial t}\right) + \frac{\partial n_\alpha}{\partial t}\right).$$

$$(8.7.40)$$

Here we have used equation (8.7.3), which, in turn, is rewritten in the form:

$$\frac{\partial u_\alpha}{\partial z} = \left(\frac{q_\alpha E_3}{m_\alpha\gamma_\alpha^3} - \frac{\partial u_\alpha}{\partial t}\right)\Big/u_\alpha.$$

Let us accept again the supposition about stationarity of amplitudes in the considered model. This means that all amplitudes in (8.7.37), (8.7.38) are depended on the longitudinal coordinate z only. Therein the expressions for partial derivatives on time t for the concentration n_α and the beam velocity u_α can be represented in the form:

$$\frac{\partial n_\alpha}{\partial t} = \omega_3\frac{\partial n_\alpha}{\partial p_3} = \frac{1}{2}\sum_{m=1}^{N}(n_{\alpha m}im\omega_3\exp\left(imp_3\right)+\text{c.\,c.}),\qquad(8.7.41)$$

$$\frac{\partial u_\alpha}{\partial t} = \omega_3\frac{\partial u_\alpha}{\partial p_3} = \frac{1}{2}\sum_{m=1}^{N}(u_{\alpha m}im\omega_3\exp\left(imp_3\right)+\text{c.\,c.}).\qquad(8.7.42)$$

Substituting expressions (8.7.41) and (8.7.42) into (8.7.40) we obtain the equations for determining the slowly varying amplitudes $n_{\alpha m}$, including the zeroth harmonics $n_{\alpha 0}$.

$$\frac{dn_{\alpha m}}{dz} = imk_3 n_{\alpha m}$$
$$-\frac{1}{\pi}\int_{-\pi}^{\pi}\left\{\frac{1}{u_\alpha}\left(\frac{n_\alpha}{u_\alpha}\left(\frac{q_\alpha E_3}{m_\alpha\gamma_\alpha^3} - \frac{\partial u_\alpha}{\partial t}\right) + \frac{\partial n_\alpha}{\partial t}\right)\right\}\exp(-imp_3)dp_3;$$

$$\frac{dn_{\alpha 0}}{dz} = -\frac{1}{2\pi}\int_{-\pi}^{\pi}\left\{\frac{1}{u_\alpha}\left(\frac{n_\alpha}{u_\alpha}\left(\frac{q_\alpha E_3}{m_\alpha\gamma_\alpha^3} - \frac{\partial u_\alpha}{\partial t}\right) + \frac{\partial n_\alpha}{\partial t}\right)\right\}dp_3, \quad(8.7.43)$$

where $m = 0,1,2,\ldots,N$, the values u_α, n_α, E_3, $\partial u_\alpha/\partial t$, $\partial n_\alpha/\partial t$ are determined by expressions (8.7.37), (8.7.38), (8.7.1), (8.7.41), (8.7.42), respectively.

Following on the analogous way it does not difficult to obtain relevant equation for the slowly varying amplitudes of wave harmonics, including the zeroth parts of the electric field:

$$\frac{dE_{3m}}{dz} = imk_3 E_{3m} + 4\pi\sum_{\alpha}q_{\alpha m}n_{\alpha m},\qquad(8.7.44)$$

where $m = 0, 1, 2, \ldots, N$, $\alpha = 1, 2, 3$.

Equations (8.7.35), (8.7.43), (8.7.44) forms a *self-consistent system* which allows to analyze *self-consistent nonlinear dynamics* of the considered model. Numerical methods of analysis are used further for accomplishing of such analysis. Some results of this analysis are shown in Figs. 8.7.1–8.7.3.

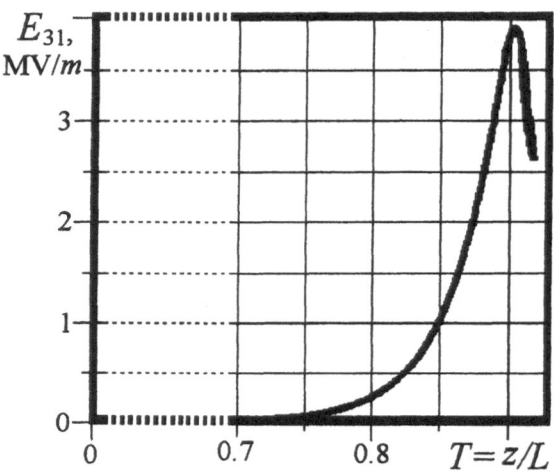

Figure 8.7.1. The dependency of amplitude of intensity E_{31} of the electric field first harmonic on non-dimensional coordinate T (where $T = z/L$, L is the system's length). This is the model where the influence of higher harmonics of the electrostatic wave does not accounted. Here the plasma frequency for each partial electron beam $\omega_b = 5 \cdot 10^{10} c^{-1}$, the length of the system $L = 160$cm; the averaged magnitude of relativistic factor for the two-velocity electron beam $\gamma = 3$; the difference between relativistic factors of the partial beam $\Delta\gamma = 0.01$; the initial magnitude of the first harmonic $E_{310} = 5, 5 \cdot 10^{-5}$V/m.

First of all, let us turn the reader attention that tree types of waves can exist in the considered two-stream system. The first are the electrostatic waves that determined by the wave part of expression (8.7.1). The second are the electron-density waves defined by expression (8.7.37). And, at last, the third are the kinematic waves (connected with particle motion) determined by definitions (8.7.38). All these waves are closely connected each with other, forming the only wave process. It should be mentioned that the understanding of intrinsic wave structure of this process is very important for its physical analysis. Unfortunately it is not a rare event in the literature that ignoring this obvious circumstance leads to some unpleasant results. Let us give one example which can obviously illustrate such a kind of situation.

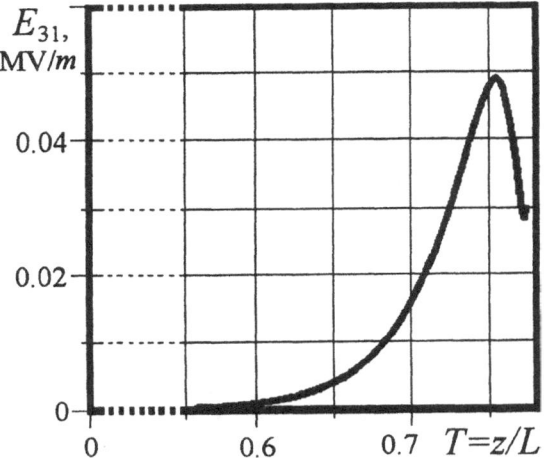

Figure 8.7.2. The dependency of amplitude of intensity of first harmonic of electric field E_{31} on the non-dimensional coordinate T (where $T = z/L$, L is the system's length) in the model, where the influence of first ten harmonics of electrostatic wave is taken into account. Here all parameters are the same with Fig. 8.7.1.

We take in view some works in references which have to do with the analysis of influence of highest wave harmonics at the general wave process. Sometimes the following erroneous calculational scheme is used. It is studied dynamics of the *first electrostatic* wave harmonics under influence of highest the kinematic wave harmonics. Relevant equations for slowly varying amplitudes are constructed by use of the simplified versions of the slowly varying amplitude method (see Section 1).

Let us to reproduce such a scheme of analysis for the case considered of the two-stream model. Namely, we analyze the nonlinear dynamics of the first harmonic of electrostatic wave amplitude (see equations (8.7.44)), accounting therein the influence of high harmonics of the kinematic waves. The result of such analysis is illustrated in Fig. 8.7.1. There the dependency of amplitude of intensity of first harmonic of electric field E_{31} on the non-dimension coordinate T(where $T = z/L$, L is the system length) is shown. It is clearly seen that saturation of amplification process occurs in the point $T \cong 0.91$ on the level ~ 3.9 MV/m.

In what follows, we turn our attention once more to such statement of the problem being not correct. The point is that the excited high harmonics of kinematic waves really should to excite (via the electron density waves) a number of high harmonic for the electrostatic waves. This means that calculating the dynamics of the first harmonics, we must take into account influence of these electrostatic waves, too. Results of

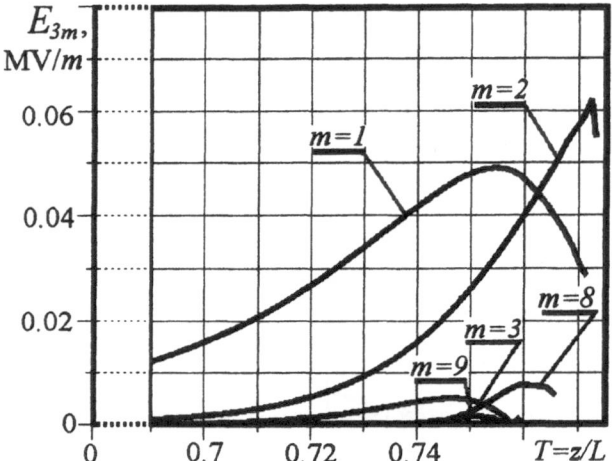

Figure 8.7.3. Dependence of amplitudes of the 1-st, 2-nd, 3-rd, 8-th, and 9-th harmonics E_{3m} on the non-dimensional longitudinal coordinate T. Here m is the harmonic number. All other parameters are the same with Fig. 8.7.1.

such 'improving' calculations are shown in Fig. 8.7.2. Comparing the results of Fig. 8.7.1 and Fig. 8.7.2, correspondingly, we see amplification properties in the second case *decrease more that 100 times*! Hence accounting the highest harmonics of electrostatic waves is always obligatory in such types of researches. Besides that, this means also that the analysis schemes of the first type (see Fig. 8.7.1) are not correct, at list in the case of two-stream or plasma-beam systems.

The conclusion formulated above is also confirmed by the results shown in Fig. 8.7.3. There the dependence of amplitudes of nine electrostatic harmonics on non-dimensional longitudinal coordinate T is shown there. It is readily seen that most of these harmonics have approximately equal amplitude order. Moreover, one of harmonics (the second one) has larger amplitude than the first harmonic even. It is obvious that the neglect of influence of higher harmonics in such situation could lead to the wrong results like that shown in Fig. 8.7.1.

References

[1] A.V. Gaponov, L.A. Ostrovsky, M.I. Rabinovich. One-dimentional waves in nonlinear dispersive media. *Izv. Vysh. Uchebn., Ser. Radiofizika (Sov. Radiophys.)*, 13(2):169–213, 1970.

[2] M.I. Rabinovich, V.I. Talanov. *Four lectures on principles of the theory of nonlinear waves and wave interactions.* Izd-vo LGU, Leningrad, 1972.

[3] M.I. Rabinovich, D.I. Trubetzkov. *Introduction in theory of oscillations and waves.* Nauka, Moscow, 1984.

[4] A.P. Sukhorukov. *Nonlinear wave interactions in optics and radiophysics.* Nauka, Moscow, 1988.

[5] N. Bloembergen. *Nonlinear optics.* Benjamin, Benjamin, 1965.

[6] A.G. Sitenko, V.M. Malnev. *Principles of the plasma theory.* Naukova Dumka, Kiev, 1994.

[7] J. Weiland, H. Wilhelmsson. *Coherent nonlinear interactions of waves in plasmas.* Pergamon Press, Oxford, 1977.

[8] L.A. Vainstein, V.A. Solntzev. *Lectures on Microwave electronics.* Sov. Radio, Moscow, 1973.

[9] V.I. Gaiduk, K.I. Palatov, D.M. Petrov. *Principles of microwave physical electronics.* Sov. Radio, Moscow, 1971.

[10] Ju.A. Mytropolskii, B.I. Moisejenkov. *Lectures on application of the asymptotic methods for solution of equations with partial derivatives.* Pub. House of Inst. of Math. Academia of Sciences of Ukraine, Kiev, 1968.

[11] Ju.A. Mytropolski, B.I. Moisejenkov. *Asymptotic methods for solution of the equations with partial derivatives.* Vyscha Shkola, Kiev, 1976.

[12] V.V. Kulish, P.B. Kosel, A.G. Kailyuk. New acceleration principle of charged particles acceleration for electronics applications. the general hierarchical approach. *International Journal of Infrared and Millimeter Waves*, 19(1):3–93, 1998.

[13] M.A. Leontovich. To the problem about propagation of electromagnetic waves in the earth atmosphere. *Izv. Akad. Nauk SSSR, ser. Fiz., (Bull. Acad. Sci. USSR, Phys. Ser.)*, 8:6–20, 1944.

[14] A.A. Ruhadze, L.S. Bogdankevich, S.E. Rosinkii, V.G. Ruhlin. *Physics of high-current relativistic beams.* Atomizdat, Moscow, 1980.

[15] R.C. Davidson. Theory of nonlinear plasmas. Mass: Benjamin, Reading, 1974.

[16] V.V. Kulish, A.V. Lysenko, V.I. Savchenko. Application of the method of averaging qusi-hydrodynamic equation for nonlinear problems of the theory of two-stream free electron lasers. *Gerald of Kyiv University, ser. Physics and Mathematics*, 4:471–480, 2000.

[17] V.V. Kulish, A.V. Lysenko, V.I. Savchenko. Method of asymptotic integration of system with particular derivatives and its application for the problems about motion of charged particles in external given electromagnetic fields. *Gerald of Sumy State University, ser. Physics and Mathematics*, 3:5–12, 2001.

[18] V.V. Kulish, A.V. Lysenko, V.I. Savchenko. Nonlinear hierarchical theory of two-stream instability in relativistic electron devices. *Gerald of Sumy State University, ser. Physics and Mathematics*, 3:12–17, 2001.

[19] V.V. Kulish. *Methods of averaging in non-linear problems of relativistic electro-dynamics*. World Federation Pub. Inc, Atlanta, 1998.

[20] V.V. Kulish, S.A. Kuleshov, A.V. Lysenko. Nonlinear self-consistent theory of superheterodyne and free electron lasers. *The International journal of infrared and millimeter waves*, 14(3):451–568, 1993.

[21] V.V. Kulish. Nonlinear self-consistent theory of free electron lasers. method of investigation. *Ukrainian Physical Journal*, 36(9):1318–1325, 1991.

[22] V.V. Kulish, A.V. Lysenko. Method of averaged kinetic equation and its use in the nonlinear problems of plasma electrodynamics. *Fizika Plasmy (sov. Plasma Physics)*, 19(2):216–227, 1993.

Appendix A
Results of calculations in the second approximation
for Chapter 8, Subsection 7.2

$$U_{u\alpha}^{(2)} = \frac{1}{2\Omega^2} \left(\frac{q_\alpha}{m_\alpha \bar\gamma_\alpha^3} \right) \times \left\{ \left(\sum_{m=1}^{N} \frac{E_{3m}}{im} e^{im\bar{p}_3} + \text{c.\,c.} \right)^2 \right.$$

$$+ \left[\left(\frac{q_\alpha}{4m_\alpha \bar\gamma_\alpha^3} \right) \left(\frac{k_3}{\Omega} - \frac{3\bar{u}_\alpha}{c^2\bar\gamma_\alpha} \right) \right]$$

$$+ \left[\left(\sum_{m=1}^{N} E_{3m} e^{im\bar{p}_3} + \text{c.\,c.} \right) \left(\sum_{m=1}^{N} \frac{E_{3m}}{m^2} e^{im\bar{p}_3} + \text{c.\,c.} \right) - \sum_{m=1}^{N} 2 \frac{E_{3m} E_{3m}^*}{m^2} \right]$$

$$\times \left[\frac{k_3}{2\Omega} \left(\frac{q_\alpha}{m_\alpha \bar\gamma_\alpha^3} \right) \right]$$

$$+ \left[\left(\sum_{m=1}^{N} \frac{1}{m^2} \frac{\partial E_{3m}}{\partial z} e^{im\bar{p}_3} + \text{c.\,c.} \right) \cdot \bar{u}_\alpha \right]$$

$$= -\frac{1}{2\Omega^2} \left(\frac{q_\alpha}{m_\alpha \bar\gamma_\alpha^3} \right) \times \left(\sum_{m=1}^{N} \frac{E_{3m}}{m^2} e^{im\bar{p}_3} + \text{c.\,c.} \right); \quad \text{(A.1)}$$

$$u_\alpha(\bar{z}, \bar{p}_3, t) = \bar{u}_\alpha(\bar{z}, t) + \frac{1}{2\Omega^2} \left(\frac{q_\alpha}{2\Omega m_\alpha \bar\gamma_\alpha^3} \right) \left(\sum_{m=1}^{N} \frac{E_{3m}}{im} e^{im\bar{p}_3} + \text{c.\,c.} \right)$$

$$+ \frac{1}{2\Omega^2} \left(\frac{q_\alpha}{m_\alpha \bar\gamma_\alpha^3} \right) \left\{ \left[\left(\sum_{m=1}^{N} \frac{E_{3m}}{im} e^{im\bar{p}_3} + \text{c.\,c.} \right)^2 - \sum_{m=1}^{N} 2 \frac{E_{3m} E_{3m}^*}{m^2} \right] \right.$$

$$\times \left[\left(\frac{q_\alpha}{4m_\alpha \bar\gamma_\alpha^3} \right) \left(\frac{k_3}{\Omega} - \frac{3\bar{u}_\alpha}{c^2\bar\gamma_\alpha} \right) \right]$$

$$+ \left[\left(\sum_{m=1}^{N} E_{3m} e^{im\bar{p}_3} + \text{c.c.} \right) \left(\sum_{m=1}^{N} \frac{E_{3m}}{m^2} e^{im\bar{p}_3} + \text{c.c.} \right) - \sum_{m=1}^{N} 2 \frac{E_{3m} E_{3m}^{*}}{m^2} \right]$$

$$\times \left[\frac{k_3}{2\Omega} \left(\frac{q_\alpha}{m_\alpha \bar{\gamma}_\alpha^3} \right) \right]$$

$$+ \left[\left(\sum_{m=1}^{N} \frac{1}{m^2} \frac{\partial E_{3m}}{\partial z} e^{im\bar{p}_3} + \text{c.c.} \right) \cdot \bar{u}_\alpha \right] \Big\} . \quad \text{(A.2)}$$

The complete solution we obtain eventually in the form:

$$u_\alpha(z,t) = \bar{u}_\alpha + \frac{1}{2\Omega} \left(\frac{q_\alpha}{m_\alpha \bar{\gamma}_\alpha^3} \right) \times \left(\sum_{m=1}^{N} \frac{E_{3m}}{im} e^{imp_3} + \text{c.c.} \right)$$

$$+ \frac{\bar{u}_\alpha}{2\Omega^2} \left(\frac{q_\alpha}{m_\alpha \bar{\gamma}_\alpha^3} \right) \left(\sum_{m=1}^{N} \frac{1}{m^2} \frac{\partial E_{3m}}{\partial z} e^{imp_3} + \text{c.c.} \right)$$

$$+ \frac{1}{8\Omega^2} \left(\frac{q_\alpha}{m_\alpha \bar{\gamma}_\alpha^3} \right)^2 \left(\frac{k_3}{\Omega} - \frac{3\bar{u}_\alpha}{c^2 \bar{\gamma}_\alpha} \right) \left(\sum_{m=1}^{N} \frac{E_{3m}}{im} e^{imp_3} + \text{c.c.} \right), \quad \text{(A.3)}$$

where the functions Ω, $\bar{\gamma}_\alpha$, E_{3m}, p_3, $\bar{u}_\alpha$ are the known functions of z and t (see Chapter 8, Section 7).

Index

averaged current–density equation, 242

ability to model principle, 38
action postulate, 41
algorithmic complexity, 45, 46
amplitude, 11, 14, 20–22, 24, 28, 91, 92,
    113, 129, 160, 164, 172, 187,
    194, 199, 201, 202, 249, 258,
    279, 284, 287–289, 291, 297,
    306, 308–310, 344–346
analysis problem, 64
analytical–numerical methods, xvii, 1, 52,
    89
angular acceleration, 9, 30
angular displacement, 30
angular velocity, 30, 34, 35
approximation of strong pumping, 288
artificial system, 60
autonomy postulate, 39
averaged current–density equation, 314
averaged equilibrium state, 315
averaged kinetic equation, 246, 348
averaged Maxwell's equation, 275
averaged quasi-hydrodynamic equation,
    237, 239, 240, 339, 340
averaged quasilinear equation, 259, 261,
    268, 270, 271, 275
averaging operator, 63, 76, 77, 92, 96
averaging operator on spatial coordinates,
    76
averaging operator on time, 76

Bhatnahar–Gross–Krook collision model,
    210
Big Bang point, 116
bucket, 179, 181, 182, 198, 201, 202
bunch, 196, 203, 204

canonical form, 208, 228, 323
canonical variable, 33, 168

case of small oscillations, 181, 188, 201
Cauchy problem, 223, 225, 226
chaotic system, 37, 60
characteristic root, 228
Cherenkov instability, 194, 195
closed hierarchical system, 59, 61, 115
closed system, 32, 36, 44
collision integral, 80, 208, 246
collision scale parameter, 80
combination oscillations, 22
comparison equation, 10, 287
comparison system, 10
complementarity postulate, 40
complex system, 37, 39–42, 44, 49, 50, 68,
    208
condition of parametric resonance, 291
conditions of compatibility, 213
conservation laws, 32
continuity equation, 211, 238, 262, 263,
    271, 342
convergence problem, 310
coupled resonance, 17
cyclic frequency, 11, 15, 20, 23, 91, 193,
    254, 287, 295

definition of resonance, 14
determined system, 36, 60
dielectric permittivity, 34
Dirichlet–Jordan theorem, 77
dispersion equation, 27, 207, 250, 278, 295
dispersion function, 27, 295, 300
dispersion law, 27, 289
dispersion relation, 27, 109, 278
distributed system, 72, 78, 82, 293
distribution function, 79, 208, 209, 211
drift motion, 182
drifting standing electromagnetic wave,
    160
Duffing's equation, 11, 146, 189, 194
Duffing's parameter, 189

351

dynamical hierarchy, 101, 115
dynamical operator, 176, 241, 296
dynamical parameter, 54
dynamical system, xix, 18, 32, 35, 36, 50,
    52, 57, 59–63, 65, 67, 68, 71,
    73, 75, 83, 111, 175, 314
dynamical variables, 50, 56, 61, 83, 296

effect of electron beam modulation, 159
effective potential, 176
electron clusters, 197
electron oscillation phase, 13, 168
electrostatic support, 329
elliptic function, 179
energy conservation law, 32, 44, 178
energy parameter, 179
equation of nonlinear pendulum, 10, 178,
    179, 186, 188
equation of stimulated nonlinear pendu-
    lum, 194
equations of comparison, 99
equations of higher hierarchical level, 4
equations of the first hierarchical level, 81,
    93, 94, 127, 139, 184, 239, 256,
    332
equivalent linear system, 81, 286, 295
essentially nonlinear system, 11
Euler variable, 35
Euler's variable, 34, 112
explicit oscillation phase, 13
extended combinative phase, 186
extended coordinate vector, 86
extended nonlinear dynamical vector-
    function, 86

fast oscillation phase, 13
fast rotating phase, 12, 102, 107, 108, 111,
    119, 130, 138, 145–147, 167,
    168, 171, 215, 331
fast variable, 90–92, 101, 102, 119, 134,
    233, 245, 295
fast varying combination phase, 16
field of forces, 31
field oscillation phase, 13
first integral of motion, 178
first law of thermodynamics, 44
force, 9, 12–14, 29, 30, 32–34, 51, 110,
    113, 167, 176, 177, 194, 199–
    202, 249, 295
four-wave parametric resonance, 16
Fourier–Bessel series, 188
free electron laser (FEL), 1, 4, 5, 13, 18,
    22, 77, 94, 161
functional operator, 64, 73, 75, 84, 88, 96,
    245

general hierarchical principle, 58, 60, 64,
    83

general standard form, 83, 285
generalized hierarchical equation of the
    first order, 73, 96, 120
generated form of solution, 286
generating equation, 10, 91, 99
gravitational field, 174, 175
gray zone, 143
Grebennikov's method, 113
group velocity, 25, 301, 307
grouping mechanism, 196, 197

Hamilton's function, 33
Hamiltonian equations, 33, 244
Hamiltonian formalism, 33
harmonic oscillations, 21, 22
Heisenberg's uncertainty principle, 41
hidden oscillation phase, 12
hierarchical analogue of the second ther-
    modynamic principle, 59, 67
hierarchical analogue of the third thermo-
    dynamic principle, 187
hierarchical   asymptotic   analytical–
    numerical methods, xvii, xix,
    xx, 4
hierarchical military unit, 114
hierarchical oscillation system, 13, 111
hierarchical principles, xx, 50, 58, 60, 63,
    73, 74, 83, 84, 113, 114, 176,
    218, 239, 302
hierarchical series, 53, 54, 57, 58, 61, 90,
    114, 117, 119, 215, 240, 242,
    246, 255, 313
hierarchical series in the normalized di-
    mensionless form, 58
hierarchical system, xviii, xix, 36, 39, 44,
    49–58, 60, 62, 64, 67, 69, 71,
    72, 78, 95, 107, 108, 111, 114,
    116, 130, 182, 192, 195, 218,
    239
hierarchical theory of oscillations and
    waves, 17
hierarchical transformations, xviii, 89,
    218, 256, 318
hierarchical tree, 51, 52, 66–69, 78, 114–
    116, 185, 196
hierarchy, xviii, xix, xxi, 4, 5, 13, 17,
    49, 52–54, 56–58, 61, 64, 73,
    84, 90, 101, 111, 114, 117, 118,
    213, 218, 241, 242, 246, 265,
    316, 317
highest hierarchical scale parameter, 4
holographic principle, 52, 59
homogeneous equation, 213

improper wave, 27
information entropy, 44, 59, 61, 115
initial phase of oscillations, 11

instability, 14, 49, 249–252, 258, 282, 315, 329, 342
instantaneous acceleration, 29
instantaneous velocity, 29, 208
intensity vector, 34
inverse problem, 64, 65, 71, 72

kinetic energy, 19, 31, 32, 115, 198–200, 203
kinetic equation, 59, 79–81, 83, 208, 236, 244, 245, 310, 311, 313, 326
klystron, 251, 252
Kramer's formula, 296, 300
Krylov–Bogolyubov's substitution, 88, 96, 117, 120, 127, 150, 153, 155, 173, 189, 215, 220, 221, 233, 234, 239, 242, 244, 245, 256, 265, 268, 275, 321, 332, 338, 340

lag-effect, 24, 26
Lagrange oscillation phase, 34
Langmuir's electron wave, 159
Lantzos' method, 112
large particle, 34, 197, 204
large scale parameter of the problem, 103
linear differential equation, xviii, 10, 91
linear frequency, 202, 203
longitudinal wave, 26
lumped system, 71, 82

magnetic permeability, 34
many-particle description, 35
mathematical pendulum, 9, 12, 13
Maxwell's equations, 33, 83, 210, 211, 236, 238, 262, 271, 309, 311, 315, 317, 326, 327, 342
method of averaged characteristics, xx, 7, 72, 89, 207, 213, 222, 225, 231, 232, 235–237, 244, 250, 254, 271, 273, 280, 283, 285, 320, 329, 339
method of averaged current–density equation, 240
method of averaged kinetic equation, 79, 236, 244
method of averaged quasi-hydrodynamic equation, 236
method of characteristics, 85, 207, 213, 219, 222, 227, 228, 261, 276, 283, 288, 340
method of exact solutions, 186
microstate, 42
Miller–Gaponov potential, 176, 177, 198, 199
moderately nonlinear system, 11
modernized standard system, 309, 317, 320

modernized version of the method of slowly varying amplitudes, 293
moment of inertia, 9, 30
moment of momentum, 30, 32
moment of momentum conservation law, 32
momentum, 5, 29, 30, 32, 33, 79, 80, 111, 164, 166, 167, 208, 212, 214, 219, 232, 236, 237, 239, 245
momentum conservation law, 32
motion integral, 32, 174, 178
motion integral of the first hierarchical level, 174
multi-frequency nonlinear resonant oscillation–wave problem, 3
multi-frequency nonlinear resonant oscillation–wave systems, 3
multi-harmonic periodic oscillations, 22
multi-resonance system, 17
multiple resonance, 16

natural hierarchical system, 51, 54, 55, 60, 69
negentropyness, 37
new fast oscillatative phase, 184
Newtonian formalism, 33
non-bound resonances, 17
non-harmonic oscillations, 21, 22
non-isochronous oscillator, 181
non-resonant case, 146, 151
non-resonant zone, 143
nonlinear dynamical system, 10, 143
nonlinear equation, xviii, 10, 91
nonlinear oscillatative system, 9
nonlinear oscillation dynamical system, 10
normalized combinative phase, 188
normalized energy, 181
normalized potential function, 179, 180
normalized time, 188

one-particle description, 35
open system, 36, 38, 60
oscillation, 21
oscillation amplitude, 12
oscillation phase, 12
overdetermined system, 212

parabolic type, 228
parametric amplification, 288, 293
parametric electron-wave lamp, 113
parametric resonance, 16, 113, 193
parametric resonant damping, 293
partial solution, 212
particle ensemble, 196, 197, 209
passage motion of particle, 198

percussive excitation of highest harmonics, 258
period of oscillations, 14
period of the oscillations, 3
perturbed model, 201, 203, 204
phase grouping, 204, 205
phase mismatch, 291
phase velocity, 25, 26, 160, 161, 164, 194, 288
physical definition of the concept of God, 68
potential energy, 31, 32, 177, 179, 198, 199
potential field, 32, 174
potential function, 32, 176, 177, 179
potential well, 174, 177, 179, 199, 202
principle of hierarchical resemblance, 59, 61, 83
principle of information compression, 58, 64, 83, 93
principle of physicality, 38
principle of quasi-stationary interaction, 200
problem small parameter, 61, 189, 309, 314
proper field of the zeroth hierarchical level, 174
proper oscillations, 13, 249
proper wave, 27, 109, 250, 290, 295, 297, 299
pumping wave, 14, 26, 161, 288
purposefulness, 37, 42, 49, 50
purposefulness principle, 38, 42

quasi-harmonic oscillations, 21
quasi-potential field, 175–177, 180
quasilinear resonance, 16, 193

Rabinovich's standard form, 85, 307, 310
relationship of characteristics, 228
relativistic factor, 15, 111, 172, 210, 241, 253, 258, 261, 278, 344
resonance, 13, 14, 18–20, 38, 108, 113, 114, 130, 134–139, 150, 161, 170, 192, 197, 200, 238, 249, 250, 284, 291, 296, 305
resonances of the same hierarchy, 17
resonant case, 143, 146, 195
resonant condition, 15, 16, 18, 20, 143, 144, 150, 195, 197, 200, 249, 250, 287, 290, 297
resonant curve, 18, 19
resonant point, 18–20, 135, 249
resonant zone, 144
rigorous version of the method of slowly varying amplitudes, 285, 294
rotating motion of the nonlinear pendulum, 198
rotating vector phase, 102

Russian matryoshka, 53

saturation of amplification, 20, 345
SCW, 28, 254
second law of dynamics, 9, 29, 31, 33, 177
second law of thermodynamics, 44
secular term, 99, 100, 120, 129, 133, 190
sefirot, 51
self-consistent system, 344
self-modeling principle, 50, 52, 56, 59, 116
self-organization, 50, 204
self-resemblance, 50, 51
separatrix, 179, 181, 182
shortened form, xvii
simplified version of slowly varying amplitude method, 284
single particle problem, 163
slow electron wave, 330
slow resonance, 17
slow variable, 102, 112, 114, 118, 136, 147, 167, 169, 170, 233, 238, 313, 331
slowest dynamical variables, 4
slowly varying amplitude, xix, xx, 7, 20, 21, 72, 82, 89, 207, 208, 250, 279, 283–285, 287–289, 291, 293, 295, 297, 302, 305, 309, 310, 312, 314, 320, 324, 326, 329, 330, 342, 343, 345
slowly varying amplitude method, 320
slowly varying combination phase, 16, 137
slowly varying cyclic frequency, 21
slowly varying initial oscillation phase, 20
small oscillations, 10, 203
small parameter of the problem, 16, 91, 186, 189, 322
space charge wave, 254, 277, 305
spatially one-dimensional model, 288, 294
standard equation of zeroth hierarchical level, 6
standard system with partial derivatives, 213
standing electromagnetic wave, 158, 164, 176, 185
static hierarchical system, 54
static system, 36
stimulated Duffing's equation, 146
stimulated oscillations, 13–15, 146, 249, 250
stochastic system, 36, 60, 79, 80, 208
straight problem, 64
strong hierarchical series, 111
strong synchronism, 160
structural complexity, 45, 46
structural hierarchical scale parameter, 53
structural hierarchy, 53, 54, 57, 62

structural operator, 62, 63, 73, 75, 84, 85, 99, 109, 111
successive approximation method, 244
successive approximations, 128
surface wave, 27
synchronous condition, 162, 164, 170
synthesis problem, 64
system definition, 35
system for optical signals transformation into microwave signals, 159
system property, 36, 39, 41, 44, 204
system with constant rotation frequency, 171
system with fast rotating phases, 102, 103, 111
system with slow and fast variables, 100

thermodynamic probability, 43, 44
third law of thermodynamics, 44
threshold, 41
transversal wave, 23, 26
tree of life, 51, 55, 66
two-level hierarchical system, 95, 101, 216
two-multiple resonance, 16
two-stream instability, 14, 29, 208, 231, 249–253, 258, 278, 329, 342

uncertainty postulate, 41
underdetermined system, 212
uniqueness, 37, 60, 249
unperturbed model, 198, 204

Van der Pol's method, xix, 93, 94, 125, 126, 284
Van der Pol's variables, 91, 146, 189, 194
vector of slow variables, 101, 117, 118, 137, 145, 146, 171, 331
vicinity of the resonant point, 18
volumetric wave, 27
vortex field, 31, 32

wave energy density, 28
wave number, 15, 24–26, 165, 193, 295, 297
wave period, 15, 23, 24
wavelength, 24, 26, 27
weak nonlinear dynamical system, 11
weak nonlinear mathematical problem, 21
weak nonlinear oscillations, 11–13, 21
weak predictability, 37
weakly nonlinear equation, 10

zero-level equation, 61

Fundamental Theories of Physics

46. P.P.J.M. Schram: *Kinetic Theory of Gases and Plasmas*. 1991   ISBN 0-7923-1392-5
47. A. Micali, R. Boudet and J. Helmstetter (eds.): *Clifford Algebras and their Applications in Mathematical Physics*. 1992   ISBN 0-7923-1623-1
48. E. Prugovečki: *Quantum Geometry*. A Framework for Quantum General Relativity. 1992   ISBN 0-7923-1640-1
49. M.H. Mac Gregor: *The Enigmatic Electron*. 1992   ISBN 0-7923-1982-6
50. C.R. Smith, G.J. Erickson and P.O. Neudorfer (eds.): *Maximum Entropy and Bayesian Methods*. Proceedings of the 11th International Workshop (Seattle, 1991). 1993   ISBN 0-7923-2031-X
51. D.J. Hoekzema: *The Quantum Labyrinth*. 1993   ISBN 0-7923-2066-2
52. Z. Oziewicz, B. Jancewicz and A. Borowiec (eds.): *Spinors, Twistors, Clifford Algebras and Quantum Deformations*. Proceedings of the Second Max Born Symposium (Wrocław, Poland, 1992). 1993   ISBN 0-7923-2251-7
53. A. Mohammad-Djafari and G. Demoment (eds.): *Maximum Entropy and Bayesian Methods*. Proceedings of the 12th International Workshop (Paris, France, 1992). 1993   ISBN 0-7923-2280-0
54. M. Riesz: *Clifford Numbers and Spinors* with Riesz' Private Lectures to E. Folke Bolinder and a Historical Review by Pertti Lounesto. E.F. Bolinder and P. Lounesto (eds.). 1993   ISBN 0-7923-2299-1
55. F. Brackx, R. Delanghe and H. Serras (eds.): *Clifford Algebras and their Applications in Mathematical Physics*. Proceedings of the Third Conference (Deinze, 1993) 1993   ISBN 0-7923-2347-5
56. J.R. Fanchi: *Parametrized Relativistic Quantum Theory*. 1993   ISBN 0-7923-2376-9
57. A. Peres: *Quantum Theory: Concepts and Methods*. 1993   ISBN 0-7923-2549-4
58. P.L. Antonelli, R.S. Ingarden and M. Matsumoto: *The Theory of Sprays and Finsler Spaces with Applications in Physics and Biology*. 1993   ISBN 0-7923-2577-X
59. R. Miron and M. Anastasiei: *The Geometry of Lagrange Spaces: Theory and Applications*. 1994   ISBN 0-7923-2591-5
60. G. Adomian: *Solving Frontier Problems of Physics: The Decomposition Method*. 1994   ISBN 0-7923-2644-X
61. B.S. Kerner and V.V. Osipov: *Autosolitons*. A New Approach to Problems of Self-Organization and Turbulence. 1994   ISBN 0-7923-2816-7
62. G.R. Heidbreder (ed.): *Maximum Entropy and Bayesian Methods*. Proceedings of the 13th International Workshop (Santa Barbara, USA, 1993) 1996   ISBN 0-7923-2851-5
63. J. Peřina, Z. Hradil and B. Jurčo: *Quantum Optics and Fundamentals of Physics*. 1994   ISBN 0-7923-3000-5
64. M. Evans and J.-P. Vigier: *The Enigmatic Photon*. Volume 1: The Field $B^{(3)}$. 1994   ISBN 0-7923-3049-8
65. C.K. Raju: *Time: Towards a Constistent Theory*. 1994   ISBN 0-7923-3103-6
66. A.K.T. Assis: *Weber's Electrodynamics*. 1994   ISBN 0-7923-3137-0
67. Yu. L. Klimontovich: *Statistical Theory of Open Systems*. Volume 1: A Unified Approach to Kinetic Description of Processes in Active Systems. 1995   ISBN 0-7923-3199-0; Pb: ISBN 0-7923-3242-3
68. M. Evans and J.-P. Vigier: *The Enigmatic Photon*. Volume 2: Non-Abelian Electrodynamics. 1995   ISBN 0-7923-3288-1
69. G. Esposito: *Complex General Relativity*. 1995   ISBN 0-7923-3340-3

Fundamental Theories of Physics

70. J. Skilling and S. Sibisi (eds.): *Maximum Entropy and Bayesian Methods*. Proceedings of the Fourteenth International Workshop on Maximum Entropy and Bayesian Methods. 1996
ISBN 0-7923-3452-3

71. C. Garola and A. Rossi (eds.): *The Foundations of Quantum Mechanics Historical Analysis and Open Questions*. 1995
ISBN 0-7923-3480-9

72. A. Peres: *Quantum Theory: Concepts and Methods*. 1995 (see for hardback edition, Vol. 57)
ISBN Pb 0-7923-3632-1

73. M. Ferrero and A. van der Merwe (eds.): *Fundamental Problems in Quantum Physics*. 1995
ISBN 0-7923-3670-4

74. F.E. Schroeck, Jr.: *Quantum Mechanics on Phase Space*. 1996
ISBN 0-7923-3794-8

75. L. de la Peña and A.M. Cetto: *The Quantum Dice*. An Introduction to Stochastic Electrodynamics. 1996
ISBN 0-7923-3818-9

76. P.L. Antonelli and R. Miron (eds.): *Lagrange and Finsler Geometry*. Applications to Physics and Biology. 1996
ISBN 0-7923-3873-1

77. M.W. Evans, J.-P. Vigier, S. Roy and S. Jeffers: *The Enigmatic Photon*. Volume 3: Theory and Practice of the $B^{(3)}$ Field. 1996
ISBN 0-7923-4044-2

78. W.G.V. Rosser: *Interpretation of Classical Electromagnetism*. 1996   ISBN 0-7923-4187-2

79. K.M. Hanson and R.N. Silver (eds.): *Maximum Entropy and Bayesian Methods*. 1996
ISBN 0-7923-4311-5

80. S. Jeffers, S. Roy, J.-P. Vigier and G. Hunter (eds.): *The Present Status of the Quantum Theory of Light*. Proceedings of a Symposium in Honour of Jean-Pierre Vigier. 1997
ISBN 0-7923-4337-9

81. M. Ferrero and A. van der Merwe (eds.): *New Developments on Fundamental Problems in Quantum Physics*. 1997
ISBN 0-7923-4374-3

82. R. Miron: *The Geometry of Higher-Order Lagrange Spaces*. Applications to Mechanics and Physics. 1997
ISBN 0-7923-4393-X

83. T. Hakioğlu and A.S. Shumovsky (eds.): *Quantum Optics and the Spectroscopy of Solids*. Concepts and Advances. 1997
ISBN 0-7923-4414-6

84. A. Sitenko and V. Tartakovskii: *Theory of Nucleus*. Nuclear Structure and Nuclear Interaction. 1997
ISBN 0-7923-4423-5

85. G. Esposito, A.Yu. Kamenshchik and G. Pollifrone: *Euclidean Quantum Gravity on Manifolds with Boundary*. 1997
ISBN 0-7923-4472-3

86. R.S. Ingarden, A. Kossakowski and M. Ohya: *Information Dynamics and Open Systems*. Classical and Quantum Approach. 1997
ISBN 0-7923-4473-1

87. K. Nakamura: *Quantum versus Chaos*. Questions Emerging from Mesoscopic Cosmos. 1997
ISBN 0-7923-4557-6

88. B.R. Iyer and C.V. Vishveshwara (eds.): *Geometry, Fields and Cosmology*. Techniques and Applications. 1997
ISBN 0-7923-4725-0

89. G.A. Martynov: *Classical Statistical Mechanics*. 1997   ISBN 0-7923-4774-9

90. M.W. Evans, J.-P. Vigier, S. Roy and G. Hunter (eds.): *The Enigmatic Photon*. Volume 4: New Directions. 1998
ISBN 0-7923-4826-5

91. M. Rédei: *Quantum Logic in Algebraic Approach*. 1998   ISBN 0-7923-4903-2

92. S. Roy: *Statistical Geometry and Applications to Microphysics and Cosmology*. 1998
ISBN 0-7923-4907-5

93. B.C. Eu: *Nonequilibrium Statistical Mechanics*. Ensembled Method. 1998
ISBN 0-7923-4980-6

94. V. Dietrich, K. Habetha and G. Jank (eds.): *Clifford Algebras and Their Application in Mathematical Physics.* Aachen 1996. 1998   ISBN 0-7923-5037-5

95. J.P. Blaizot, X. Campi and M. Ploszajczak (eds.): *Nuclear Matter in Different Phases and Transitions.* 1999   ISBN 0-7923-5660-8

96. V.P. Frolov and I.D. Novikov: *Black Hole Physics.* Basic Concepts and New Developments. 1998   ISBN 0-7923-5145-2; Pb 0-7923-5146

97. G. Hunter, S. Jeffers and J-P. Vigier (eds.): *Causality and Locality in Modern Physics.* 1998   ISBN 0-7923-5227-0

98. G.J. Erickson, J.T. Rychert and C.R. Smith (eds.): *Maximum Entropy and Bayesian Methods.* 1998   ISBN 0-7923-5047-2

99. D. Hestenes: *New Foundations for Classical Mechanics (Second Edition).* 1999   ISBN 0-7923-5302-1; Pb ISBN 0-7923-5514-8

100. B.R. Iyer and B. Bhawal (eds.): *Black Holes, Gravitational Radiation and the Universe.* Essays in Honor of C. V. Vishveshwara. 1999   ISBN 0-7923-5308-0

101. P.L. Antonelli and T.J. Zastawniak: *Fundamentals of Finslerian Diffusion with Applications.* 1998   ISBN 0-7923-5511-3

102. H. Atmanspacher, A. Amann and U. Müller-Herold: *On Quanta, Mind and Matter Hans Primas in Context.* 1999   ISBN 0-7923-5696-9

103. M.A. Trump and W.C. Schieve: *Classical Relativistic Many-Body Dynamics.* 1999   ISBN 0-7923-5737-X

104. A.I. Maimistov and A.M. Basharov: *Nonlinear Optical Waves.* 1999   ISBN 0-7923-5752-3

105. W. von der Linden, V. Dose, R. Fischer and R. Preuss (eds.): *Maximum Entropy and Bayesian Methods Garching, Germany 1998.* 1999   ISBN 0-7923-5766-3

106. M.W. Evans: *The Enigmatic Photon Volume 5: O(3) Electrodynamics.* 1999   ISBN 0-7923-5792-2

107. G.N. Afanasiev: *Topological Effects in Quantum Mecvhanics.* 1999   ISBN 0-7923-5800-7

108. V. Devanathan: *Angular Momentum Techniques in Quantum Mechanics.* 1999   ISBN 0-7923-5866-X

109. P.L. Antonelli (ed.): *Finslerian Geometries A Meeting of Minds.* 1999   ISBN 0-7923-6115-6

110. M.B. Mensky: *Quantum Measurements and Decoherence Models and Phenomenology.* 2000   ISBN 0-7923-6227-6

111. B. Coecke, D. Moore and A. Wilce (eds.): *Current Research in Operation Quantum Logic.* Algebras, Categories, Languages. 2000   ISBN 0-7923-6258-6

112. G. Jumarie: *Maximum Entropy, Information Without Probability and Complex Fractals.* Classical and Quantum Approach. 2000   ISBN 0-7923-6330-2

113. B. Fain: *Irreversibilities in Quantum Mechanics.* 2000   ISBN 0-7923-6581-X

114. T. Borne, G. Lochak and H. Stumpf: *Nonperturbative Quantum Field Theory and the Structure of Matter.* 2001   ISBN 0-7923-6803-7

115. J. Keller: *Theory of the Electron.* A Theory of Matter from START. 2001   ISBN 0-7923-6819-3

116. M. Rivas: *Kinematical Theory of Spinning Particles.* Classical and Quantum Mechanical Formalism of Elementary Particles. 2001   ISBN 0-7923-6824-X

117. A.A. Ungar: *Beyond the Einstein Addition Law and its Gyroscopic Thomas Precession.* The Theory of Gyrogroups and Gyrovector Spaces. 2001   ISBN 0-7923-6909-2

118. R. Miron, D. Hrimiuc, H. Shimada and S.V. Sabau: *The Geometry of Hamilton and Lagrange Spaces.* 2001   ISBN 0-7923-6926-2

Fundamental Theories of Physics

119. M. Pavšič: *The Landscape of Theoretical Physics: A Global View*. From Point Particles to the Brane World and Beyond in Search of a Unifying Principle. 2001   ISBN 0-7923-7006-6
120. R.M. Santilli: *Foundations of Hadronic Chemistry*. With Applications to New Clean Energies and Fuels. 2001   ISBN 1-4020-0087-1
121. S. Fujita and S. Godoy: *Theory of High Temperature Superconductivity*. 2001
      ISBN 1-4020-0149-5
122. R. Luzzi, A.R. Vasconcellos and J. Galvão Ramos: *Predictive Statitical Mechanics*. A Nonequilibrium Ensemble Formalism. 2002   ISBN 1-4020-0482-6

KLUWER ACADEMIC PUBLISHERS – DORDRECHT / BOSTON / LONDON